METHODS IN MOLECULAR BIOLOGY

Series Editor
John M. Walker
School of Life and Medical Sciences
University of Hertfordshire
Hatfield, Hertfordshire, UK

For further volumes:
http://www.springer.com/series/7651

For over 35 years, biological scientists have come to rely on the research protocols and methodologies in the critically acclaimed *Methods in Molecular Biology* series. The series was the first to introduce the step-by-step protocols approach that has become the standard in all biomedical protocol publishing. Each protocol is provided in readily-reproducible step-by-step fashion, opening with an introductory overview, a list of the materials and reagents needed to complete the experiment, and followed by a detailed procedure that is supported with a helpful notes section offering tips and tricks of the trade as well as troubleshooting advice. These hallmark features were introduced by series editor Dr. John Walker and constitute the key ingredient in each and every volume of the *Methods in Molecular Biology* series. Tested and trusted, comprehensive and reliable, all protocols from the series are indexed in PubMed.

Quantum Mechanics in Drug Discovery

Edited by

Alexander Heifetz

Evotec (UK) Ltd., Abingdon, Oxfordshire, UK

Humana Press

Editor
Alexander Heifetz
Evotec (UK) Ltd.
Abingdon, Oxfordshire, UK

ISSN 1064-3745 ISSN 1940-6029 (electronic)
Methods in Molecular Biology
ISBN 978-1-0716-0281-2 ISBN 978-1-0716-0282-9 (eBook)
https://doi.org/10.1007/978-1-0716-0282-9

This Humana imprint is published by the registered company Springer Science+Business Media, LLC, part of Springer Nature.
The registered company address is: 1 New York Plaza, New York, NY 10004, U.S.A.

Preface

The application of quantum mechanical (QM) methods in drug discovery is becoming increasingly popular. This is largely a consequence of significant improvements in computing power/speed, which have led to advances in QM algorithm development and new applications of QM methods. Such methods are now routinely applied to many computer-aided drug discovery problems, including characterizing protein–water–ligand and protein–protein interactions, providing estimates of binding affinities, determining ligand energies and bioactive conformations, refinement of molecular geometries, scoring docked protein–ligand poses, describing molecular similarity, structure–activity relationship (SAR) analysis, and ADMET prediction. QM applications are rapidly replacing classical molecular mechanics (MM) methods because there is no need to use MM approximations when you can perform QM calculations with much higher accuracy and similar speed.

This book provides a unique overview of the QM methods, protocols, and applications used to deal with various challenges in drug discovery and structure exploration. It compares the performance of QM methods with alternative computational approaches such as molecular mechanics. Finally, the book provides an overall perspective on QM methods and their role in drug discovery. A review of these techniques will allow a diverse audience, including structural and molecular biologists, computational and medicinal chemists, pharmacologists and drug designers, to navigate through and effectively deploy these advances.

Oxfordshire, UK *Alexander Heifetz*

The most intelligent picture ever taken

Participants of the 5th Solvay Conference on Quantum Mechanics, 1927; 17 of the 29 attendees were or became Nobel Prize winners, including Marie Curie, who, alone among them, had won Nobel Prizes in two separate scientific disciplines.

From back to front and from left to right: Auguste Piccard, Émile Henriot, Paul Ehrenfest, Édouard Herzen, Théophile de Donder, Erwin Schrödinger, Jules-Émile Verschaffelt, Wolfgang Pauli, Werner Heisenberg, Ralph Howard Fowler, Léon Brillouin, Peter Debye, Martin Knudsen, William Lawrence Bragg, Hendrik Anthony Kramers, Paul Dirac, Arthur Compton, Louis de Broglie, Max Born, Niels Bohr, Irving Langmuir, Max Planck, Marie Skłodowska Curie, Hendrik Lorentz, Albert Einstein, Paul Langevin, Charles Eugène Guye, Charles Thomson Rees Wilson, Owen Willans Richardson.

Contents

Contributors

STEVEN M. ABEL • *Department of Chemical and Biomolecular Engineering, The University of Tennessee, Knoxville, TN, USA*

YURI ALEXEEV • *Argonne Leadership Computing Facility and Computational Science Division, Argonne National Laboratory, Lemont, IL, USA*

ANDREW ANIGHORO • *Evotec (UK) Ltd., Oxfordshire, UK*

M. GABRIELA AUCAR • *Computational Drug Design and Biomedical Informatics Laboratory, Translational Medicine Research Institute (IIMT), CONICET-Universidad Austral, Pilar-Derqui, Buenos Aires, Argentina*

MIKE J. BODKIN • *Evotec (UK) Ltd., Abingdon, Oxfordshire, UK*

STEVEN BROMIDGE • *Evotec (UK) Ltd., Oxfordshire, UK*

RICHARD A. BRYCE • *Division of Pharmacy and Optometry, School of Health Sciences, University of Manchester, Manchester, UK*

DONATELLA CALLEGARI • *Drug Design and Discovery Group, Department of Food and Drug, University of Parma, Parma, Italy*

CLAUDIO N. CAVASOTTO • *Computational Drug Design and Biomedical Informatics Laboratory, Translational Medicine Research Institute (IIMT), CONICET-Universidad Austral, Derqui-Pilar, Buenos Aires, Argentina; Austral Institute for Applied Artificial Intelligence, Universidad Austral, Derqui-Pilar, Buenos Aires, Argentina; Facultad de Ciencias Biomédicas, Universidad Austral, Derqui-Pilar, Buenos Aires, Argentina; Facultad de Ingeniería, Universidad Austral, Derqui-Pilar, Buenos Aires, Argentina*

MOUHMAD H. ELAYYAN • *Department of Chemical Engineering, Tennessee Tech University, Cookeville, TN, USA*

DMITRI G. FEDOROV • *Research Center for Computational Design of Advanced Functional Materials (CD-FMat), National Institute of Advanced Industrial Science and Technology (AIST), Tsukuba, Ibaraki, Japan*

KAORI FUKUZAWA • *School of Pharmacy and Pharmaceutical Sciences, Hoshi University, Tokyo, Japan*

CONSTANTINO DIAZ GONZALEZ • *Global Research Informatics/Cheminformatics and Drug Design, Evotec (France) SAS, Toulouse, France*

MATTHEW HABGOOD • *Evotec (UK) Ltd., Oxfordshire, UK*

XIAO HE • *Shanghai Engineering Research Center of Molecular Therapeutics and New Drug Development, School of Chemistry and Molecular Engineering, East China Normal University, Shanghai, China; NYU-ECNU Center for Computational Chemistry at NYU Shanghai, Shanghai, China*

ALEXANDER HEIFETZ • *Evotec (UK) Ltd., Abingdon, Oxfordshire, UK*

STEPHAN IRLE • *Computational Sciences and Engineering Division & Chemical Sciences Division, Oak Ridge National Laboratory, Oak Ridge, TN, USA; Bredesen Center for Interdisciplinary Research and Graduate Education, The University of Tennessee, Knoxville, TN, USA*

TIM JAMES • *Evotec (UK) Ltd., Abingdon, Oxfordshire, UK*

YUTO KOMEIJI • *Biomedical Research Institute, National Institute of Advanced Industrial Science and Technology, Ibaraki, Japan*

MARTIN KOTEV • *Global Research Informatics/Cheminformatics and Drug Design, Evotec (France) SAS, Toulouse, France*
HUI LI • *Department of Chemistry, University of Nebraska-Lincoln, Lincoln, NE, USA*
JINFENG LIU • *Department of Basic Medicine and Clinical Pharmacy, China Pharmaceutical University, Nanjing, China*
ALESSIO LODOLA • *Drug Design and Discovery Group, Department of Food and Drug, University of Parma, Parma, Italy*
VLADIMIR MIRONOV • *Department of Chemistry, Lomonosov Moscow State University, Moscow, Russian Federation*
MARCO MOR • *Drug Design and Discovery Group, Department of Food and Drug, University of Parma, Parma, Italy*
INAKI MORAO • *Evotec (UK) Ltd., Abingdon, Oxfordshire, UK*
HIROYA NAKATA • *Kyocera Corporation, Research Institute for Advanced Materials and Devices, Seika-cho, Soraku-gun, Kyoto, Japan*
YOSHIO OKIYAMA • *Division of Medicinal Safety Science, National Institute of Health Sciences, Kawasaki, Kanagawa, Japan*
ALFONSO POZZAN • *Research Informatics, Computational Chemistry, Aptuit SRL, an Evotec Company, Verona, Italy*
SILVIA RIVARA • *Drug Design and Discovery Group, Department of Food and Drug, University of Parma, Parma, Italy*
LAURIE SARRAT • *Global Research Informatics/Cheminformatics and Drug Design, Evotec (France) SAS, Toulouse, France*
LAURA SCALVINI • *Drug Design and Discovery Group, Department of Food and Drug, University of Parma, Parma, Italy*
VLADIMIR SLADEK • *Institute of Chemistry, Centre for Glycomics, Slovak Academy of Sciences, Bratislava, Slovakia*
MICHELLE SOUTHEY • *Evotec (UK) Ltd., Abingdon, Oxfordshire, UK*
MARAT R. TALIPOV • *Department of Chemistry & Biochemistry, New Mexico State University, Las Cruces, NM, USA*
SHIGENORI TANAKA • *Department of Computational Science, Graduate School of System Informatics, Kobe University, Kobe, Japan*
CHRISTOFER S. TAUTERMANN • *Department of Medicinal Chemistry, Boehringer Ingelheim Pharma GmbH & Co. KG, Biberach, Germany*
ANDREA TOWNSEND-NICHOLSON • *Division of Biosciences, Research Department of Structural & Molecular Biology, Institute of Structural & Molecular Biology, University College London, London, UK*
VAN Q. VUONG • *Computational Sciences and Engineering Division & Chemical Sciences Division, Oak Ridge National Laboratory, Oak Ridge, TN, USA; Bredesen Center for Interdisciplinary Research and Graduate Education, The University of Tennessee, Knoxville, TN, USA*

Chapter 1

Current and Future Challenges in Modern Drug Discovery

Christofer S. Tautermann

Abstract

Drug discovery is an expensive, time-consuming, and risky business. To avoid late-stage failure, learnings from past projects and the development of new approaches are crucial. New modalities and emerging new target spaces allow the exploration of unprecedented indications or to address so far undrugable targets. Late-stage attrition is usually attributed to the lack of efficacy or to compound-related safety issues. Efficacy has been shown to be related to a strong genetic link to human disease, a better understanding of the target biology, and the availability of biomarkers to bridge from animals to humans. Compound safety can be improved by ligand optimization, which is becoming increasingly demanding for difficult targets. Therefore, new strategies include the design of allosteric ligands, covalent binders, and other modalities. Design methods currently heavily rely on artificial intelligence and advanced computational methods such as free energy calculations and quantum chemistry. Especially for quantum chemical methods, a more detailed overview is given in this chapter.

Key words Drug design, Drug discovery, R&D spending, R&D efficiency, New chemical entities, Quantum mechanics in drug design

1 Introduction

Drug discovery is a very challenging, cost-intensive, and in terms of investment risky endeavor. On average, it takes 10–15 years and an investment of more than 2.5 billion dollars to get a new drug to the market [1]. Especially the clinical phases cause extremely high costs [2] and late-stage failures, especially in phase II studies, are quite common. Whereas in the 1990s the pharmacokinetic (PK) properties were often causing clinical drug failure, nowadays the lack of efficacy and safety findings cause late attrition [3]. Therefore pharmaceutical companies have developed strategies to overcome these issues, and several key factors for success have been identified [3, 4]. To tackle the issue of lacking efficacy, the drug target requires a thorough validation, if possible with translational biomarkers as well as a relevant patient selection. Safety issues are often compound related rather than target dependent and can be overcome by the selection of the right compound, be it small

Alexander Heifetz (ed.), *Quantum Mechanics in Drug Discovery*, Methods in Molecular Biology, vol. 2114,
https://doi.org/10.1007/978-1-0716-0282-9_1,

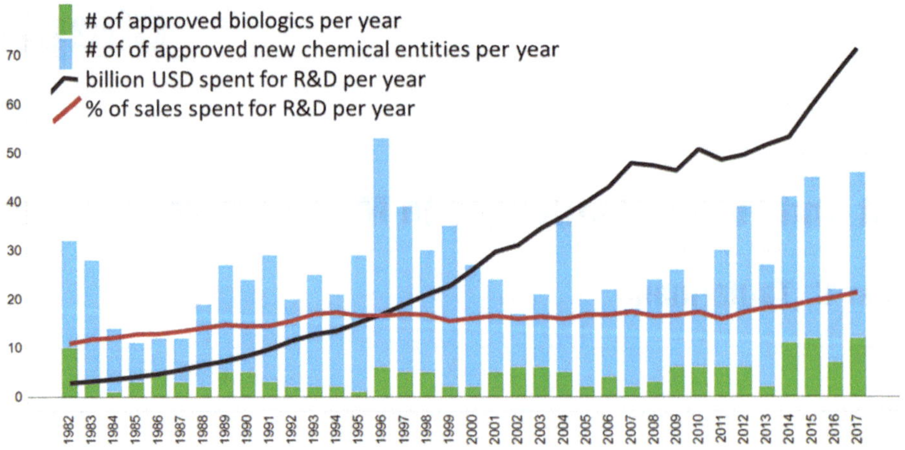

Fig. 1 Drug approvals and R&D spendings over the last decades. The absolute values of R&D spendings incline much steeper than the number of approved drugs

molecule, antibody, or any other modality [3]. This observation shows that the quest for the best compound for a certain target is crucial.

Although the hurdles are high to get new drugs to the market, there is a trend that the number of drug approvals increases over the last 15 years [5]. On average 43 drugs were approved by the FDA in the years 2014–2018, whereas the average in the first decade of the century has been 23 drugs per year (*see* Fig. 1). 2018 has been an exceptionally successful year with 59 FDA-approved drugs [5]. Although these numbers are impressive, they have to be seen in context of the overall spending in research and development (R&D) in pharmaceutical industry. In the last few years, more than 20% of total sales are invested in R&D, but in the beginning of the 1980s, it was merely 10% (according to the PhRMA membership survey on https://www.phrma.org/). Therefore, the efforts that are put into the finding and development of new drugs have doubled in 35 years. Figure 1 gives a good overview over the last 4 decades of drug discovery and this demonstrates that innovation is still ongoing in the field. Forecasts of R&D spending for the next decades predict a further increase; most notably there will also be a geographic shift of R&D towards Asiatic countries [6].

From a high-level view, the drug discovery business looks like a healthy, stable market. However, it is a very risky business, where late-stage failures are causing immense costs. Thus, companies do their best to identify the very best compounds for the expensive clinical phases in order to reduce attrition. In the next few paragraphs, I will discuss the general challenges in drug discovery. Because this article is part of a book on quantum mechanics (QM) in drug design, I will clearly focus on the discovery of new molecular entities (NMEs) and the currently most promising new methods will be discussed in more detail.

2 General Challenges in Drug Design

2.1 Commercial Potential

As already pointed out, in 2018 the FDA has approved 59 new drugs, which is very impressive just from the sheer amount. When analyzing the commercial potential of these drugs, only two out of 59 are predicted to achieve annual sales of more than 2 billion USD per year [5]. Generally, the average projected peak sales for new drugs are declining [5]. This may have several reasons, but one simple explanation is that many drugs are aiming for quite small indications. For example, in the last few years, a large fraction of new approvals has been for oncology with more than 50 new approvals over the last 5 years [5, 7, 8]. However, cancer drugs are very often tissue and cancer specific and therefore they are only suitable for a quite small subpopulation of patients. Having said this, there is still a huge commercial potential in cancer drugs, if benefit for additional subpopulations can be demonstrated and additional approvals can be gained. Beyond the cancer drugs, the so-called orphan drugs account for the highest fraction of new approvals in the last 5 years [5]. Orphan drugs are defined as drugs which target rare diseases and thus per definition aim for a small patient population. Often the awareness of such diseases is low and (depending on the patient population) the commercial potential is often not given [9]. However, to foster drug development for orphan diseases, authorities grant incentives (such as tax reduction or market exclusivity) for the approval of orphan drugs. Another reason for the lack of potential new blockbusters is that generic drugs are already available for many indications with huge patient populations. For example, atorvastatin had revenues of more than ten billion dollar per year in 2004–2010. However, since the end of its patent life in 2011, the revenue dropped significantly [10]. This means that there is no need and no commercial potential for additional lipid-lowering medications with a statin-like mechanism. Or to say it in other words, the low-hanging fruits are taken.

2.2 Translation into Clinics

The largest attrition in clinical phases is observed in phase II studies. In this stage of drug discovery, the short-term safety in human has already been tested in phase I. The main goal is the assessment of the efficacy of the compound yielding a clinical proof of concept. A recent analysis from four major pharma companies [3] revealed that the cause for attrition in phase II is mainly caused by lack of efficacy followed by safety issues. Efficacy for a certain indication is usually preclinically tested in animal models. However, the translatability from animal models to the human situation is not always given. There are several reasons for this. Often the disease condition cannot adequately be induced in animals. For example, psychiatric disorders such as depression are induced in mice by injection of LPS causing severe inflammation and sick-like behavior [11]. Whether such mechanisms are comparable to the human situation of depression may be questioned. This means that in many cases, behavioral animal models are not the first choice for target validation and better predictors for clinical efficacy are required. In the best case, the target can be validated by human genetic evidence, where subpopulations carry mutations in alleles that cause changes in the target function compared to wild type. Based on the phenotypes of this subpopulation and correlations to target function, genetic evidence can be demonstrated statistically. This method is also used for target identification in GWAS (genome-wide association) studies, where genetic variants in the human population are linked to disease [12]. The lack of human genetic evidence increases the risk of a failure in the clinical proof of concept [4]. In addition to a good genetic evidence, the availability of relevant biomarkers decreases the attrition significantly [3]. The translatability between species is much higher on a biomarker basis rather than based on a behavioral readout. The second main reason for attrition in clinics is safety, where attrition based on compound-related safety issues has been shown to be much higher than target-related issues [4]. Large-scale analysis investigating the dependence of safety issues on compound properties [3] revealed that further focus on physicochemical parameters won't lead to a significant reduction of attrition. The authors suggest that a lot of work on compound properties has to be done to understand safety-related features. The well-known rule of five [13] has provided a good framework to establish a drug-like space enabling good PK properties, and indeed, very few clinical failures are nowadays caused by PK issues. There are currently no such general rules available which allow the minimization of safety-related effects which go beyond the obvious endpoints such as hERG or CYP inhibition. The development of methods that are able to predict compound toxicity in vivo would be huge step in the direction of the reduction of safety-related attrition. Obviously, this is an endeavor where computational chemistry will play a key role.

To sum up, genetic evidence, relevant biomarkers, and prediction of toxicity will be key to reduce clinical attrition and have to be worked on in the future.

2.3 Unexplored Opportunities and Different Modalities

Expiration of patent lifetime forces pharmaceutical industry into a process of continuous innovation. Starting with the identification of an unmet medical need, targets are identified and drugs are developed to fill again the pipeline. As pointed out before, the low-hanging fruits are already taken, where a blockbuster indication target could be more or less straightforwardly addressed by developing a small molecule. Since several years, new drug modalities become much more important and nowadays the majority of the best selling drugs are biologics [5]. Ten years ago, the picture was just the opposite, where the majority of the best selling drugs were small molecules [14]. This is also reflected in the new drug approvals—the fraction of biologics is steadily increasing over the past few years [5] (*see* Fig. 1). There are several reasons for this steady incline of biologics: First, some targets are not easily drugable by small molecules. For example, the CGRP receptor has been worked on for decades as a target for migraine [15], always with the goal to find a small-molecule drug, but finally in 2018 an antibody targeting CGRP has been approved [5]. Generally speaking, about 1/3 of the potentially drugable genome is unexplored [16] and biologic entities will definitely play a major role in its exploration. The largest disadvantage of antibodies and other biologics for the patient is the way of administration, as they have to be injected. However, targets that are currently deemed undrugable by small molecules can often be targeted by biologics. In addition, the repertoire of biologics is huge—most prominent are the well-known antibody therapeutics (often found in oncology), but also therapeutic peptides (such as insulin analogs) and proteins are marketed. Recently also siRNA-based therapeutics and even therapeutics based on viruses are approved [5]. Such modalities are the basis for gene therapy, which aims for the correction of genetic defects in patients. In addition to biologics, also other new drug modalities are under development, which exploit physiologic pathways and help in drug design for so far undrugable targets [17]. One prominent example is PROTACs, which attach a target to a ligase in the ubiquitin pathway and thereby lead to target depletion in the body [18]. One common disadvantage that all new modalities have in common is the higher cost of goods. Usually the production of small molecules on large scale is much easier and less expensive [19], rendering them to a preferred modality for pharmaceutical industry.

The legacy from decades of small-molecule drug discovery is a huge portfolio of methods for drug design. With these methods and new developments in the field, we have a solid foundation to overcome recent challenges in small-molecule drug design.

3 Small-Molecule Drug Design Challenges

As eluded above, small molecules have several advantages, such as being cell and brain permeable, the lower cost of goods in their production, and oral administration. However, they are not always the easiest path forward for challenging targets. Especially in the case of difficult or so far undrugable targets, new design strategies are required to be successful [17].

3.1 Selectivity

Targets can be difficult due to different reasons. In many cases, selectivity against closely related homologs can be decisive for a good safety profile. Also for so-called easily drugable targets, selectivity may be a showstopper. Most prominently for aminergic GPCRs [20–22] and kinases [23], several drugs are on the market; however very often these drugs display polypharmacology [24]. For tyrosine phosphatases the active site is highly conserved [25] and therefore reaching selectivity combined with good physicochemical parameters hampers drug development. A way to circumvent selectivity problems within protein families is the development of allosteric drugs. For example, for GPCRs, allosteric modulators are well known for several different families [26]. For the metabotropic glutamate receptors, the orthosteric binding site is highly conserved because all family members bind glutamate as agonist in an extracellular flytrap domain. The only chance to achieve selectivity is targeting of an allosteric site that is embedded in the transmembrane helical bundle. For amine receptors, several locations of allosteric modulators have been elucidated by X-ray [26], and all the allosteric binding pockets show lower sequence conservation than the orthosteric sites. Even for tyrosine phosphatases, recently the first allosteric activator has been reported [27] which binds far away from the active site and is communicating through allosteric long-range interactions with the active site. The drawback of allosteric sites is that they are often small and hydrophobic, because often they have not evolved to host small-molecule ligands. In addition, the communication to the active/orthosteric site is not always clear, and the effort to characterize allosteric modulators pharmacologically is much more demanding than for simple active site binders [28, 29].

Another way to tackle selectivity emerged quite recently and it involves the water network surrounding the ligand binding site [30]. Several studies in the literature demonstrate that ligand affinity can be modulated by displacing water molecules [31, 32]. This can efficiently be used to selectively modulate ligand affinities towards different proteins and thereby gain selectivity. The prediction of such effects is quite difficult, but free energy calculations are one promising way forward [33].

3.2 Flexibility

Another difficult case is the development of small-molecule drugs for highly flexible targets. The usual key and lock picture of ligand and target is not applicable for cases where the ligand induces the shape of the binding pocket. In extreme cases, intrinsically disordered regions of proteins are targeted and become ordered by ligand binding. For these targets, conventional drug design approaches fail and often these molecules are found by pharmacological screening [34]. If small molecules induce an ordering of some regions, entropic loss is expected from the protein, and this has to be compensated by a strong gain in enthalpy or by a favorable desolvation entropy upon ligand binding [35]. This explains why the molecules are often hydrophobic and weak binders. For the more conventional cases of protein flexibility, where ordered regions fluctuate and cryptic pockets are opening up, several methods for drug design are reported [36, 37]. Especially in the case of small fluctuations and local flexibility, molecular dynamics (MD) simulations help to understand protein flexibility upon ligand binding, which can display a very complex behavior [38]. The detection of induced or cryptic pockets is often achieved by so-called co-solvent MD simulations. In this case molecular fragments (such as benzene or methanol) are added to the simulation in quite high concentration and the preferred binding sites on the proteins are identified [39]. Often these fragments are sufficiently potent to induce rotamer changes of amino acids in the protein and thereby induce the opening of (sub)pockets or other small-scale movements.

3.3 Non-protein Targets

Proteins are not the only targets where small-molecule drugs can act on. Several drugs are known which bind DNA, but in many cases this is in the field of oncology, where a drug kills tumor cells by causing quite unspecific DNA damage [40]. The side effects of DNA damage, however, are not desirable, and therefore drugs binding, but not damaging, DNA are the clear goal. For example, the antimalarial drug quinacrine intercalates into DNA and also shows antitumor activity [41]. The quite homogenous structure of the DNA makes it a difficult target for a specific interaction with small molecules. On the other hand, RNA is now coming into focus of drug discovery [42]. RNA shows a high diversity in biological functions ranging from being the genetic carrier for translation to modulation of gene expression [43]. This makes it a highly interesting drug target, and it has been demonstrated that RNA is drugable [42]. RNA forms a distinct 3-dimensional structure; therefore, the approaches to drug design for RNA (especially for riboswitches) are similar to the common protein-targeting approaches. Just for small RNA sequences such as micro-RNA, other methods like library vs. library screens help to identify starting points for optimization [44].

3.4 Difficult Pockets

Often, the failure of small-molecule drug finding is attributed to the difficulty to find potent ligands for a potential binding site. Especially if it comes to protein–protein interfaces (PPI) or DNA/RNA–protein interfaces, the topology of the interaction site is decisive [45]. The interaction region is often very large and quite flat and it contains several hotspots (i.e., amino acids which strongly contribute to the binding affinity of the proteins) [46]. To be highly potent, the small molecule needs to address several of the hotspots, which may be spatially separated. High potency is required, because the binding affinity of two proteins of a PPI is quite often in the sub-nM range. In addition, if there is no defined pocket for ligand binding, the binding affinity does not benefit from terms like desolvation entropy [47] because it stays on the surface. Undoubtedly, there are success stories of small molecules targeting PPIs, but the resulting molecules are often quite large and complex because they are addressing several spread out regions on the proteins [45]. Quite impressively, small-molecule PPI inhibitors can exploit inducible pockets on the PPI surface and thereby gain affinity. Computational methods can substantially contribute to PPI inhibitor design. Methods like free energy calculations are applied to detect PPI hotspots [48] and mixed solvent methods are used to detect inducible pockets [39]. Aromatic interactions are very common in PPIs [49] and their strength can be estimated by QM methods [50]. With this knowledge, ligand design can be supported very well by computational chemistry. Another approach for targeting large PPI interfaces is by covalent ligands. In this case, the binding affinity to the protein is not the only decisive parameter, but also the reactivity of the ligand's warhead towards the protein side chain is of high importance. Too reactive molecules will label too many other proteins unselectively, resulting in toxicity. Therefore the optimal profile would comprise a strong binding affinity together with a medium reactivity, in order to react only in the desired binding site [51]. To estimate warhead reactivity, usually QM methods are indispensable, either for calculating the electrophilicity index [52] or for determining the transition state of the reaction [53]. An energetically proper description of a chemical reaction always requires QM treatment; therefore the resurgence of covalent approaches leads to the increased usage of QM methods in computational chemistry.

Another kind of difficult pocket may arise from non-advantageous physicochemical properties, e.g., found in phosphatases, where the active site is highly polar. A potent ligand accordingly also has to be polar to complement the binding site, but then its properties lead to inacceptable ADME properties. A common way out of this problem is the identification and targeting of allosteric sites. For tyrosine phosphatases this approach has been successful [27, 54]; however, such an approach often requires elaborate methods. Straightforward screening for biochemically active hits may identify allosteric binders, but in most cases they are much less active than active site binders and are missed in hit triaging. Quite frequently allosteric binders are identified by fragment-based screens, where either NMR or X-ray methods reveal the actual binding site. Fragments usually bind with a very low affinity and have to be optimized to yield functional effects. Therefore, the identification of a remote pocket is only the first step, and the proof of its communication to the active site is a more complex task. In this case, computational methods help to identify allosteric communication pathways to the active site and can help to prioritize fragments binding to potentially functional pockets [55].

4 State of the Art and Trends in Small-Molecule Drug Design

Several challenges in the discovery and design of small molecules have been discussed in the previous chapters. Some of the challenges cannot be solved by drug design itself, for example, the commercial potential is only indirectly dependent on the quality of a drug. For all typical challenges in drug design, there are currently state-of-the-art solutions available, which clearly contribute to end up with safer and better small-molecule drugs. To start with, the affinity and the functional effect of a ligand on a target are of highest importance. Therefore, methods are required, which are able to assess the binding affinity accurately and in the best case also predict how to strengthen the interaction of the ligand to the target. Free energy calculations allow the estimation of changes in affinity when comparing ligands and they can be applied in a medium throughput [33]. Sometimes these calculations encounter difficulties due to inappropriate underlying force fields. To overcome this issue, recently methods have been established to perform free energy calculations including QM assessment at various snapshots [56]. The new QM-corrected calculations come at a high computational cost, but they are designed to have a higher accuracy. Another approach deals with the energy decomposition of the ligand/target interaction energy. Based on QM calculations of the amino acids surrounding the ligand, the fragments in molecules (FMO) approach evaluate the interaction energy of ligands with individual amino acids [57–59]. This allows the identification of regions where the ligand interacts very well with the protein and regions, where additional binding energy can theoretically be gained. In cases where the binding mode of the ligand to the target is unknown, docking procedures have to be employed to predict the binding geometry, but conventional methods have severe shortcomings [60] which can be partly overcome by QM [61]. Huge progress has been made by employing MD methods, especially Markov State models to predict ligand binding to a target [62]. All such approaches support the optimization of binding affinity, but this is only *one* parameter, which has to be optimized during drug development. The ADME properties also have to be in an acceptable range, and to optimize these, machine learning is very often applied. Recently neuronal networks encounter a renaissance and basically every endpoint in the ADME field has been predicted by deep learning networks [63]. Machine learning is a branch of artificial intelligence (AI) methods, which are currently highly popular in drug design [64]. AI in drug design covers many conceptually highly different methods, which have in common that a network or more general a machine gets trained on experimental data and makes predictions on new molecules. With every experimental data point, the quality of the model improves and therefore

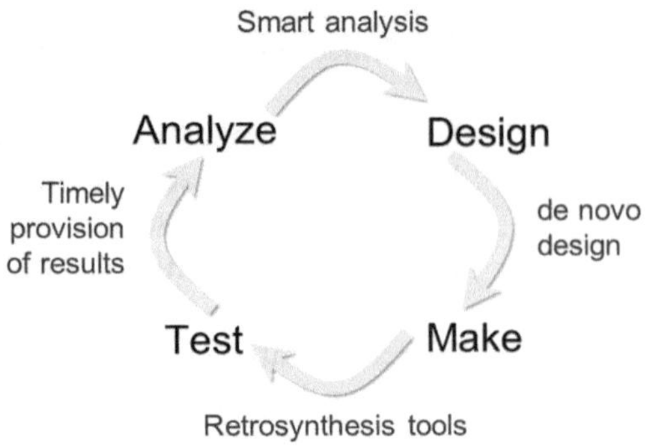

Fig. 2 The design-make-test-analyze cycle and the contributions of AI and virtual assistants

a large set of experiments to start with is desirable. What makes AI methods so special is the ability not only to predict properties of molecules but also to suggest new molecules themselves, which are a product of a training on existing molecules [65]. This has fundamental impact on the design-make-test-analyze (DMTA) cycle, because already the design may be massively supported by AI, as shown in Fig. 2. Recently Vidler and Baumgartner have described a DMTA cycle as they implemented it at Lilly [66]. On one hand, computational methods have been employed to analyze data and to provide the new test data timely to the medicinal chemists. On the other hand, they also have implemented a basic AI method to design new molecules based on known SAR and recent test data. In general, the use of virtual assistants to provide and analyze data is well established, and it enables the quick and thorough understanding of SAR. Also for design, virtual assistants are valuable to predict various endpoints of newly designed molecules. The use of AI in the design phase of the cycle is relatively new, but extremely powerful. Based on various concepts (ranging from pure matched molecular pair analysis to generative networks), a large number of new molecules is generated and the properties (including potency on the target of interest) are predicted by machine learning [67]. By automated rules (or human intervention), the best molecules are selected for synthesis and also here AI plays a major role. Recently a network-based method has been introduced that supports retrosynthesis of molecules [68]. Thus, AI also heavily supports the make step in the DMTA cycle. Last, but not least virtual assistants help to decide which assays should be tested for every molecule and which compounds have failed at some point in the testing cascade. To sum up, the enhancement of the DMTA cycle by virtual assistants and AI (*see* Fig. 2) leads to a general speedup of research projects. This is caused by shortening cycle times and reducing the number of compounds necessary for successfully driving a project. The most crucial challenge is to get potency and ADME predictions of high quality. Quite surprisingly, also in this case QM methods play an important role. Several studies demonstrate an increase in quality when adding QM descriptors in the machine learning models [69, 70], especially if it comes to prediction of reactivity [52]. A more thorough overview of the application of QM methods in drug discovery is provided in Table 1 as well as in several other reviews [71, 72].

Table 1
Typical problems in drug design which benefit from application of QM methods

Category	Problem	Methods	Benefits through QM	Example references
Ligand/target interactions	Improvement of docking scores	QM on entire proteins, QM/MM, FMO	Improved pose ranking, better enrichments	[61, 73]
	Enhance free energy calculations	QM/PBSA, QM/TI, LIE approaches	Higher accuracy by including QM energies	[56, 74, 75]
	Interpretation of ligand/target interactions to guide ligand optimization	FMO and energy decomposition approaches	Not possible by force fields	[57, 76]
	Investigation of isolated ligand/target interactions (isolated hydrogen bonds, π-stacking, cation/π)	QM treatment of model systems	Higher accuracy through QM, capture electron correlation effects	[50, 77–79]
	Optimization of ligand/target geometries, e.g., for X-ray refinement	QM/MM calculations and linear scaling QM	Correct descriptions of complex interactions	[80]
	Decomposition of interaction energies into single contributions	Energy decomposition analyses with linear scaling DFT or FMO	Not possible by force fields	[58, 81]
	Elucidation of ligand and protein flexibility	QM/MM MD simulations	Breaking and forming of bonds possible	[82]
Protein/protein interactions	Identification of hotspots	FMO method to estimate local energetics	Improved accuracy on interaction energetics	[83]
Ligand based methods	Determination of bioactive conformations	QM optimization of conformational ensembles	Boltzmann-weighted ensemble based on accurate energies	[84, 85]
	Torsional angle energy profiles	Constrained QM optimizations	Improved accuracy on energy profiles	[86]
QSAR/AI	ADME(T) predictions	QM descriptors to improve prediction quality (often semiempirical)	Improved model performance, especially for toxicity endpoints	[69, 70, 87–89]

(continued)

Table 1 (continued)

Category	Problem	Methods	Benefits through QM	Example references
Reaction mechanisms	Investigation of enzymatic reactions	QM/MM or FMO and reaction path following	Not possible with force fields	[90, 91]
Covalent ligand design	Reactivity predictions	Warhead reactivity prediction and transition state finding	Not possible with force fields	[52, 53, 92]
Spectroscopic methods	Identification of chemical species	Prediction of IR, VCD, and NMR spectra	Not possible with force fields—Important for reaction product identification	[93, 94]

5 Summary

This chapter has given a very condensed overview on the current challenges in drug discovery. From a high-level point of view, drug design is a highly risky and extremely expensive endeavor, where the commercial potential for newly introduced drugs decreases on average every year. As a risk mitigation strategy, new therapeutic modalities (e.g., biologics) have been successfully introduced which allow targeting proteins that are not drugable by small molecules. However, the quest for small molecules as new drugs is still ongoing due to the advantage of oral bioavailability, brain and cell penetration, and comparable low cost of goods in production. The research phase in drug design is currently revolutionized by the use of virtual assistants and artificial intelligence. These systems ensure a comprehensive view on existing SAR, analyze trends, and even may enhance the design of the next round of new molecules. In addition, the synthesis planning is supported by retrosynthesis tools. This all allows the project team to shorten timelines in the design-make-test-analyze cycle significantly and thereby to reduce the costs in the research phase of a drug design program.

From a molecular modeling standpoint, there has been a constant evolution of new methods in small-molecule drug design over the last years. Free energy calculations have become state of the art and are applied on a regular basis to prioritize compounds before synthesis. QM methods are commonly used for a more thorough analysis of protein/ligand interactions and allow the identification of structural features for improvement on the ligand. Last but not least, several strategies have evolved to target difficult pockets:

Covalent inhibitors are developed where the free energy of binding to a difficult pocket or a selectivity issue requires enhancement of binding by the formation of a covalent bond to the target. On the other hand, allosteric pockets can be identified and targeted, thereby avoiding to address the orthosteric site. This has especially advantages for gaining selectivity to phylogenetic neighbors.

Beyond that, new target classes (such as RNA) are emerging, where all the design principles need to be amended. However, in the end the existing and the upcoming powerful design methods will enable a steady flow of new drug approvals to tackle unmet medical need.

References

1. DiMasi JA, Grabowski HG, Hansen RW (2016) Innovation in the pharmaceutical industry: new estimates of R&D costs. J Health Econ 47:20–33
2. Sertkaya A, Wong H-H, Jessup A, Beleche T (2016) Key cost drivers of pharmaceutical clinical trials in the United States. Clin Trials 13 (2):117–126
3. Waring MJ, Arrowsmith J, Leach AR, Leeson PD, Mandrell S, Owen RM, Pairaudeau G, Pennie WD, Pickett SD, Wang J, Wallace O, Weir A (2015) An analysis of the attrition of drug candidates from four major pharmaceutical companies. Nat Rev Drug Discov 14:475
4. Cook D, Brown D, Alexander R, March R, Morgan P, Satterthwaite G, Pangalos MN (2014) Lessons learned from the fate of AstraZeneca's drug pipeline: a five-dimensional framework. Nat Rev Drug Discov 13:419
5. Mullard A (2019) 2018 FDA drug approvals. Nat Rev Drug Discov 18(2):85–89
6. Dehmer SP, Pardey PG, Beddow JM, Chai Y (2019) Reshuffling the global R&D deck, 1980–2050. PLoS One 14(3):e0213801
7. Mullard A (2017) 2016 FDA drug approvals. Nat Rev Drug Discov 16:73
8. Mullard A (2018) 2017 FDA drug approvals. Nat Rev Drug Discov 17:81
9. Mariz S, Reese JH, Westermark K, Greene L, Goto T, Hoshino T, Llinares-Garcia J, Sepodes B (2016) Worldwide collaboration for orphan drug designation. Nat Rev Drug Discov 15:440
10. Harrison C (2010) Patent watch: the patent cliff steepens. Nat Rev Drug Discov 10:12
11. Remus JL, Dantzer R (2016) Inflammation models of depression in rodents: relevance to psychotropic drug discovery. Int J Neuropsychopharmacol 19(9):pyw028
12. Manolio TA (2010) Genomewide association studies and assessment of the risk of disease. New Engl J Med 363(2):166–176
13. Lipinski CA, Lombardo F, Dominy BW, Feeney PJ (2001) Experimental and computational approaches to estimate solubility and permeability in drug discovery and development settings. Adv Drug Deliv Rev 46(1):3–26
14. McGrath NA, Brichacek M, Njardarson JT (2010) A graphical journey of innovative organic architectures that have improved our lives. J Chem Educ 87(12):1348–1349
15. Urits I, Jones MR, Gress K, Charipova K, Fiocchi J, Kaye AD, Viswanath O (2019) CGRP antagonists for the treatment of chronic migraines: a comprehensive review. Curr Pain Headache Rep 23(5):29
16. Oprea TI, Bologa CG, Brunak S, Campbell A, Gan GN, Gaulton A, Gomez SM, Guha R, Hersey A, Holmes J, Jadhav A, Jensen LJ, Johnson GL, Karlson A, Leach AR, Ma'ayan A, Malovannaya A, Mani S, Mathias SL, McManus MT, Meehan TF, von Mering C, Muthas D, Nguyen D-T, Overington JP, Papadatos G, Qin J, Reich C, Roth BL, Schürer SC, Simeonov A, Sklar LA, Southall N, Tomita S, Tudose I, Ursu O, Vidović D, Waller A, Westergaard D, Yang JJ, Zahoránszky-Köhalmi G (2018) Unexplored therapeutic opportunities in the human genome. Nat Rev Drug Discov 17:317
17. Wang J, Li S, Li H (2019) Drug design of "Undruggable" targets. Chin. J Chem 37 (5):501–512
18. Toure M, Crews CM (2016) Small-molecule PROTACS: new approaches to protein degradation. Angew Chem Int Ed 55(6):1966–1973
19. Schnorrenberg G (2019) New trends in drug discovery. In: Fischer J, Klein C, Childers WE (eds) Successful drug discovery, vol 3, pp 1–39

20. Michino M, Beuming T, Donthamsetti P, Newman AH, Javitch JA, Shi L (2015) What can crystal structures of aminergic receptors tell us about designing subtype-selective ligands? Pharmacol Rev 67(1):198–213
21. Vass M, Podlewska S, de Esch IJP, Bojarski AJ, Leurs R, Kooistra AJ, de Graaf C (2019) Aminergic GPCR–ligand interactions: a chemical and structural map of receptor mutation data. J Med Chem 62(8):3784–3839
22. Hauser AS, Attwood MM, Rask-Andersen M, Schiöth HB, Gloriam DE (2017) Trends in GPCR drug discovery: new agents, targets and indications. Nat Rev Drug Discov 16:829
23. Ferguson FM, Gray NS (2018) Kinase inhibitors: the road ahead. Nat Rev Drug Discov 17 (5):353–377
24. Reddy AS, Zhang S (2013) Polypharmacology: drug discovery for the future. Expert Rev Clin Pharmacol 6(1):41–47
25. Barr AJ, Ugochukwu E, Lee WH, King ONF, Filippakopoulos P, Alfano I, Savitsky P, Burgess-Brown NA, Müller S, Knapp S (2009) Large-scale structural analysis of the classical human protein tyrosine phosphatome. Cell 136(2):352–363
26. Wold EA, Chen J, Cunningham KA, Zhou J (2019) Allosteric modulation of class A GPCRs: targets, agents, and emerging concepts. J Med Chem 62(1):88–127
27. Tautermann CS, Binder F, Büttner FH, Eickmeier C, Fiegen D, Gross U, Grundl MA, Heilker R, Hobson S, Hoerer S, Luippold A, Mack V, Montel F, Peters S, Bhattacharya S, Vaidehi N, Schnapp G, Thamm S, Zeeb M (2019) Allosteric activation of striatal-enriched protein tyrosine phosphatase (STEP, PTPN5) by a fragment-like molecule. J Med Chem 62 (1):306–316
28. Wenthur CJ, Gentry PR, Mathews TP, Lindsley CW (2014) Drugs for allosteric sites on receptors. Annu Rev Pharmacol Toxicol 54 (1):165–184
29. Changeux J-P, Christopoulos A (2017) Allosteric modulation as a unifying mechanism for receptor function and regulation. Diabetes Obes Metab 19(S1):4–21
30. Betz M, Wulsdorf T, Krimmer SG, Klebe G (2016) Impact of surface water layers on protein–ligand binding: how well are experimental data reproduced by molecular dynamics simulations in a thermolysin test case? J Chem Inf Model 56(1):223–233
31. Robinson D, Bertrand T, Carry J-C, Halley F, Karlsson A, Mathieu M, Minoux H, Perrin M-A, Robert B, Schio L, Sherman W (2016) Differential water thermodynamics determine PI3K-beta/delta selectivity for solvent-exposed ligand modifications. J Chem Inf Model 56(5):886–894
32. Schiebel J, Gaspari R, Wulsdorf T, Ngo K, Sohn C, Schrader TE, Cavalli A, Ostermann A, Heine A, Klebe G (2018) Intriguing role of water in protein-ligand binding studied by neutron crystallography on trypsin complexes. Nat Commun 9(1):3559
33. Christ CD, Fox T (2014) Accuracy assessment and automation of free energy calculations for drug design. J Chem Inf Model 54 (1):108–120
34. Ruan H, Sun Q, Zhang W, Liu Y, Lai L (2019) Targeting intrinsically disordered proteins at the edge of chaos. Drug Discov Today 24 (1):217–227
35. Neira JL, Bintz J, Arruebo M, Rizzuti B, Bonacci T, Vega S, Lanas A, Velázquez-Campoy A, Iovanna JL, Abián O (2017) Identification of a drug targeting an intrinsically disordered protein involved in pancreatic adenocarcinoma. Sci Rep 7:39732
36. Antunes DA, Devaurs D, Kavraki LE (2015) Understanding the challenges of protein flexibility in drug design. Expert Opin Drug Discovery 10(12):1301–1313
37. Buonfiglio R, Recanatini M, Masetti M (2015) Protein flexibility in drug discovery: from theory to computation. ChemMedChem 10 (7):1141–1148
38. Amaral M, Kokh DB, Bomke J, Wegener A, Buchstaller HP, Eggenweiler HM, Matias P, Sirrenberg C, Wade RC, Frech M (2017) Protein conformational flexibility modulates kinetics and thermodynamics of drug binding. Nat Commun 8(1):2276
39. Ustach VD, Lakkaraju SK, Jo S, Yu W, Jiang W, MacKerell AD (2019) Optimization and evaluation of site-identification by ligand competitive saturation (SILCS) as a tool for target-based ligand optimization. J Chem Inf Model 59(6):3018–3035
40. Gurova K (2009) New hopes from old drugs: revisiting DNA-binding small molecules as anticancer agents. Future Oncol 5 (10):1685–1704
41. Hossain M, Giri P, Kumar GS (2008) DNA intercalation by Quinacrine and methylene blue: a comparative binding and thermodynamic characterization study. DNA Cell Biol 27(2):81–90
42. Connelly Colleen M, Moon Michelle H, Schneekloth John S Jr (2016) The emerging role of RNA as a therapeutic target for small molecules. Cell Chem Biol 23(9):1077–1090

43. Breaker Ronald R, Joyce Gerald F (2014) The expanding view of RNA and DNA function. Chem Biol 21(9):1059–1065
44. Disney MD (2019) Targeting RNA with small molecules to capture opportunities at the intersection of chemistry, biology, and medicine. J Am Chem Soc 141(17):6776–6790
45. Arkin Michelle R, Tang Y, Wells James A (2014) Small-molecule inhibitors of protein--protein interactions: progressing toward the reality. Chem Biol 21(9):1102–1114
46. Guo W, Wisniewski JA, Ji H (2014) Hot spot-based design of small-molecule inhibitors for protein–protein interactions. Bioorg Med Chem Lett 24(11):2546–2554
47. Tang Z, Chang C-eA (2018) Binding thermodynamics and kinetics calculations using chemical host and guest: a comprehensive picture of molecular recognition. J Chem Theory Comput 14(1):303–318
48. John Kenneth M, Shuxing Z (2012) Computational prediction of protein hot spot residues. Curr Pharm Des 18(9):1255–1265
49. Daze KD, Hof F (2013) The Cation–π interaction at protein–protein interaction interfaces: developing and learning from synthetic mimics of proteins that bind methylated Lysines. Acc Chem Res 46(4):937–945
50. Pinheiro S, Soteras I, Gelpí JL, Dehez F, Chipot C, Luque FJ, Curutchet C (2017) Structural and energetic study of cation–π–cation interactions in proteins. Phys Chem Chem Phys 19(15):9849–9861
51. Lonsdale R, Ward RA (2018) Structure-based design of targeted covalent inhibitors. Chem Soc Rev 47(11):3816–3830
52. Palazzesi F, Grundl MA, Pautsch A, Weber A, Tautermann CS (2019) A fast Ab initio predictor tool for covalent reactivity estimation of acrylamides. J Chem Inf Model 59 (8):3565–3571
53. Flanagan ME, Abramite JA, Anderson DP, Aulabaugh A, Dahal UP, Gilbert AM, Li C, Montgomery J, Oppenheimer SR, Ryder T, Schuff BP, Uccello DP, Walker GS, Wu Y, Brown MF, Chen JM, Hayward MM, Noe MC, Obach RS, Philippe L, Shanmugasundaram V, Shapiro MJ, Starr J, Stroh J, Che Y (2014) Chemical and computational methods for the characterization of covalent reactive groups for the prospective design of Irreversible Inhibitors. J Med Chem 57(23):10072–10079
54. Salamoun JM, Wipf P (2016) Allosteric modulation of phosphatase activity may redefine therapeutic value. J Med Chem 59 (17):7771–7772
55. Vaidehi N, Bhattacharya S (2016) Allosteric communication pipelines in G-protein-coupled receptors. Curr Opin Pharmacol 30:76–83
56. Fox SJ, Dziedzic J, Fox T, Tautermann CS, Skylaris C-K (2014) Density functional theory calculations on entire proteins for free energies of binding: application to a model polar binding site. Proteins: struct. Funct Bioinform 82 (12):3335–3346
57. Heifetz A, Trani G, Aldeghi M, MacKinnon CH, McEwan PA, Brookfield FA, Chudyk EI, Bodkin M, Pei Z, Burch JD, Ortwine DF (2016) Fragment molecular orbital method applied to Lead optimization of novel Interleukin-2 inducible T-cell kinase (ITK) inhibitors. J Med Chem 59(9):4352–4363
58. Morao I, Fedorov DG, Robinson R, Southey M, Townsend-Nicholson A, Bodkin MJ, Heifetz A (2017) Rapid and accurate assessment of GPCR–ligand interactions using the fragment molecular orbital-based density-functional tight-binding method. J Comput Chem 38(23):1987–1990
59. Choi J, Kim H-J, Jin X, Lim H, Kim S, Roh I-S, Kang H-E, No KT, Sohn H-J (2018) Application of the fragment molecular orbital method to discover novel natural products for prion disease. Sci Rep 8(1):13063
60. Pantsar T, Poso A (2018) Binding affinity via docking: fact and fiction. Molecules 23 (8):1899
61. Crespo A, Rodriguez-Granillo A, Lim VT (2017) Quantum-mechanics methodologies in drug discovery: applications of docking and scoring in lead optimization. Curr Top Med Chem 17(23):2663–2680
62. Chodera JD, Noé F (2014) Markov state models of biomolecular conformational dynamics. Curr Opin Struct Biol 25:135–144
63. Wenzel J, Matter H, Schmidt F (2019) Predictive multitask deep neural network models for ADME-Tox properties: learning from large data sets. J Chem Inf Model 59(3):1253–1268
64. Hessler G, Baringhaus K-H (2018) Artificial Intelligence in Drug Design. Molecules 23 (10):2520
65. Skalic M, Jiménez J, Sabbadin D, De Fabritiis G (2019) Shape-based generative modeling for de novo drug design. J Chem Inf Model 59 (3):1205–1214
66. Vidler LR, Baumgartner MP (2019) Creating a virtual assistant for medicinal chemistry. ACS Med Chem Lett 10(7):1051–1055
67. Kadurin A, Nikolenko S, Khrabrov K, Aliper A, Zhavoronkov A (2017) druGAN: an advanced generative adversarial autoencoder model for de novo generation of new molecules with

desired molecular properties in Silico. Mol Pharm 14(9):3098–3104

68. Segler MHS, Preuss M, Waller MP (2018) Planning chemical syntheses with deep neural networks and symbolic AI. Nature 555:604
69. El Kerdawy A, Güssregen S, Matter H, Hennemann M, Clark T (2013) Quantum mechanics-based properties for 3D-QSAR. J Chem Inf Model 53(6):1486–1502
70. Hennemann M, Friedl A, Lobell M, Keldenich J, Hillisch A, Clark T, Göller AH (2009) CypScore: quantitative prediction of reactivity toward cytochromes P450 based on Semiempirical molecular orbital theory. ChemMedChem 4(4):657–669
71. Arodola OA, Soliman MES (2017) Quantum mechanics implementation in drug-design workflows: does it really help? Drug Des Devel Ther 11:2551–2564
72. Cavasotto CN, Adler NS, Aucar MG (2018) Quantum chemical approaches in structure-based virtual screening and Lead optimization. Front Chem 6:188
73. Raha K, Merz KM (2005) Large-scale validation of a quantum mechanics based scoring function: predicting the binding affinity and the binding mode of a diverse set of protein–ligand complexes. J Med Chem 48 (14):4558–4575
74. Ryde U, Söderhjelm P (2016) Ligand-binding affinity estimates supported by quantum-mechanical methods. Chem Rev 116 (9):5520–5566
75. Cavasotto CN, Aucar MG, Adler NS (2019) Computational chemistry in drug lead discovery and design. Int J Quantum Chem 119(2): e25678
76. Chudyk EI, Sarrat L, Aldeghi M, Fedorov DG, Bodkin MJ, James T, Southey M, Robinson R, Morao I, Heifetz A (2018) Exploring GPCR-ligand interactions with the fragment molecular orbital (FMO) method. In: Heifetz A (ed) Computational methods for GPCR drug discovery. Springer New York, New York, NY, pp 179–195. https://doi.org/10.1007/978-1-4939-7465-8_8
77. Bootsma AN, Doney AC, Wheeler SE (2019) Predicting the strength of stacking interactions between Heterocycles and aromatic amino acid side chains. J Am Chem Soc 141 (28):11027–11035
78. Huber RG, Margreiter MA, Fuchs JE, von Grafenstein S, Tautermann CS, Liedl KR, Fox T (2014) Heteroaromatic π-stacking energy landscapes. J Chem Inf Model 54 (5):1371–1379
79. El Kerdawy A, Tautermann CS, Clark T, Fox T (2013) Economical and accurate protocol for calculating hydrogen-bond-acceptor strengths. J Chem Inf Model 53(12):3262–3272
80. Zheng M, Reimers JR, Waller MP, Afonine PV (2017) Q|R: quantum-based refinement. Acta Crystallogr Sect D 73(1):45–52
81. Phipps MJS, Fox T, Tautermann CS, Skylaris C-K (2015) Energy decomposition analysis approaches and their evaluation on prototypical protein–drug interaction patterns. Chem Soc Rev 44(10):3177–3211
82. Barbault F, Maurel F (2015) Simulation with quantum mechanics/molecular mechanics for drug discovery. Expert Opin Drug Discovery 10(10):1047–1057
83. Lim H, Jin X, Kim J, Hwang S, Shin KB, Choi J, No KT (2019) Investigation of hot spot region in XIAP inhibitor binding site by fragment molecular orbital method. Comput Struct Biotechnol J 17:1217–1225
84. Butler KT, Luque FJ, Barril X (2009) Toward accurate relative energy predictions of the bioactive conformation of drugs. J Comput Chem 30(4):601–610
85. Kanal IY, Keith JA, Hutchison GR (2018) A sobering assessment of small-molecule force field methods for low energy conformer predictions. Int J Quantum Chem 118(5):e25512
86. Hao M-H, Haq O, Muegge I (2007) Torsion angle preference and energetics of small-molecule ligands bound to proteins. J Chem Inf Model 47(6):2242–2252
87. Reenu V (2015) Exploring the role of quantum chemical descriptors in modeling acute toxicity of diverse chemicals to *Daphnia magna*. J Mol Graph Model 61:89–101
88. Choi J-S, Ha MK, Trinh TX, Yoon TH, Byun H-G (2018) Towards a generalized toxicity prediction model for oxide nanomaterials using integrated data from different sources. Sci Rep 8(1):6110
89. Olsen L, Montefiori M, Tran KP, Jørgensen FS (2019) SMARTCyp 3.0: enhanced cytochrome P450 site-of-metabolism prediction server. Bioinformatics 35(17):3174–3175
90. Lonsdale R, Harvey JN, Mulholland AJ (2012) A practical guide to modelling enzyme-catalysed reactions. Chem Soc Rev 41 (8):3025–3038
91. Steinmann C, Fedorov DG, Jensen JH (2013) Mapping enzymatic catalysis using the effective fragment molecular orbital method: towards all ab initio biochemistry. PLoS One 8(4):e60602
92. Lonsdale R, Burgess J, Colclough N, Davies NL, Lenz EM, Orton AL, Ward RA (2017) Expanding the armory: predicting and tuning covalent warhead reactivity. J Chem Inf Model 57(12):3124–3137

93. Merten C, Golub TP, Kreienborg NM (2019) Absolute configurations of synthetic molecular scaffolds from vibrational CD spectroscopy. J Org Chem 84(14):8797–8814

94. Xin D, Sader CA, Chaudhary O, Jones P-J, Wagner K, Tautermann CS, Yang Z, Busacca CA, Saraceno RA, Fandrick KR, Gonnella NC, Horspool K, Hansen G, Senanayake CH (2017) Development of a 13C NMR chemical shift prediction procedure using B3LYP/cc-pVDZ and empirically derived systematic error correction terms: a computational small molecule structure elucidation method. J Org Chem 82(10):5135–5145

Chapter 2

QM Implementation in Drug Design: Does It Really Help?

Jinfeng Liu and Xiao He

Abstract

Computational chemistry allows one to characterize the structure, dynamics, and energetics of protein–ligand interactions, which makes it a valuable tool in drug discovery in both academic research and pharmaceutical industry. Molecular mechanics (MM)-based approaches are widely utilized to assist the discovery of new drug candidates. However, the complexity of protein–ligand interactions challenges the accuracy and efficiency of the commonly used empirical methods. Aiming to provide better accuracy in the description of protein–ligand interactions, quantum mechanics (QM)-based approaches are becoming increasingly explored. In principle, QM calculation includes all contributions to the energy, accounting for terms usually missing in empirical force fields, and provides a greater degree of transferability. The usefulness of QM in drug design cannot be overemphasized. In this chapter, we present recent developments and applications of fragment-based QM method in studying the protein–ligand and protein–protein interactions. We critically discuss the performance of the fragment-based QM method at different ab initio levels while trying to answer a critical question: do QM-based methods really help in drug design?

Key words Drug design, Protein–ligand interaction, Protein–protein interaction, Binding affinity, Quantum mechanics, EE-GMFCC

1 Introduction

The pharmacological activity of the drug is ultimately due to the spatial arrangement and electronic nature of its atoms and the way these atoms interact with the target biomacromolecule [1, 2]. Computational chemistry, which allows one to characterize the structure, dynamics, and energetics of such interactions, is a well-established and valuable tool in the drug discovery process [3–5]. Many molecular mechanics (MM)-based computational tools are developed to efficiently assist the discovery of new drug candidates, and these computationally inexpensive methods are nowadays routinely used in drug design [2, 6–19]. However, the complexity of protein–ligand interactions challenges the accuracy and efficiency of these commonly used empirical methods [20–22]. Empirical force fields usually adopt the mean field approximation and lack many important effects, such as electronic

Alexander Heifetz (ed.), *Quantum Mechanics in Drug Discovery*, Methods in Molecular Biology, vol. 2114,
https://doi.org/10.1007/978-1-0716-0282-9_2,

polarization, charge transfer, and halogen bonding [23, 24]. This empirical approximation is known to be problematic when studying protein–ligand interaction [23, 25–30].

Quantum mechanics (QM) accounts for all contributions to the energy and thus is, in principle, theoretically approaching to the exact result [5, 31–35]. Moreover, it provides a greater degree of transferability, avoiding system-dependent parameterizations [5, 33, 36]. Moreover, the first principles nature of QM allows it to systematically improve the accuracy of the description of the interactions between molecules [34, 37, 38]. Therefore, there has been a great interest in improving ligand binding estimates by using QM methods. A series of reviews have been published on the use of QM methods for the calculation of protein–ligand binding affinities [2, 34, 37, 39–42]. However, the practical use of conventional QM calculation in drug design applications is still to be realized due, in large part, to computational limitations. Accurate and efficient QM calculation on the macromolecule presents a grand challenge in computational chemistry.

In the past decade, considerable effort has been made to the development of various linear-scaling and/or fragmentation methods to extend the applicability of QM on large molecular systems [43–49]. Due to the computational efficiency and easy implementation, the fragmentation QM methods have attracted significant interest, and a range of fragmentation QM methods for large systems have been proposed [38, 43–45, 50–63]. According to the fragmentation scheme, the macromolecule can be divided into a series of subsystems (fragments), which could be feasibly calculated by QM at diverse ab initio levels. Then molecular properties of the whole system can be obtained by taking a proper linear combination of the corresponding terms of individual fragments.

The electrostatically embedded generalized molecular fractionation with conjugate caps (EE-GMFCC) method was a fragment-based approach, which was developed in our group for accurate calculation of the protein energy [43, 64]. The accuracy and efficiency of EE-GMFCC have been validated through numerical tests by comparison with the corresponding full-system QM calculations [64–68]. Recently, the EE-GMFCC method has been successfully utilized for protein structure optimization [69], ab initio molecular dynamics simulation [70–73], and protein–ligand and protein–protein binding calculations [24, 74]. In this chapter, we mainly describe the use of EE-GMFCC method for the calculation of protein–ligand and protein–protein binding affinities. The performance of this fragment-based QM method at different ab initio levels is presented. The necessity and validity of application of QM method in drug design are discussed, and we are trying to give some concluding remarks in response to the question: do QM-based methods really help in drug design?

2 Theory

In this section, we give a brief review of the EE-GMFCC method used when calculating protein–ligand and protein–protein binding affinities. For a more detailed description of the method, the interested reader is referred to our previously published articles [24, 64, 74].

2.1 The EE-GMFCC Method

The illustration of the fragmentation scheme in EE-GMFCC is shown in Fig. 1. A protein with N residues is divided into N-2 individual fragments by cutting through the peptide bond (excluding the first and the last peptide bonds), and a pair of conjugate caps is used to saturate each fragment to protect its chemical environment (*see* Fig. 1a). Hydrogen atoms are added to terminate the molecular caps to avoid dangling bonds. Each individual fragment can be calculated by QM independently. For the non-neighboring residues but spatially in close contact (the minimal distance between two residues falls within a defined distance threshold λ) (*see* Fig. 1b), we also calculated these two-body interactions by QM explicitly. All of the fragment calculations are embedded in the electrostatic field of the point charges representing the remaining fragments of the protein. The total energy of protein (or protein–protein complex) can then be expressed by the following formula: [24].

$$
\begin{aligned}
E_{\text{EE-GMFCC}} = & \sum_{i=2}^{N-1} \widetilde{E}\left(\text{Cap}_{i-1}^{*} A_i \text{Cap}_{i+1}\right) \\
& - \sum_{i=2}^{N-2} \widetilde{E}\left(\text{Cap}_i^{*} \text{Cap}_{i+1}\right) \\
& + \sum_{\substack{i,j > i+2 \\ |R_i - R_j| \le \lambda}} \left(\widetilde{E}_{ij} - \widetilde{E}_i - \widetilde{E}_j\right)_{\text{QM}} - E_{\text{DC}} \qquad (1)
\end{aligned}
$$

where $\widetilde{E}\left(\text{Cap}_{i-1}^{*} A_i \text{Cap}_{i+1}\right)$ represents the sum of the self-energy of fragment (the ith residue A_i capped with a left cap Cap_{i-1}^{*} and a right cap Cap_{i+1}) and the interaction energy between fragment and background charges. $\widetilde{E}\left(\text{Cap}_i^{*} \text{Cap}_{i+1}\right)$ is the self-energy of concap along with the interaction energy between the concap and background charges, which needs to be deducted. The third term is the two-body QM interaction between two non-neighboring residues (i and j) in close contact (within the distance threshold λ). Because of the electrostatic embedding scheme, the interactions between distant non-neighboring residues are doubly counted and thus are approximately deducted by Coulomb interactions in the last term E_{DC}. In our previous studies [24, 64, 74], the Amber94

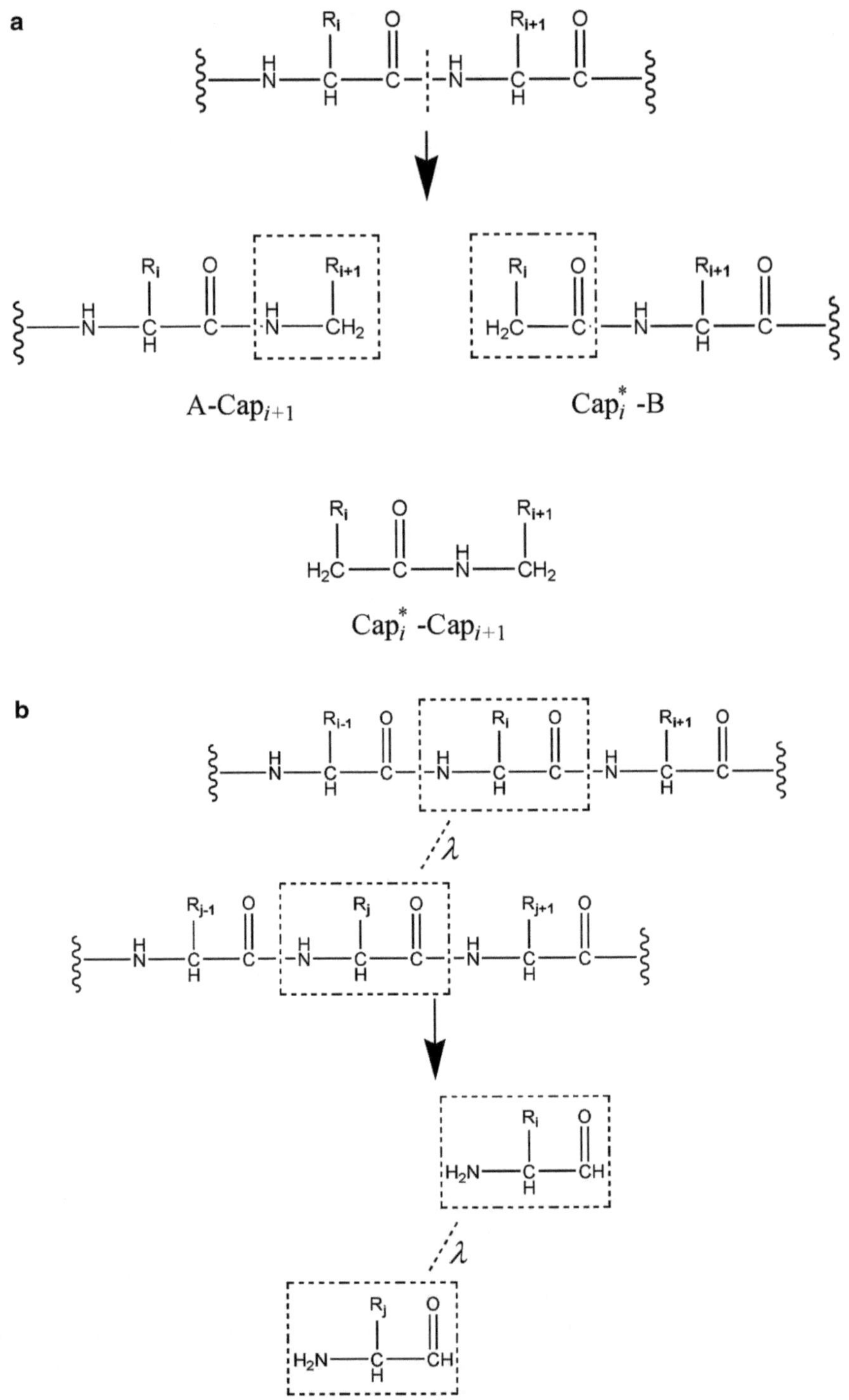

Fig. 1 (**a**) EE-GMFCC scheme in which the peptide bond is cut and the fragments are capped with Cap_{i+1} and its conjugate Cap_i^*, where subscript *i* denotes the *i*th amino acid in the given protein. The concap is defined as the fused molecular species of $\mathrm{Cap}_i^* - \mathrm{Cap}_{i+1}$. (**b**) The generalized concap (Gconcap) scheme and the atomic structure of the Gconcap. (Liu, J. F.; Wang, X. W.; Zhang, J. Z. H.; He, X. Calculation of protein-ligand binding affinities based on a fragment quantum mechanical method. *Rsc Adv.* **2015**, *5*, 107020–107030—Adapted by permission of The Royal Society of Chemistry)

charge model [75] is validated to be a good choice for describing the embedding field.

In a similar way, the total energy of the protein-ligand binding complex can be derived using the following formula: [24].

$$
\begin{aligned}
E_{\text{EE-GMFCC}} = & \sum_{i=2}^{N-1} \widetilde{E}(\text{Cap}_{i-1}^{*} A_i \text{Cap}_{i+1}) - \sum_{i=2}^{N-2} \widetilde{E}(\text{Cap}_{i}^{*} \text{Cap}_{i+1}) \\
& + \sum_{\substack{i,j > i+2 \\ |R_i - R_j| \leq \lambda}} (\widetilde{E}_{ij} - \widetilde{E}_i - \widetilde{E}_j)_{\text{QM}} \\
& + \widetilde{E}(\text{lig}) + \sum_{|R_{\text{lig}} - R_i| \leq \lambda} (\widetilde{E}_{\text{lig},i} - \widetilde{E}_{\text{lig}} - \widetilde{E}_i) - E_{\text{DC}}
\end{aligned}
\tag{2}
$$

where the ligand is treated as one fragment and $\widetilde{E}(\text{lig})$ represents the self-energy of ligand and the interaction energy between the ligand and its environment represented by the background charges. The two-body QM interactions between the ligand and its nearby residues, namely, $\widetilde{E}_{\text{lig},i} - \widetilde{E}_{\text{lig}} - \widetilde{E}_i$, are also included.

2.2 The EE-GMFCC-CPCM Method

The conductor-like polarized continuum model (CPCM) [76] is one of the most widely used solvent models in QM calculation. The EE-GMFCC-CPCM method was developed for ab initio calculation of the electrostatic solvation energy of proteins by combining the EE-GMFCC and CPCM methods [65]. Normally, the electrostatic solvation energy (G(ele)) can be expressed as the sum of the wave function distortion energy (E(wfd)) and the electrostatic solute-solvent interaction energy (G(es)):

$$G(\text{ele}) = E(\text{wfd}) + G(\text{es}) \tag{3}$$

Based on the EE-GMFCC approach, the E(wfd) can be approximately obtained through

$$E(\text{wfd}) = \sum_{k=1}^{N_{\text{F}}} \Delta\widetilde{E}_k - \sum_{k=1}^{N_{\text{C}}} \Delta\widetilde{E}_k^{c} + \sum_{k=1}^{N_{\text{GC}}} \Delta\left(\widetilde{E}_k^{ij} - \widetilde{E}_k^{i} - \widetilde{E}_k^{j}\right) \tag{4}$$

where $\Delta\widetilde{E}$ represents the difference of the fragment energy (including the self-energy of each fragment and the interaction energy between the fragment and its corresponding background charges) in gas phase and solution phase. N_{F}, N_{C}, and N_{GC} denote the number of capped fragments (including the single ligand), concaps, and Gconcaps, respectively. Owing to the used fixed charge model, the doubly counted charge–charge interaction energy (the last term in Eqs. 1 and 2) is exactly canceled out between the gas phase and solution phase in the current EE-GMFCC-CPCM calculations.

The electrostatic solute–solvent interaction energy $G(\mathrm{es})$ is given by.

$$G(\mathrm{es}) = \frac{1}{2}\sum_{\mu} q_{\mu}\left[\sum_{\alpha}\frac{Z_{\alpha}}{|r_{\mu} - \mathbf{R}_{\alpha}|} - \int\frac{\rho(\mathbf{r})}{|\mathbf{r}_{\mu} - \mathbf{r}|}\mathrm{d}\mathbf{r}\right] = \frac{1}{2}\sum_{\mu} q_{\mu}\varphi(\mathbf{r}_{\mu}) \quad (5)$$

where q_{μ} and $\mathbf{r}_{\mu}$ represent the induced surface charges and their positions, respectively, which are determined through the CPCM method. Z_{α} and $\mathbf{R}_{\alpha}$ are the nuclear charges and their corresponding coordinates. $\rho(\mathbf{r})$ is the electron density at position $\mathbf{r}$. $\varphi(\mathbf{r}_{\mu})$ is the electrostatic potential on cavity surface site μ generated by both the nuclei and the electrons of the solute, which can be derived from the EE-GMFCC approach:

$$\varphi(\mathbf{r}_{\mu}) = \sum_{k=1}^{N_{\mathrm{F}}}\varphi_{k}(\mathbf{r}_{\mu}) - \sum_{k=1}^{N_{\mathrm{C}}}\varphi_{k}^{\mathrm{C}}(\mathbf{r}_{\mu}) + \sum_{k=1}^{N_{\mathrm{GC}}} \times \left[\varphi_{k}^{ij}(\mathbf{r}_{\mu}) - \varphi_{k}^{i}(\mathbf{r}_{\mu}) - \varphi_{k}^{j}(\mathbf{r}_{\mu})\right] \quad (6)$$

In the EE-GMFCC-CPCM approach, the QM calculation of each fragment is conducted in the presence of the surface charges. The newly obtained wave function is then employed to determine the surface charges, and this procedure is repeated in a self-consistent fashion until the final electrostatic solute-solvent interaction energy and the surface charges reach convergence [76].

3 Applications and Discussion

3.1 Protein–Ligand Binding Affinity Calculation

Based on the thermodynamic cycle [15], the protein-ligand binding affinity ($\Delta G_{\mathrm{P-L}}$) can be formulated as

$$\Delta G_{\mathrm{P-L}} = G_{\mathrm{PL}} - G_{\mathrm{P}} - G_{\mathrm{L}} \quad (7)$$

where G_{PL}, G_{P}, and G_{L} are the free energies of the protein–ligand complex, the protein monomer, and the ligand, respectively. The free energy of each molecule is estimated by

$$G = E_{\mathrm{gas}} + G_{\mathrm{solv}} + G_{\mathrm{np}} - \mathrm{TS} \quad (8)$$

where $E_{\mathrm{gas}} = E_{\mathrm{EE-GMFCC}} + E_{\mathrm{dispersion}}$ is the energy of the molecule in gas phase, which includes the molecular electronic energy calculated by the EE-GMFCC method and the dispersion correction

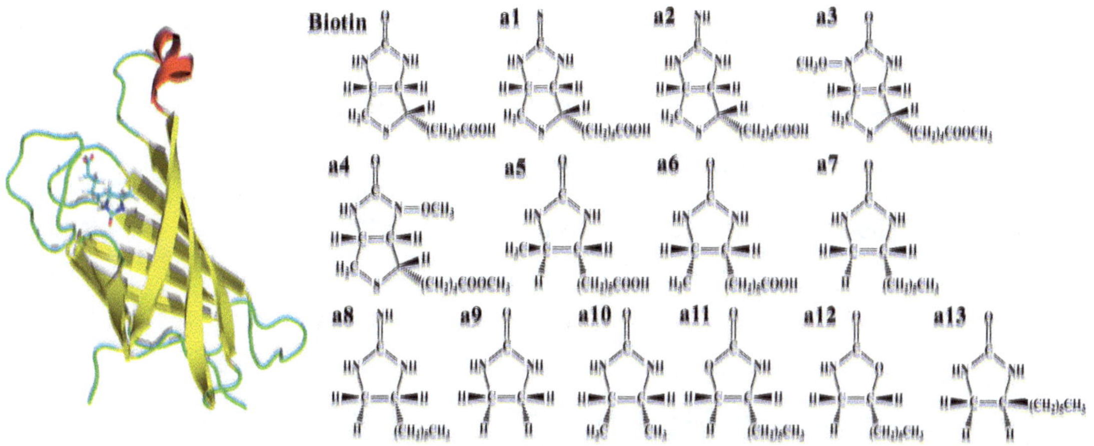

Fig. 2 The structures of avidin-biotin binding complex and the biotin analogues. (Liu, J. F.; Wang, X. W.; Zhang, J. Z. H.; He, X. Calculation of protein-ligand binding affinities based on a fragment quantum mechanical method. *Rsc Adv.* **2015**, *5*, 107020–107030—Adapted by permission of The Royal Society of Chemistry)

energy obtained from the density functional dispersion correction (DFT-D3) method [77]; G_{solv} is the EE-GMFCC-CPCM calculated electrostatic solvation energy; G_{np} is the non-electrostatic solvation energy, which is calculated by the non-electrostatic term in the CPCM model; and S is the conformational entropy. Here we did not take into account the entropic contribution because of the calculation inaccuracy and expensive computation for sampling the conformations. We computed the binding affinities of biotin and 13 biotin analogues bound to avidin (*see* Fig. 2) using the EE-GMFCC method [24, 64, 65] at the HF/6-31G∗-D and B3LYP/6-31G∗-D levels, respectively, and compared to the experimental values [78–81]. All the fragment QM calculations were carried out using the Gaussian09 program [82].

We first used one single snapshot to calculate the binding affinity for each system. At the HF/6-31G∗-D level, the calculated affinities are shown in Table 1, and the correlation coefficient R between the experimental and calculated binding affinities is 0.75 (*see* Fig. 3). The results obtained at the B3LYP/6-31G∗-D level are shown in Table 1 and Fig. 4, and the correlation coefficient R is 0.85. We also calculated the binding affinities by using the force field, and the obtained correlation coefficient R is 0.73. Hence, the QM method outperforms empirical approaches in predicting the protein–ligand binding affinity for these binding complexes.

Binding affinity predictions usually require extensive sampling of protein-ligand conformations through molecular dynamics (MD) simulation. However, it is computationally too expensive in the QM methods, even when the fragmentation or linear-scaling QM method is utilized. For better including the effect of conformational sampling within affordable QM computational time,

Table 1
The calculated binding affinities (in kcal/mol) of biotin and 13 biotin analogues bound to avidin using the EE-GMFCC-CPCM method in comparison with the experiment (the entropic contribution is not included). The basis set utilized was 6-31G∗ and the empirical dispersion correction was added

	Single snapshot		Multiple snapshots	
Biotin and analogues	HF	B3LYP	B3LYP	ΔG_{exp}
Biotin	−35.55	−54.45	−72.51	−20.40
a1	−27.98	−49.77	−48.95	−16.90
a2	−44.61	−62.41	−65.29	−14.30
a3	2.14	−36.66	−41.21	−8.80
a4	−20.69	−34.17	−47.77	−12.20
a5	−32.51	−51.05	−52.93	−16.50
a6	−36.16	−63.95	−59.85	−14.00
a7	−22.28	−30.98	−27.47	−11.10
a8	−8.34	−16.91	−24.78	−7.40
a9	−11.36	−17.17	−25.06	−4.50
a10	−21.02	−28.5	−32.13	−6.40
a11	−8.87	−16.33	−14.93	−5.00
a12	−7.32	−15.87	−16.88	−7.40
a13	−27.74	−34.57	−34.02	−9.10

three representative snapshots were selected from the MD simulation for each system and used for binding affinity calculations. The average binding affinities obtained from B3LYP/6-31G∗-D level are shown in Table 1 and Fig. 4. Importantly, with three representative protein–ligand structures, we achieved a better correlation of $R = 0.88$ between the experimental and calculated binding affinities, as compared to the result ($R = 0.85$) when the single structure was used. Therefore, the ensemble-averaged binding affinity over multiple snapshots generally gives more reliable result than that calculated on the single structure.

3.2 Protein–Protein Binding Affinity Calculation

The protein–protein binding affinity is calculated by

$$\Delta G_{\text{P-P}} = G_{\text{PP}} - G_{\text{P1}} - G_{\text{P2}} \quad (9)$$

where the free energy of the complex G_{PP} and the free energies of the protein monomers G_{P1} and G_{P2} can be obtained from Eq. 8. Here the polar contribution to the solvation free energy was computed by solving the Poisson-Boltzmann (PB) equation which

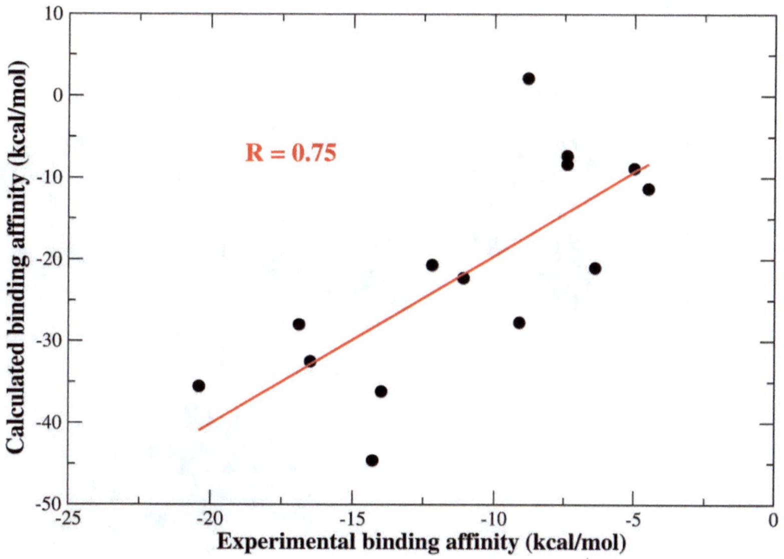

Fig. 3 The correlation between the experimental and calculated binding affinities of biotin and 13 biotin analogues bound to avidin based on the single protein–ligand structure at the HF/6-31G∗-D level using EE-GMFCC-CPCM. (Liu, J. F.; Wang, X. W.; Zhang, J. Z. H.; He, X. Calculation of protein–ligand binding affinities based on a fragment quantum mechanical method. *Rsc Adv.* **2015**, *5*, 107020–107030—Reproduced by permission of The Royal Society of Chemistry)

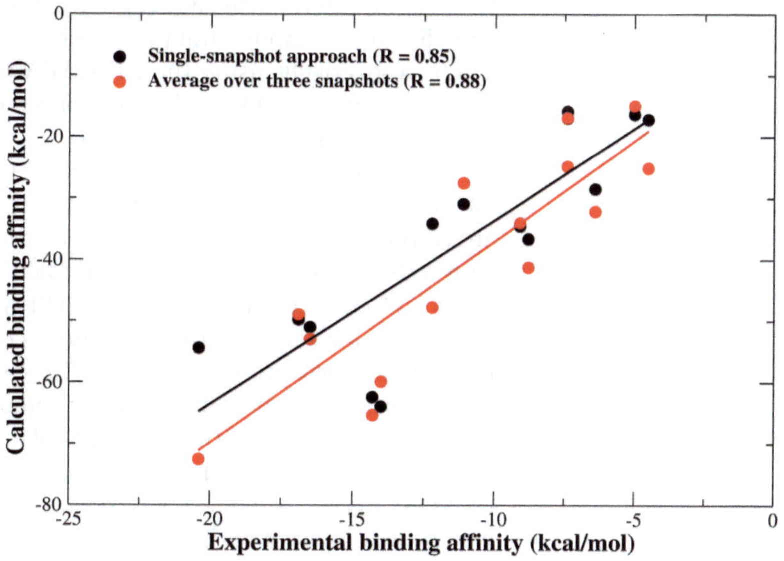

Fig. 4 The correlation between the experimental and calculated binding affinities of biotin and 13 biotin analogues bound to avidin at the B3LYP/6-31G∗-D level using EE-GMFCC-CPCM. (Liu, J. F.; Wang, X. W.; Zhang, J. Z. H.; He,X. Calculation of protein–ligand binding affinities based on a fragment quantum mechanical method. *Rsc Adv.* **2015**, *5*, 107020–107030—Reproduced by permission of The Royal Society of Chemistry)

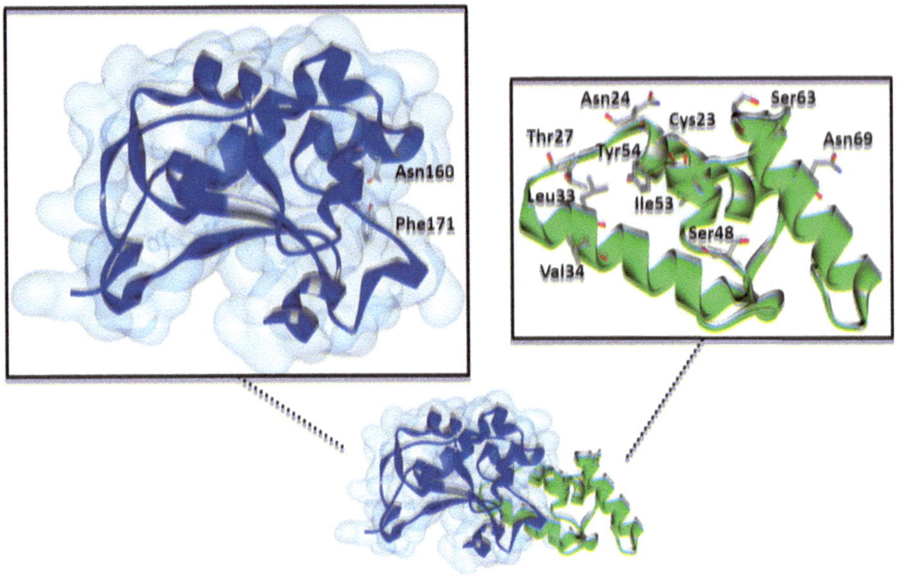

Fig. 5 The crystal structure of E9-Im9 complex. The mutated residues studied are indicated using the stick model (E9 in blue and Im9 in green). (Reproduced from Ref. 74 with permission from John Wiley and Sons)

could be evaluated using the DelPhi program [83] based on the atomic charges from the force field or the EE-GMFCC method. We computed the binding affinity changes $\Delta\Delta G$ of the cytotoxic endonuclease domain from the bacterial toxin colicin 9 enzyme (E9) complexed with its cognate immunity protein 9 (Im9) (Fig. 5) for ten alanine mutations in Im9 and two alanine mutations in E9, respectively, using the EE-GMFCC method [64, 74] at the B3LYP/6-31G$*$-D3 level, and compared the calculated results with the experimental data [84–86].

According to our calculated results (*see* Table 2), seven mutants affect the binding free energies by 0–2 kcal/mol and five mutants affect the binding affinities by more than 2 kcal/mol. Figure 6 shows the correlation between the calculated relative binding affinities ($\Delta\Delta G$, with reference to the wild type) and the experimental values. As one can see from the figure, comparing with the Amber force field (correlation coefficient $R = -0.053$), the QM method significantly improves the relative binding affinity predictions. Moreover, the interaction energy calculated by EE-GMFCC greatly improves the correlation between calculated free energy changes ($\Delta\Delta E_{EE-GMFCC} + \Delta\Delta E_{D3} + \Delta\Delta G_{PB\_EE-GMFCC}$) and experimental values, as compared to the results using the electrostatic energy from Amber force field ($\Delta\Delta E_{ele\_Amber} + \Delta\Delta E_{D3} + \Delta\Delta G_{PB\_EE-GMFCC}$). In addition, the absolute errors are also clearly decreased, especially for the mutants of V34A, N160A, and F171A. As a result, the mean unsigned error (MUE) was declined to 0.95 from 1.62 kcal/mol. Moreover, the correlation between the energy of $\Delta\Delta E_{EE-GMFCC} + \Delta\Delta E_{D3} + \Delta\Delta G_{PB\_EE-GMFCC}$ and experimental

Table 2
The calculated binding free energy change (in kcal/mol, $\Delta\Delta G_{bind\_cal} = \Delta\Delta E_{EE-GMFCC} + \Delta\Delta E_{D3} + \Delta\Delta G_{PB\_EE-GMFCC}$) between the wild-type E9–Im9 complex and twelve mutants using the EE-GMFCC method at the B3LYP/6-31G*-D3 level (the entropic contribution is not included)

System	$\Delta\Delta G_{cal}$	$\Delta\Delta G_{exp}$
C23A	0.96	0.92
N24A	0.53	0.14
T27A	1.40	0.73
L33A	2.86	3.42
V34A	3.10	2.58
S48A	−0.30	0.01
I53A	1.79	0.85
Y54A	9.77	4.84
S63A	−0.16	0.87
N69A	0.66	0.28
N160A	2.66	2.30
F171A	5.22	3.90
MUE[a]	0.95	

[a]Mean unsigned error of calculated $\Delta\Delta G_{cal}$ with respect to experimental values

results also improved from 0.775 to 0.906, as compared to the energy of $\Delta\Delta E_{EE-GMFCC} + \Delta\Delta E_{VDW} + \Delta\Delta G_{PB\_EE-GMFCC}$ (*see* Fig. 6a). By comparison of the results given by $\Delta\Delta E_{EE-GMFCC} + \Delta\Delta E_{D3} + \Delta\Delta G_{PB\_EE-GMFCC}$ and $\Delta\Delta E_{EE-GMFCC} + \Delta\Delta E_{D3} + \Delta\Delta G_{PB\_Amber}$, the atomic charges derived from EE-GMFCC for the solvation energy calculation greatly improve the correlation with the experimental results, where the correlation coefficient given by $\Delta\Delta E_{EE-GMFCC} + \Delta\Delta E_{D3} + \Delta\Delta G_{PB\_EE-GMFCC}$ is increased from 0.813 to 0.906, and the MUE also falls to 0.95 from 1.03 kcal/mol, as compared to the results given by $\Delta\Delta E_{EE-GMFCC} + \Delta\Delta E_{D3} + \Delta\Delta G_{PB\_Amber}$. Therefore, we conclude that the inclusion of the QM effects is critical for accurate prediction of the protein–protein binding affinities.

Partial charges of the protein and the PB surface charges could polarize each other until convergence is reached. The polar contribution to the solvation free energy was evaluated using the PB method. Figure 6a and b shows the results for these two binding free energy changes (namely, $\Delta\Delta E_{EE-GMFCC} + \Delta\Delta E_{D3} + \Delta\Delta G_{PB\_EE-GMFCC}$ and $\Delta\Delta E_{EE-GMFCC} + \Delta\Delta E_{D3} + \Delta\Delta G_{PB\_Amber}$), respectively. The result in Fig. 6a combines the interaction energy of EE-GMFCC, the dispersion interaction energy from the DFT-D3

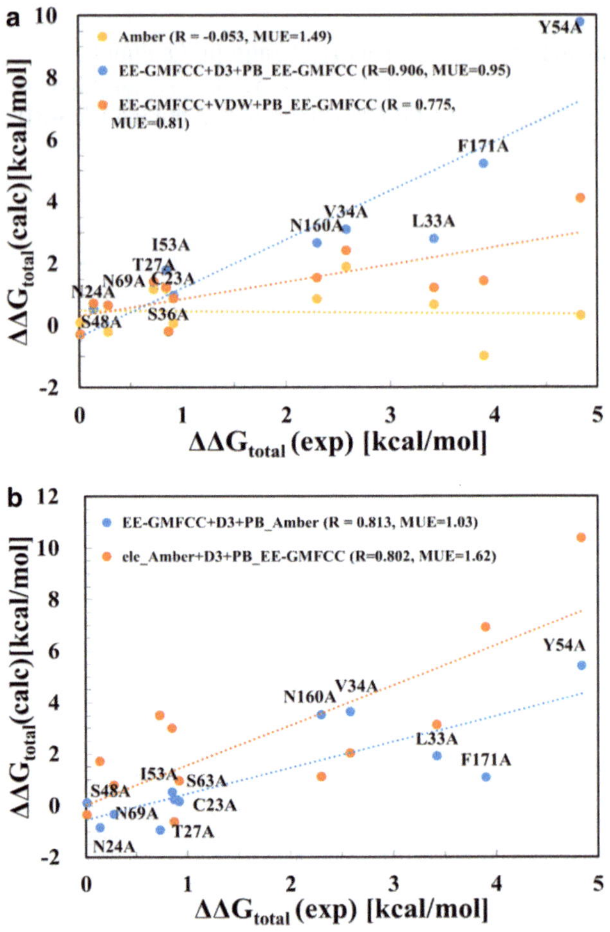

Fig. 6 The correlation between the experimental binding affinity changes and calculated results (in kcal/mol) between the wild-type E9–Im9 complex and twelve mutants using the EE-GMFCC method (the entropic contribution is not included). MUE: mean unsigned error of calculated $\Delta\Delta G$ with respect to experimental values. (**a**) $\Delta\Delta G_{bind\_cal} = \Delta\Delta E_{ele\_Amber} + \Delta\Delta E_{VDW} + \Delta\Delta G_{PB\_Amber}$ (points colored in yellow) $\Delta\Delta G_{bind\_cal} = \Delta\Delta E_{EE\text{–}GMFCC} + \Delta\Delta E_{D3} + \Delta\Delta G_{PB\_EE\text{–}GMFCC}$ (points colored in blue) $\Delta\Delta G_{bind\_cal} = \Delta\Delta E_{EE\text{–}GMFCC} + \Delta\Delta E_{VDW} + \Delta\Delta G_{PB\_EE\text{–}GMFCC}$ (points colored in orange). (**b**) $\Delta\Delta G_{bind\_cal} = \Delta\Delta E_{EE\text{–}GMFCC} + \Delta\Delta E_{D3} + \Delta\Delta G_{PB\_Amber}$ (points colored in blue) $\Delta\Delta G_{bind\_cal} = \Delta\Delta E_{ele\_Amber} + \Delta\Delta E_{D3} + \Delta\Delta\Delta G_{PB\_EE\text{–}GMFCC}$ (points colored in orange). (Reproduced from Ref. 74 with permission from John Wiley and Sons)

method, nonpolar solvation energy, and polar solvation energy with the polarized charges ($R = 0.906$), while the result in Fig. 6b has the same energy combinations except that the polar solvation energy was obtained from the Amber force field ($R = 0.813$). The MUE falls from 1.03 to 0.95 kcal/mol when the PB polar solvation energy with the polarized charges was included. Hence, the self-consistent calculation of PB solvation energy (or mutual polarization between the protein and surface charges) is required for accurate prediction of protein-protein binding free energy. However, to obtain the polarized charges through the EE-GMFCC method, it takes about 13.5 h (wall time) for one iteration on the 20-core Intel Xeon E7-8860 2.27GHz processor at the B3LYP/6-31G*-D3 level, and it usually takes 4–5 iterations to reach convergence.

4 Conclusions

This chapter has illustrated the use of fragment-based QM calculations to improve estimates of protein–ligand and protein–protein binding affinities. Considering the increasing awareness of the shortcomings of standard MM force fields, QM calculations automatically include effects of polarization, charge transfer, charge penetration, and the coupling of the various terms. Therefore, current QM approaches could offer more accurate and detailed information on the nature and relative strengths of protein–ligand and protein–protein interactions, which is valuable in drug design. The QM method will become a more prominent tool in the drug discovery field with the development of more efficient QM methods. The use of QM approaches, especially the linear-scaling and/or fragment-based QM methods, in drug design will certainly experience tremendous growth in the coming years.

At present, however, we have to strike a compromise between the attained accuracy and computational efficiency by applying QM methods in drug discovery, as the entropy calculation and conformational sampling are still major hurdles to effective drug design methods in both the classical and QM approaches. Nowadays, QM could be applied to "lead" compounds to provide useful insights into the protein–ligand binding interactions.

Acknowledgments

This work was supported by the National Key R&D Program of China (Grant No. 2016YFA0501700), National Natural Science Foundation of China (Grant Nos. 21703289, 21673074, 21922301, and 21761132022), Shanghai Municipal Natural Science Foundation (Grant No. 18ZR1412600), Young Top-Notch Talent Support Program of Shanghai, NYU-ECNU Center for Computational Chemistry at NYU Shanghai, "Double First-Class" University project (CPU2018GY09), and the Fundamental Research Funds for China Pharmaceutical University (2632019FY01). We also thank the Supercomputer Center of East China Normal University (ECNU Multifunctional Platform for Innovation 001) for providing computational time.

References

1. Bohm HJ, Klebe G (1996) What can we learn from molecular recognition in protein-ligand complexes for the design of new drugs? Angew Chem Int Ed 35:2589–2614
2. Cavalli A, Carloni P, Recanatini M (2006) Target-related applications of first principles quantum chemical methods in drug design. Chem Rev 106:3497–3519
3. Jorgensen WL (2009) Efficient drug lead discovery and optimization. Acc Chem Res 42:724–733
4. Cavasotto CN, Orry AJW (2007) Ligand docking and structure-based virtual screening in drug discovery. Curr Top Med Chem 7:1006–1014
5. Cavasotto CN, Adler NS, Aucar MG (2018) Quantum chemical approaches in structure-based virtual screening and lead optimization. Front Chem 6:188
6. Jorgensen WL (2004) The many roles of computation in drug discovery. Science 303:1813–1818
7. Liu J, He X, Zhang JZH (2013) Improving the scoring of protein-ligand binding affinity by including the effects of structural water and electronic polarization. J Chem Inf Model 53:1306–1314
8. Friesner RA, Banks JL, Murphy RB, Halgren TA, Klicic JJ, Mainz DT, Repasky MP, Knoll EH, Shelley M, Perry JK, Shaw DE, Francis P, Shenkin PS (2004) Glide: a new approach for rapid, accurate docking and scoring. 1. Method and assessment of docking accuracy. J Med Chem 47:1739–1749
9. Morris GM, Goodsell DS, Halliday RS, Huey R, Hart WE, Belew RK, Olson AJ (1998) Automated docking using a Lamarckian genetic algorithm and an empirical binding free energy function. J Comput Chem 19:1639–1662
10. Rarey M, Kramer B, Lengauer T, Klebe G (1996) A fast flexible docking method using an incremental construction algorithm. J Mol Biol 261:470–489
11. Jones G, Willett P, Glen RC, Leach AR, Taylor R (1997) Development and validation of a genetic algorithm for flexible docking. J Mol Biol 267:727–748
12. Paul N, Rognan D (2002) ConsDock: a new program for the consensus analysis of protein-ligand interactions. Proteins 47:521–533
13. Weis A, Katebzadeh K, Soderhjelm P, Nilsson I, Ryde U (2006) Ligand affinities predicted with the MM/PBSA method: dependence on the simulation method and the force field. J Med Chem 49:6596–6606
14. Wang JM, Hou TJ, Xu XJ (2006) Recent advances in free energy calculations with a combination of molecular mechanics and continuum models. Curr Comput Aided Drug Des 2:287–306
15. Kollman PA, Massova I, Reyes C, Kuhn B, Huo SH, Chong L, Lee M, Lee T, Duan Y, Wang W, Donini O, Cieplak P, Srinivasan J, Case DA, Cheatham TE (2000) Calculating structures and free energies of complex molecules: combining molecular mechanics and continuum models. Acc Chem Res 33:889–897
16. Wang W, Donini O, Reyes CM, Kollman PA (2001) Biomolecular simulations: recent developments in force fields, simulations of enzyme catalysis, protein-ligand, protein-protein, and protein-nucleic acid noncovalent interactions. Annu Rev Biophys Biomol Struct 30:211–243
17. Muegge I, Martin YC (1999) A general and fast scoring function for protein-ligand interactions: a simplified potential approach. J Med Chem 42:791–804
18. Gohlke H, Klebe G (2002) Approaches to the description and prediction of the binding affinity of small-molecule ligands to macromolecular receptors. Angew Chem Int Ed 41:2645–2676
19. Lyne PD, Lamb ML, Saeh JC (2006) Accurate prediction of the relative potencies of members of a series of kinase inhibitors using molecular docking and MM-GBSA scoring. J Med Chem 49:4805–4808
20. Arodola OA, Soliman MES (2017) Quantum mechanics implementation in drug-design workflows: does it really help? Drug Des Devel Ther 11:2551–2564
21. Soderhjelm P, Kongsted J, Ryde U (2010) Ligand affinities estimated by quantum chemical calculations. J Chem Theory Comput 6:1726–1737
22. Pearlman DA, Charifson PS (2001) Are free energy calculations useful in practice? A comparison with rapid scoring functions for the p38 MAP kinase protein system. J Med Chem 44:3417–3423
23. Ji CG, Mei Y (2014) Some practical approaches to treating electrostatic polarization of proteins. Acc Chem Res 47:2795–2803
24. Liu JF, Wang XW, Zhang JZH, He X (2015) Calculation of protein-ligand binding affinities based on a fragment quantum mechanical method. RSC Adv 5:107020–107030

25. Duan LL, Mei Y, Zhang DW, Zhang QG, Zhang JZH (2010) Folding of a helix at room temperature is critically aided by electrostatic polarization of Intraprotein hydrogen bonds. J Am Chem Soc 132:11159–11164
26. Tong Y, Mei Y, Li YL, Ji CG, Zhang JZH (2010) Electrostatic polarization makes a substantial contribution to the free energy of Avidin-biotin binding. J Am Chem Soc 132:5137–5142
27. Kumbhar S, Fischer FD, Waller MP (2012) Assessment of weak intermolecular interactions across QM/MM noncovalent boundaries. J Chem Inf Model 52:93–98
28. Parker TM, Sherrill CD (2015) Assessment of empirical models versus high-accuracy Ab initio methods for Nucleobase stacking: evaluating the importance of charge penetration. J Chem Theory Comput 11:4197–4204
29. Hansen N, van Gunsteren WF (2014) Practical aspects of free-energy calculations: a review. J Chem Theory Comput 10:2632–2647
30. Sherrill CD, Sumpter BG, Sinnokrot MO, Marshall MS, Hohenstein EG, Walker RC, Gould IR (2009) Assessment of standard force field models against high-quality Ab initio potential curves for prototypes of pi-pi, CH/pi, and SH/pi interactions. J Comput Chem 30:2187–2193
31. Zhou T, Huang D, Caflisch A (2008) Is quantum mechanics necessary for predicting binding free energy? J Med Chem 51:4280–4288
32. Vondrasek J, Bendova L, Klusak V, Hobza P (2005) Unexpectedly strong energy stabilization inside the hydrophobic core of small protein rubredoxin mediated by aromatic residues: correlated ab initio quantum chemical calculations. J Am Chem Soc 127:2615–2619
33. Raha K, Merz KM (2004) A quantum mechanics-based scoring function: study of zinc ion-mediated ligand binding. J Am Chem Soc 126:1020–1021
34. Ryde U, Soderhjelm P (2016) Ligand-binding affinity estimates supported by quantum-mechanical methods. Chem Rev 116:5520–5566
35. Sooderhjelm P, Kongsted J, Genheden S, Ryde U (2010) Estimates of ligand-binding affinities supported by quantum mechanical methods. Interdiscip Sci 2:21–37
36. Raha K, Merz KM (2005) Large-scale validation of a quantum mechanics based scoring function: predicting the binding affinity and the binding mode of a diverse set of protein-ligand complexes. J Med Chem 48:4558–4575
37. Raha K, Peters MB, Wang B, Yu N, WollaCott AM, Westerhoff LM, Merz KM (2007) The role of quantum mechanics in structure-based drug design. Drug Discov Today 12:725–731
38. Zhang DW, Zhang JZH (2003) Molecular fractionation with conjugate caps for full quantum mechanical calculation of protein-molecule interaction energy. J Chem Phys 119:3599–3605
39. Gleeson MP, Gleeson DQM (2009) MM calculations in drug discovery: a useful method for studying binding phenomena? J Chem Inf Model 49:670–677
40. Zhou T, Huang DZ, Caflisch A (2010) Quantum mechanical methods for drug design. Curr Top Med Chem 10:33–45
41. Ilatovskiy AV, Abagyan R, Kufareva I (2013) Quantum mechanics approaches to drug research in the era of structural Chemogenomics. Int J Quantum Chem 113:1669–1675
42. Jensen JH (2015) Predicting accurate absolute binding energies in aqueous solution: thermodynamic considerations for electronic structure methods. Phys Chem Chem Phys 17:12441–12451
43. He X, Zhu T, Wang XW, Liu JF, Zhang JZH (2014) Fragment quantum mechanical calculation of proteins and its applications. Acc Chem Res 47:2748–2757
44. Gordon MS, Fedorov DG, Pruitt SR, Slipchenko LV (2012) Fragmentation methods: a route to accurate calculations on large systems. Chem Rev 112:632–672
45. Collins MA, Bettens RPA (2015) Energy-based molecular fragmentation methods. Chem Rev 115:5607–5642
46. Goedecker S (1999) Linear scaling electronic structure methods. Rev Mod Phys 71:1085–1123
47. He X, Merz KM Jr (2010) Divide and conquer Hartree-Fock calculations on proteins. J Chem Theory Comput 6:405–411
48. Strain MC, Scuseria GE, Frisch MJ (1996) Achieving linear scaling for the electronic quantum coulomb problem. Science 271:51–53
49. Kohn W (1996) Density functional and density matrix method scaling linearly with the number of atoms. Phys Rev Lett 76:3168–3171
50. Kitaura K, Ikeo E, Asada T, Nakano T, Uebayasi M (1999) Fragment molecular orbital method: an approximate computational method for large molecules. Chem Phys Lett 313:701–706
51. Nakano T, Kaminuma T, Sato T, Fukuzawa K, Akiyama Y, Uebayasi M, Kitaura K (2002) Fragment molecular orbital method: use of approximate electrostatic potential. Chem Phys Lett 351:475–480

52. Fedorov DG, Kitaura K (2007) Extending the power of quantum chemistry to large systems with the fragment molecular orbital method. J Phys Chem A 111:6904–6914
53. He X, Zhang JZH (2005) A new method for direct calculation of total energy of protein. J Chem Phys 122:031103
54. He X, Zhang JZH (2006) The generalized molecular fractionation with conjugate caps/ molecular mechanics method for direct calculation of protein energy. J Chem Phys 124:184703
55. Deev V, Collins MA (2005) Approximate ab initio energies by systematic molecular fragmentation. J Chem Phys 122:154102
56. Collins MA, Deev VA (2006) Accuracy and efficiency of electronic energies from systematic molecular fragmentation. J Chem Phys 125:104104
57. Exner TE, Mezey PG (2003) Ab initio quality properties for macromolecules using the ADMA approach. J Comput Chem 24:1980–1986
58. Isegawa M, Wang B, Truhlar DG (2013) Electrostatically embedded molecular tailoring approach and validation for peptides. J Chem Theory Comput 9:1381–1393
59. Li SH, Li W, Ma J (2014) Generalized energy-based fragmentation approach and its applications to macromolecules and molecular aggregates. Acc Chem Res 47:2712–2720
60. Richard RM, Herbert JM (2012) A generalized many-body expansion and a unified view of fragment-based methods in electronic structure theory. J Chem Phys 137:064113
61. Dahlke EE, Truhlar DG (2007) Electrostatically embedded many-body expansion for large systems, with applications to water clusters. J Chem Theory Comput 3:46–53
62. Dahlke EE, Truhlar DG (2007) Electrostatically embedded many-body correlation energy, with applications to the calculation of accurate second-order Moller-Plesset perturbation theory energies for large water clusters. J Chem Theory Comput 3:1342–1348
63. Xie WS, Gao JL (2007) Design of a next generation force field: the X-POL potential. J Chem Theory Comput 3:1890–1900
64. Wang XW, Liu JF, Zhang JZH, He X (2013) Electrostatically embedded generalized molecular fractionation with conjugate caps method for full quantum mechanical calculation of protein energy. J Phys Chem A 117:7149–7161
65. Jia XY, Wang XW, Liu JF, Zhang JZH, Mei Y, He X (2013) An improved fragment-based quantum mechanical method for calculation of electrostatic solvation energy of proteins. J Chem Phys 139:214104
66. Liu JF, He X (2017) Accurate prediction of energetic properties of ionic liquid clusters using a fragment-based quantum mechanical method. Phys Chem Chem Phys 19:20657–20666
67. Jin XS, Zhang JZH, He X (2017) Full QM calculation of RNA energy using electrostatically embedded generalized molecular fractionation with conjugate caps method. J Phys Chem A 121:2503–2514
68. Liu JF, Qi LW, Zhang JZH, He X (2017) Fragment quantum mechanical method for large-sized ion-water clusters. J Chem Theory Comput 13:2021–2034
69. Liu JF, Zhang JZH, He X (2016) Fragment quantum chemical approach to geometry optimization and vibrational spectrum calculation of proteins. Phys Chem Chem Phys 18:1864–1875
70. Liu JF, Zhu T, Wang XW, He X, Zhang JZH (2015) Quantum fragment based ab initio molecular dynamics for proteins. J Chem Theory Comput 11:5897–5905
71. Liu JF, He X, Zhang JZH (2017) Structure of liquid water—a dynamical mixture of tetrahedral and 'ring-and-chain' like structures. Phys Chem Chem Phys 19:11931–11936
72. Liu JF, He X, Zhang JZH, Qi LW (2018) Hydrogen-bond structure dynamics in bulk water: insights from ab initio simulations with coupled cluster theory. Chem Sci 9:2065–2073
73. Liu JF, Zhang JZH, He X (2018) Probing the ion-specific effects at the water/air Interface and water-mediated ion pairing in sodium halide solution with Ab initio molecular dynamics. J Phys Chem B 122:10202–10209
74. Wang YQ, Liu JF, Li JJ, He X (2018) Fragment-based quantum mechanical calculation of protein-protein binding affinities. J Comput Chem 39:1617–1628
75. Cornell WD, Cieplak P, Bayly CI, Gould IR, Merz KM, Ferguson DM, Spellmeyer DC, Fox T, Caldwell JW, Kollman PA (1996) A second generation force field for the simulation of proteins, nucleic acids, and organic molecules. J Am Chem Soc 118:2309–2309
76. Barone V, Cossi M (1998) Quantum calculation of molecular energies and energy gradients in solution by a conductor solvent model. J Phys Chem A 102:1995–2001
77. Grimme S, Antony J, Ehrlich S, Krieg H (2010) A consistent and accurate ab initio parametrization of density functional

dispersion correction (DFT-D) for the 94 elements H-Pu. J Chem Phys 132:154104

78. Livnah O, Bayer EA, Wilchek M, Sussman JL (1993) 3-dimensional structures of avidin and the avidin-biotin complex. Proc Natl Acad Sci U S A 90:5076–5080
79. Kuhn B, Kollman PA (2000) Binding of a diverse set of ligands to avidin and streptavidin: an accurate quantitative prediction of their relative affinities by a combination of molecular mechanics and continuum solvent models. J Med Chem 43:3786–3791
80. Kuhn B, Kollman PA (2000) A ligand that is predicted to bind better to avidin than biotin: insights from computational fluorine scanning. J Am Chem Soc 122:3909–3916
81. Jia X, Zeng J, Zhang JZH, Mei Y (2014) Accessing the applicability of polarized protein-specific charge in linear interaction energy analysis. J Comput Chem 35:737–747
82. Frisch, M. J.; Trucks, G. W.; Schlegel, H. B.; Scuseria, G. E.; Robb, M. A.; Cheeseman, J. R.; Scalmani, G.; Barone V.; Mennucci, B.; Petersson, G. A.; Nakatsuji, H.; Caricato, M.; Li, X.; Hratchian, H. P.; Izmaylov, A. F.; Bloino, J.; Zheng, G.; Sonnenberg, J. L.; Hada, M.; Ehara, M.; Toyota, K.; Fukuda, R.; Hasegawa, J.; Ishida, M.; Nakajima, T.; Honda, Y.; Kitao, O.; Nakai, H.; Vreven, T.; J. A. Montgomery, Peralta, J. E.; Ogliaro, F.; Bearpark, M.; Heyd, J. J.; Brothers, E.; Kudin, K. N.; Staroverov, V. N.; Keith, T.; Kobayashi, R.; Normand, J.; Raghavachari, K.; Rendell, A.; Burant, J. C.; Iyengar, S. S.; Tomasi, J.; Cossi, M.; Rega, N.; Millam, J. M.; Klene, M.; Knox, J. E.; Cross, J. B.; Bakken, V.; Adamo, C.; Jaramillo, J.; Gomperts, R.; Stratmann, R. E.; Yazyev, O.; Austin, A. J.; Cammi, R.; Pomelli, C.; Ochterski, J. W.; Martin, R. L.; Morokuma, K.; Zakrzewski, V. G.; Voth, G. A.; Salvador, P.; Dannenberg, J. J.; Dapprich, S.; Daniels, A. D.; Farkas, O.; Foresman, J. B.; Ortiz, J. V.; Cioslowski, J.; Fox, D. J. Gaussian, Inc., Wallingford CT 2010
83. Honig B, Nicholls A (1995) Classical electrostatics in biology and chemistry. Science 268:1144–1149
84. Keeble AH, Joachimiak LA, Mate MJ, Meenan N, Kirkpatrick N, Baker D, Kleanthous C (2008) Experimental and computational analyses of the energetic basis for dual recognition of immunity proteins by colicin endonucleases. J Mol Biol 379:745–759
85. Luitz MP, Zacharias M (2013) Role of tyrosine hot-spot residues at the interface of colicin E9 and immunity protein 9: a comparative free energy simulation study. Proteins 81:461–468
86. Kuhlmann UC, Pommer AJ, Moore GR, James R, Kleanthous C (2000) Specificity in protein-protein interactions: the structural basis for dual recognition in endonuclease colicin-immunity protein complexes. J Mol Biol 301:1163–1178

Chapter 3

Guiding Medicinal Chemistry with Fragment Molecular Orbital (FMO) Method

Alexander Heifetz, Tim James, Michelle Southey, Mike J. Bodkin, and Steven Bromidge

Abstract

The understanding of binding interactions between a protein and a small molecule plays a key role in the rationalization of potency and selectivity and in design of new ideas. However, even when a target of interest is structurally enabled, visual inspection and force field-based molecular mechanics calculations cannot always explain the full complexity of the molecular interactions that are critical in drug design. Quantum mechanical methods have the potential to address this shortcoming, but traditionally, computational expense has made the application of these calculations impractical. The fragment molecular orbital (FMO) method offers a solution that combines accuracy, speed, and the ability to characterize important interactions (i.e. its strength in kcal/mol and chemical nature: hydrophobic, electrostatic, etc) that would otherwise be hard to detect. In this chapter, we describe the FMO method and illustrate its application in the discovery of the benzothiazole (BZT) series as novel tyrosine kinase ITK inhibitors for treatment of allergic asthma.

Key words *FMO* fragment molecular orbitals method, *ITK* interleukin-2 inducible T-cell kinase, *PIE* pair interaction energy, PIEDA pair interaction energy decomposition analysis, SAR, BZT, Benzothiazole ITK inhibitors, IND indazole ITK inhibitors

1 Introduction

The understanding of binding interactions between a protein and a small molecule plays a key role in the rationalization of potency and selectivity [1] and in design of new targets for synthesis. The efficiency and cost-effectiveness of the drug-discovery process can be accelerated by the availability of structural data regarding the target protein and by the reliability of the computational tools for data exploration [2–4]. However, even with a structurally enabled target, "visual inspection" and force field-based molecular mechanics (MM) calculations often used for the rationalization of ligand–protein potency cannot always explain the full complexity of the molecular interactions [1]. There is increasing evidence [1, 5–7]

Alexander Heifetz (ed.), *Quantum Mechanics in Drug Discovery*, Methods in Molecular Biology, vol. 2114,
https://doi.org/10.1007/978-1-0716-0282-9_3,

that there are a number of underappreciated interactions such as CH/π [8, 9], halogen/π [10], cation/π [11], non-classical H-bonds [12], and others, which play important roles in protein–ligand binding that are not adequately parameterized in the most popular force fields (FF) [6]. The use of quantum mechanical (QM) methods has traditionally been employed to improve the reliability of the exploration of protein–ligand interactions [13, 14]. Despite their many advantages, QM methods have not been tractable until recently for large biological molecules such as proteins, due to high computational expense [15].

The fragment molecular orbital (FMO) method [9, 14, 16] offers a substantial computational time saving over traditional QM methods [17]. This is achieved by partitioning the larger system into smaller portions called fragments (Fig. 1). For example, in proteins, each residue can be represented by a fragment. Similarly, the ligand can be represented by single or multiple fragments as necessary. By performing QM calculations on fragments, one can make the computational cost scaling with respect to system size nearly linear [17].

FMO can be applied to the majority of ligand–protein complex. It requires a protein–ligand complex as input and provides a comprehensive list of interactions and reveals the composition of those interactions related to their electronic structure [15]. The strength of these interactions is quantified in terms of energy stabilization of the complex (in kcal/mol). By summing up the pair interaction energies (PIEs) calculated for individual fragment interactions, one obtains an estimate of the overall binding energy gain that can be related to the resultant K_i of the compound as measured in a biological assay. A typical FMO calculation on a protein–ligand complex takes approximately 4 h on a 36 CPU core to complete, which is a significant cost improvement over classical QM calculations [9].

Pair interaction energy (PIE) between any two fragments calculated by FMO is a sum of four energy terms, electrostatics, exchange repulsion, charge transfer, and dispersion, and is provided by a pair interaction energy decomposition analysis (PIEDA [18], Fig. 1). The electrostatic and charge transfer terms are dominant in H-bonds (classical and nonclassical [1]), polar interactions, and salt bridges, while the dispersion term is more prominent in hydrophobic interactions. The role of hydrophobic interactions is vital for biomolecular recognition, but there is still no reliable, predictive method for its quantification [1]. The exchange repulsion term describes steric repulsion between electrons [15] that prevents atoms from clashing with each other. Such information is essential for medicinal chemists to be able to rationally approach modification of lead compounds in order to increase favourable interactions.

FMO is a highly useful tool for rational structure-based drug design (SBDD) [9, 20, 21], as it provides an accurate and comprehensive list of strong, weak, or repulsive interactions between the

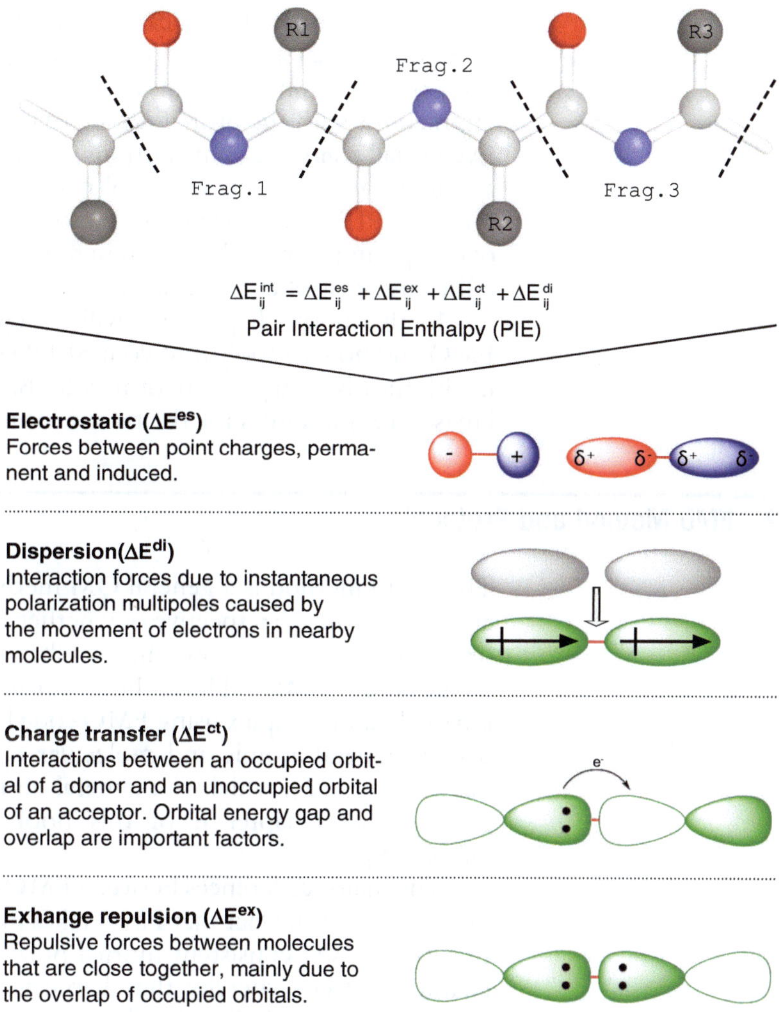

Fig. 1 Workflow for PIEDA calculations and details on each of the PIE terms that are computed. The electrostatic component arises from the Coulomb interaction between polarized charge distributions of fragments. The dispersion term arises from the interaction between instantaneous dipole moments of two fragments. It is hydrophobic (nonpolar) in nature and is obtained in PIEDA from the correlation energy of electrons. The charge transfer term arises from the interaction between occupied orbitals of a donor and unoccupied orbitals of an acceptor. The exchange repulsion term is derived from the interaction between fragments situated in close proximity and is always repulsive; it is due to the Pauli repulsion and is related to the overlap of two occupied orbitals. (Reprinted with permission from Journal Of Med Chem. 2016 May 12;59(9):4352–63 [19]. Copyright (2019) American Chemical Society)

ligand and its surrounding residues. Chemists can more readily relate to QM results as the method quantifies chemically interesting interactions, in particular, nonpolar interactions which historically have been hard to quantify. Such information is highly useful in

SBDD in terms of modifications, scaffold replacement (scaffold hoping) linking (specifically in case of fragment-based drug discovery), or extension of chemical moieties to form stronger or new interactions with the protein or alternatively to remove repulsions. Even small energetic contributions, when summed, may give rise to a large total value. The chemical nature of such interactions computed by PIEDA can be important for the design of the next round of compounds for synthesis that in return significantly decreases the effort and cost of chemical synthesis.

In this book chapter, we will review how technically to use FMO and how to apply it to your SBDD project and will exemplify it with the discovery of benzothiazole (BZT) series as novel tyrosine kinase ITK inhibitors for the treatment of allergic asthma [19].

2 FMO Method and Protocol

The FMO method is a general QM method that can be applied to any set of atoms, irrespective of whether they are a part of a soluble or a membrane-bound protein (*see* Chapter 4 of this book for more details about FMO). Here, the FMO method [9] was applied to kinase–ligand complex using FMO code [17] version 5.0 as embedded in general atomic and molecular electronic structure system (GAMESS) [22], which is a general ab initio quantum chemistry package (*see* Chapter 8 of this book for more details about GAMESS).

The major differences between FMO and MM methods mainly originate from the fact that FMO takes into consideration polarization (in the self-consistent mutual polarization of fragments) and charge transfer (whereby the charge is allowed to flow between fragments) [9, 23]. The detailed description of the FMO strategy and methodology can be found in the published reviews [9, 14, 18], including a detailed mathematical formulation that is beyond the scope of this manuscript. In this work, we used a well-established FMO protocol that was previously described and tested [8, 9, 24–26].

The FMO calculation consists of the following steps:

1. Fragmentation (i.e., assigning atoms in a system to a fragment). Each residue and ligand are characterized as a fragment.
2. Fragment self-consistent field (SCF) calculations in the embedding polarizable potential, so that fragments mutually polarize each other in a self-consistent fashion whereby intra-fragment charge transfer and other quantum effects are accounted for.
3. Fragment pair SCF calculations, bringing in inter-fragment charge transfer.
4. Total property (energy, gradient, etc.) evaluation.

By performing QM computations on fragments, one can achieve high efficiency, often resulting from linear scaling and computational speed. The FMO method has been parallelized for PC clusters [17]. In this work, we used the MP2 method (second-order Møller-Plesset perturbation theory [27]) with the 6-31G∗ basis set. This basis set is most commonly used and is often considered the best compromise between speed and accuracy [24]. Residues and water molecules within a radius of ≤4.5 Å around the ligand atoms were included in the FMO calculations since previous work demonstrated [28] that including these atoms significantly increases the speed of the calculation without compromising the results.

PIE—interaction energy (ΔE^{int}) between fragments *i* and *j* is a sum of 4 PIE terms: electrostatics (ΔE^{es}), exchange repulsion (ΔE^{ex}), charge transfer (ΔE^{ct}), and dispersion (ΔE^{di}) (Eq. 1). The chemical definition of each term is described in Fig. 1:

$$\Delta E_{ij}^{\text{int}} = \Delta E_{ij}^{\text{es}} + \Delta E_{ij}^{\text{ex}} + \Delta E_{ij}^{\text{ct}} + \Delta E_{ij}^{\text{di}} \tag{1}$$

The PIE is not a difference between energies of the protein–ligand complex and the sum of the "free" protein and ligand, but rather represents the "strength" of the interaction between the ligand and protein residues in the complex. The Δ signs refer to the differences in the total QM energy of a fragment pair *ij* and two individual fragments *i* and *j*, both computed in the receptor–ligand complex. In the present work, fragment *i* was the ligand and the other *n* fragments were receptor residues and water molecules. In the equations above there is no self-interaction (the sums exclude $j = i$). The total PIE calculated by the FMO method describes the stability of the receptor–ligand complex. This stability correlates to, but is not the same as, the binding energy [24]. The difference lies in the polarization factors—the ligand is polarized by the protein and vice versa [17]. Based on previous reports [9], we consider any interaction with an absolute PIE greater than or equal to 3.0 kcal/mol to be significant.

3 Discovery of ITK BZT Inhibitors

We would like to demonstrate how FMO has been exploited to guide the discovery of benzothiazole [29] (BZT—Table 1) ITK inhibitors [19]. ITK is a tyrosine kinase involved in T cell development, differentiation, and effector function [30, 31]. It has been shown to play an important role in the development of T-cell-dependent late-phase responses of allergic asthma. Studies using an ITK knockout transgenic mouse model suggest that ITK kinase activity is required for the control of Th2 responses and the development of allergic asthma [32]. The wealth of preclinical evidence

Table 1
Benzothiazole ITK inhibitors (BZT) [29]

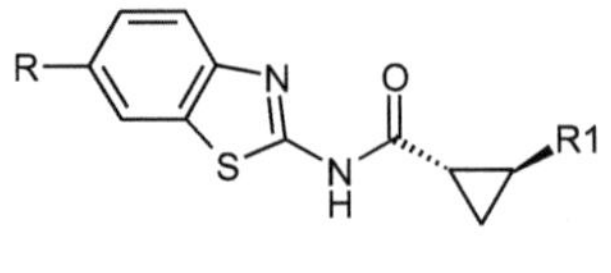

Compound	R	R1	ITK Ki (nM)	Total PIE (kcal/mol)
2a	MeO	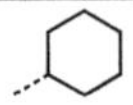	3,500	-51.33
2b	MeO	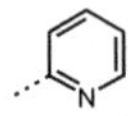	2,000	-51.87
2c	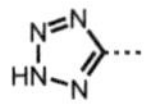	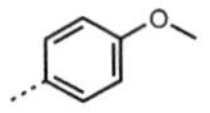	1,100	-55.70
2d	MeO	Ph	460	-53.64
2e	MeO	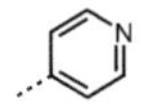	220	-53.64
2f	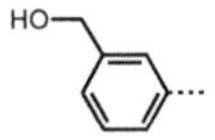	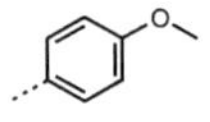	160	-75.35
2g	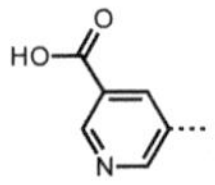	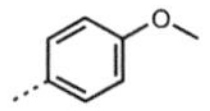	130	-72.41
2h	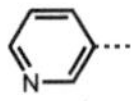	Ph	85	-70.94
2i	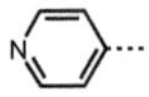	Ph	50	-68.03
2j	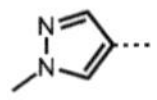	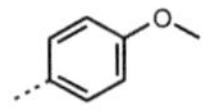	43	-69.29
2k	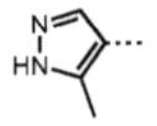	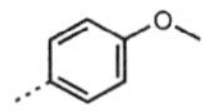	30	-65.12
2l	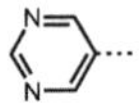	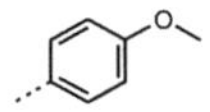	25[b]	-67.49
2m	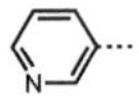	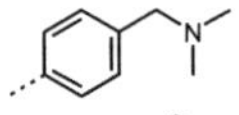	8.5	-84.62
2n	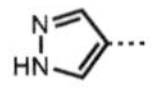	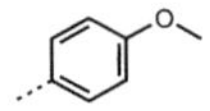	6.4	-91.70
2o	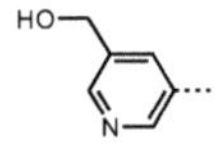	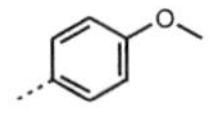	5.3	-82.89
2p	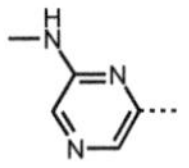	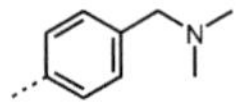	4.1[b]	-91.44
2q	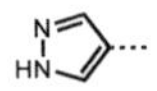	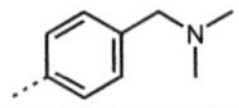	0.69	-98.92

supporting the role of ITK in allergic asthma and other inflammatory disorders has prompted substantial research effort from the pharmaceutical community toward the discovery of selective inhibitors of this kinase.

A key challenge early on of our ITK project was to improve in vitro potency of the primary hit (compound **2d**) while either maintaining or enhancing other attributes such as selectivity, solubility, and DMPK properties. This required a detailed exploration of the interactions between ITK protein residues and compound **2d**, aided by the availability of crystal structures or homology models of protein/inhibitor complexes. Classical hydrogen bonds (H-bonds) with kinase active site hinge residues were readily identified by visual inspection; however, other key interactions were less straightforward to detect by eye. FMO detected that initial hit, compound **2d** (Fig. 2a), forms 6 key interactions with ITK. Two H-bonds are formed between hinge residue Met438 and the two nitrogen atoms of the benzothiazole amide of **2d**. A nonclassical H-bond is formed by the backbone carbonyl of Glu436 and the hydrogen atom on C4 of the benzothiazole ring. Similar to other ITK ATP-site binders, the inhibitor is positioned in a narrow lipophilic cleft between the N and C lobes of ITK and forms a CH–π interaction with Ile369 and a hydrophobic interaction with Leu489. Finally, the cyclopropyl moiety directs the attached benzene ring to form a face-to-edge, π–stacking interaction with Phe437. Analogues where this benzene ring was replaced with a cyclohexyl (**2a**, Table 1) or 2-pyridyl (**2b**) group displayed decreased potency, which was consistent with observations from the FMO analysis.

A rational SBDD process resulted in the design of **2p** and **2q** (Table 1). It was suggested that an aromatic substituent off the 6-position of the benzothiazole core might form an interaction with the gatekeeper residue Phe435 and facilitate additional interactions with residues in the DFG/αC helix area of ITK. FMO analyses of **2p** (Fig. 2b) and **2q** (Fig. 2c) detected many new interactions beyond the original 6 observed for the parent **2d** (Fig. 2a). Compound **2q** displayed 10 interactions with ITK compared to 9 in the case of **2p**. A face-to-face π–stacking interaction with Phe435 and a nonclassical H-bond between a cyclopropyl hydrogen and the carbonyl backbone oxygen of Glu439 were observed for both **2p** and **2q**. Interestingly, a strong H-bond interaction with Glu406 (αC helix residue) was present solely with ligand **2p**, while two significant interactions with Ser499 and DFG Asp500 were predicted solely in the case of **2q**. According to the total PIE calculation, **2q** was predicted to be more potent than **2p**, and indeed, the experimental binding affinities confirmed this prediction (Table 1). Compounds **2p** and **2q** demonstrated 112- and 657-fold increases in ITK potency, respectively, over the original HTS hit (**2d**) [29].

In summary, FMO methodology was useful for exploring the interactions of **2d** with ITK and for identifying potential interactions that this compound lacked with ITK. This subsequently enabled the use of FMO in the design process of **2p** and **2q**, two

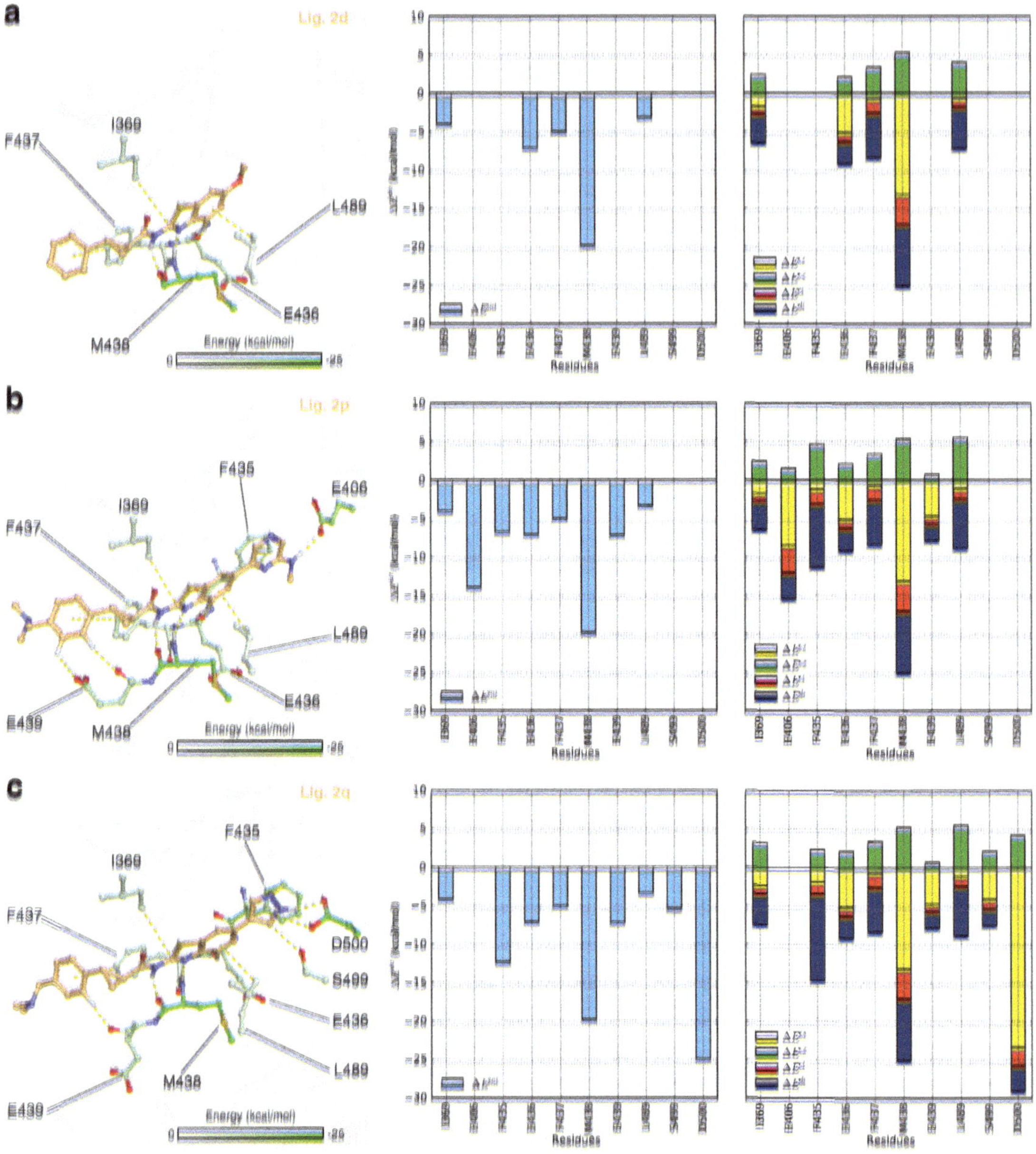

Fig. 2 FMO results for the BZT series: compounds **2d** (**a**), **2p** (**b**), and **2q** (**c**). The carbon atoms of the ligand are shown in light orange and those of the receptor according to PIE values calculated by FMO (the scale is from dark green for the strongest interactions to white for the weakest ones). Key interactions as calculated by FMO are marked as yellow dashed lines. The left-hand bar plots describe the PIE of the most significant residues, and the right-hand plots describe the PIEDA of these key interactions. PIE terms: electrostatics, dispersion, charge transfer, and exchange repulsion are color coded in yellow, blue, red, and green, respectively. We consider any interaction with an absolute PIE $\geq$3.0 kcal/mol to be significant. (Reprinted with permission from Journal Of Med Chem. 2016 May 12;59(9):4352–63 [19]. Copyright (2019) American Chemical Society)

compounds that made several additional interactions with ITK and benefitted from the increased potency derived from these interactions.

To assess how accurately the total pair interaction energy (PIE) described the interactions between ITK and our BZT compounds (Table 1), we performed a correlation between total PIE energy calculated by FMO and the experimentally measured binding affinities (pK_i). The correlation plot is shown in Fig. 3a [19]. We observed a significant correlation ($r^2 = 0.83$) between the experimental values and FMO calculated PIEs. This observation provided us with confidence that the PIE energy calculated by FMO can be used as a predictor ITK–ligand potency. It is important to emphasize that we did not expect to see a perfect correlation between pK_i and PIE since the binding affinity is not only dependent on specific interactions calculated by the FMO technique but also on additional energy terms such as entropy, desolvation, and strain energy present in the ligand's bioactive conformation that is not accounted for by FMO. Also noteworthy is the fact that for the ITK protein–ligand inhibitor complexes studied, the active site residues did not significantly change conformation, making this lock and key relationship well suited to a QM analysis where the protein is held rigid.

Performance of the FMO method was also compared [19] with binding energy values calculated using the MM/GBSA method (Fig. 3). The FMO method clearly outperformed the force field-based approach, demonstrating a high correlation ($r^2 = 0.839$) with experimentally measured ITK pK_i values relative to that obtained by the MM/GBSA method ($r^2 = 0.061$).

4 Notes

1. We described how FMO was successfully applied to the SBDD of novel and selective ITK inhibitors [19]. FMO provided detailed information on the individual contribution and chemical nature of each ITK residue to the ligand binding that normally would be hard to detect without QM.
2. This information was useful for the design of subsequent generations of BZT analogues. It helped to rationalize differences in ITK potencies when compounds appeared to be very similar to visual inspection. The correlation between experimentally measured affinity and FMO calculated pairwise interaction energy gave us the confidence to use it in design and as a filter of new targets for synthesis.
3. FMO is efficient in cases of being applied to well-modeled or crystal structures. We have demonstrated [19] that FMO can be useful in cases where at least one crystal structure per chemical series is available, provided that binding poses can be

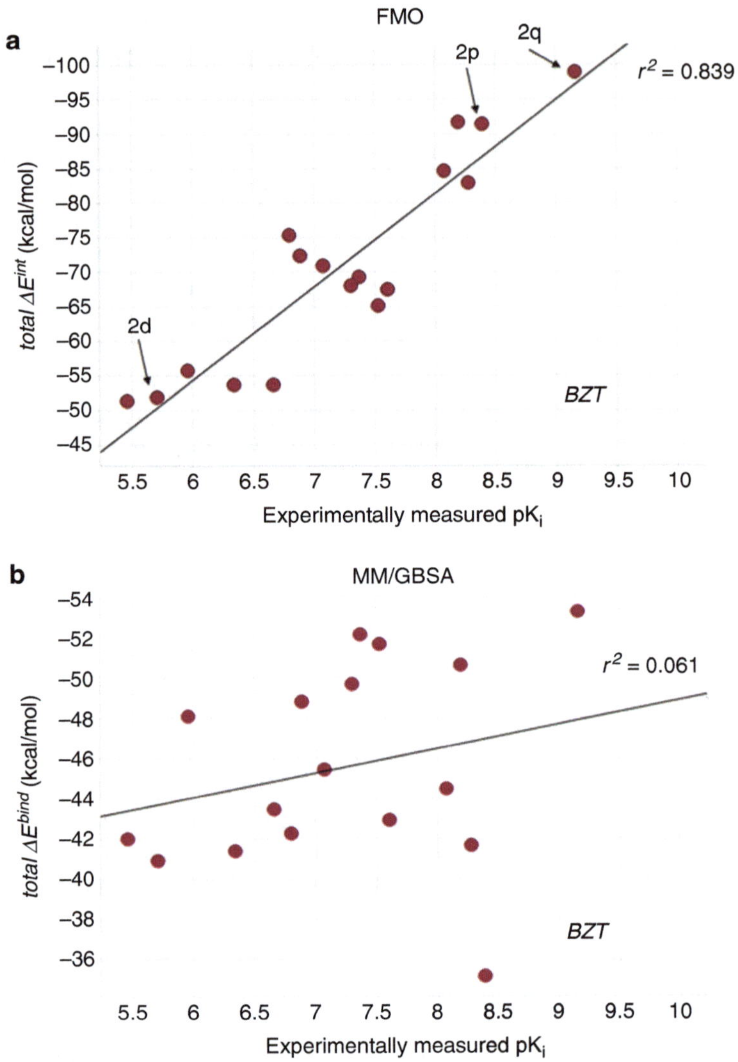

Fig. 3 Correlation plots between experimentally measured affinity (ITK pK_i) and FMO (**a**) and MM/GBSA (**b**) calculated ITK–ligand binding energy (*total* ΔE^{bind}) using the Amber12EHT force field as implemented MOE (version 2015.10) for BZT. (Adapted with permission from Journal Of Med Chem. 2016 May 12;59 (9):4352–63 [19]. Copyright (2019) American Chemical Society)

generated for the remaining ligands by overlaying them onto the appropriate parent compounds. The FMO version used in this project performed its calculations in a vacuum, without taking into account the desolvation effect. The most recent version of FMO now includes a polarizable continuum solvation model (PCM) [33] that allows the inclusion of a desolvation term into the FMO calculation.

4. An advantage of the FMO method is quantitation of the effect if carefully applied to specific series in a lead optimization project where the protein is reasonably rigid. Another benefit

is the detection of more subtle nontraditional H-bonding/electrostatic stabilization that can contribute to potency or selectivity differences. Such subtle interactions can be exploited in the examination of multiple candidate analogues, since the technique is computationally efficient and can be rapidly run in a time frame soon enough to impact fast-moving projects in a lead optimization stage. Like many other SBDD computational approaches such as docking and de novo design, it is important to know the conformational preferences of the complex under study before undertaking these calculations, as significant protein motion will undermine the results. Nevertheless, in the case of ITK compound design, these calculations proved useful in revealing subtle interactions that rationalized potency differences and provided suggestions for synthesis prioritization. The in silico methods that were applied in this project, including FMO, complemented each other and provided a powerful, efficient, and cost-effective platform for successful SBDD against this target.

Acknowledgments

Some sections of this chapter are reprinted (adapted) with permission from Journal of Med Chem. 2016 May 12;59(9):4352–63 [19]. Copyright (2019) American Chemical Society.

References

1. Bissantz C, Kuhn B, Stahl M (2010) A medicinal chemist's guide to molecular interactions. J Med Chem 53:5061–5084
2. Tautermann CS (2014) GPCR structures in drug design, emerging opportunities with new structures. Bioorg Med Chem Lett 24:4073–4079
3. Shonberg J, Kling RC, Gmeiner P, Lober S (2015) GPCR crystal structures: medicinal chemistry in the pocket. Bioorg Med Chem 23:3880–3906
4. Jazayeri A, Dias JM, Marshall FH (2015) From G protein-coupled receptor structure resolution to rational drug design. J Biol Chem 290:19489–19495
5. Tong Y, Mei Y, Li YL, Ji CG, Zhang JZ (2010) Electrostatic polarization makes a substantial contribution to the free energy of avidin-biotin binding. J Am Chem Soc 132:5137–5142
6. Raha K, Peters MB, Wang B, Yu N, Wollacott AM, Westerhoff LM, Merz KM Jr (2007) The role of quantum mechanics in structure-based drug design. Drug Discov Today 12:725–731
7. Beratan DN, Liu C, Migliore A, Polizzi NF, Skourtis SS, Zhang P, Zhang Y (2015) Charge transfer in dynamical biosystems, or the treachery of (static) images. Acc Chem Res 48:474–481
8. Ozawa T, Okazaki K, Kitaura K (2011) CH/pi hydrogen bonds play a role in ligand recognition and equilibrium between active and inactive states of the beta2 adrenergic receptor: an ab initio fragment molecular orbital (FMO) study. Bioorg Med Chem 19:5231–5237
9. Fedorov DG, Nagata T, Kitaura K (2012) Exploring chemistry with the fragment molecular orbital method. Phys Chem Chem Phys 14:7562–7577
10. Lu Y-X, Zou J-W, Wang Y-H, Yu Q-S (2007) Substituent effects on noncovalent halogen/π interactions: theoretical study. Int J Quantum Chem 107:1479–1486
11. Gallivan JP, Dougherty DA (1999) Cation-pi interactions in structural biology. Proc Natl Acad Sci U S A 96:9459–9464
12. Johnston RC, Cheong PH (2013) C-H...O non-classical hydrogen bonding in the stereomechanics of organic transformations: theory and recognition. Org Biomol Chem 11:5057–5064

13. Yu N, Li X, Cui G, Hayik SA, Merz KM 2nd (2006) Critical assessment of quantum mechanics based energy restraints in protein crystal structure refinement. Protein Sci 15:2773–2784
14. Fedorov DG, Kitaura K (2007) Extending the power of quantum chemistry to large systems with the fragment molecular orbital method. J Phys Chem A 111:6904–6914
15. Phipps MJ, Fox T, Tautermann CS, Skylaris CK (2015) Energy decomposition analysis approaches and their evaluation on prototypical protein-drug interaction patterns. Chem Soc Rev 44:3177–3211
16. Kitaura K, Ikeo E, Asada T, Nakano T, Uebayasi M (1999) Fragment molecular orbital method: an approximate computational method for large molecules. Chem Phys Lett 313:701–706
17. Alexeev Y, Mazanetz MP, Ichihara O, Fedorov DG (2012) GAMESS as a free quantum-mechanical platform for drug research. Curr Top Med Chem 12:2013–2033
18. Fedorov DG, Kitaura K (2007) Pair interaction energy decomposition analysis. J Comput Chem 28:222–237
19. Heifetz A, Trani G, Aldeghi M, MacKinnon CH, McEwan PA, Brookfield FA, Chudyk EI, Bodkin M, Pei Z, Burch JD, Ortwine DF (2016) Fragment molecular orbital method applied to lead optimization of novel Interleukin-2 inducible T-cell kinase (ITK) inhibitors. J Med Chem 59:4352–4363
20. Barker JJ, Barker O, Courtney SM, Gardiner M, Hesterkamp T, Ichihara O, Mather O, Montalbetti CA, Muller A, Varasi M, Whittaker M, Yarnold CJ (2010) Discovery of a novel Hsp90 inhibitor by fragment linking. ChemMedChem 5:1697–1700
21. Sawada T, Fedorov DG, Kitaura K (2010) Binding of influenza a virus hemagglutinin to the sialoside receptor is not controlled by the homotropic allosteric effect. J Phys Chem B 114:15700–15705
22. Schmidt MW, Baldridge KK, Boatz JA, Elbert ST, Gordon MS, Jensen JH, Koseki S, Matsunaga N, Nguyen KA, Su S, Windus TL, Dupuis M, Montgomery JA (1993) General atomic and molecular electronic structure system. J Comput Chem 14:1347–1363
23. Fedorov DG, Kitaura K (2012) Energy decomposition analysis in solution based on the fragment molecular orbital method. J Phys Chem A 116:704–719
24. Mazanetz MP, Ichihara O, Law RJ, Whittaker M (2011) Prediction of cyclin-dependent kinase 2 inhibitor potency using the fragment molecular orbital method. J Cheminform 3:2
25. Yoshino R, Yasuo N, Inaoka DK, Hagiwara Y, Ohno K, Orita M, Inoue M, Shiba T, Harada S, Honma T, Balogun EO, da Rocha JR, Montanari CA, Kita K, Sekijima M (2015) Pharmacophore modeling for anti-Chagas drug design using the fragment molecular orbital method. PLoS One 10:e0125829
26. Hitaoka S, Chuman H, Yoshizawa K (2015) A QSAR study on the inhibition mechanism of matrix metalloproteinase-12 by arylsulfone analogs based on molecular orbital calculations. Org Biomol Chem 13:793–806
27. Fedorov DG, Kitaura K (2004) Second order Moller-Plesset perturbation theory based upon the fragment molecular orbital method. J Chem Phys 121:2483–2490
28. Heifetz A, Chudyk EI, Gleave L, Aldeghi M, Cherezov V, Fedorov DG, Biggin PC, Bodkin MJ (2016) The fragment molecular orbital method reveals new insight into the chemical nature of GPCR-ligand interactions. J Chem Inf Model 56:159–172
29. MacKinnon CH, Lau K, Burch JD, Chen Y, Dines J, Ding X, Eigenbrot C, Heifetz A, Jaochico A, Johnson A, Kraemer J, Kruger S, Krulle TM, Liimatta M, Ly J, Maghames R, Montalbetti CA, Ortwine DF, Perez-Fuertes Y, Shia S, Stein DB, Trani G, Vaidya DG, Wang X, Bromidge SM, Wu LC, Pei Z (2013) Structure-based design and synthesis of potent benzothiazole inhibitors of interleukin-2-inducible T cell kinase (ITK). Bioorg Med Chem Lett 23:6331–6335
30. Schaeffer EM, Debnath J, Yap G, McVicar D, Liao XC, Littman DR, Sher A, Varmus HE, Lenardo MJ, Schwartzberg PL (1999) Requirement for Tec kinases Rlk and Itk in T cell receptor signaling and immunity. Science 284:638–641
31. Fowell DJ, Shinkai K, Liao XC, Beebe AM, Coffman RL, Littman DR, Locksley RM (1999) Impaired NFATc translocation and failure of Th2 development in Itk-deficient CD4+ T cells. Immunity 11:399–409
32. Mueller C, August A (2003) Attenuation of immunological symptoms of allergic asthma in mice lacking the tyrosine kinase ITK. J Immunol 170:5056–5063
33. Li H, Fedorov DG, Nagata T, Kitaura K, Jensen JH, Gordon MS (2010) Energy gradients in combined fragment molecular orbital and polarizable continuum model (FMO/PCM) calculation. J Comput Chem 31:778–790

Chapter 4

Analyzing Interactions with the Fragment Molecular Orbital Method

Dmitri G. Fedorov

Abstract

Basic concepts in the analysis of binding using the fragment molecular orbital method are discussed at length: polarization, desolvation, and interaction. The components in the pair interaction energy decomposition analysis are introduced, and the analysis is illustrated for a water dimer and a protein–ligand complex.

Key words Interaction, Binding, Energy decomposition analysis, FMO, PIEDA, Polarization, Charge transfer, Solvent screening, Dispersion, Protein–ligand complex, Water dimer, Desolvation

1 Introduction

Biochemical systems, such as DNA or proteins, can consist of many thousands of atoms. Although quantum-mechanical (QM) methods are accurate and general, especially ab initio QM methods considering electron correlation, their cost steeply increases with the system size. Various routes to reducing this scaling have been explored [1, 2], and a promising direction is to use fragmentation [3–6]. A molecular system can be divided into pieces called fragments, which are calculated using QM methods, so that at least formally, the scaling of computational cost is proportional to the number of fragments. Fragment-based methods are suitable for fragment-based drug discovery [7].

In chemistry functional groups form a key concept, and molecular environment perturbs their properties (e.g., the acidity of carboxyls COO^-). Likewise, one can consider molecular fragments. Fragments can be computed taking the environment into consideration, so a water molecule near a negative amino acid residue is not the same as another water near a neutral residue. The environment is accounted for via an *embedding potential*, which changes the properties (the electronic state) of each

Alexander Heifetz (ed.), *Quantum Mechanics in Drug Discovery*, Methods in Molecular Biology, vol. 2114,
https://doi.org/10.1007/978-1-0716-0282-9_4,

fragment, which is known as the *polarization*. Another key concept is the *interaction* between fragments, which is the main subject of this chapter. Adding an embedding and considering interactions tends to increase the computational cost, and much effort has been invested in reducing the cost of the fragmentation method by using judicious approximations [3].

Fragments are typically larger than functional groups, although there are some rare examples when an atom can be made into a fragment, for example, for inert gases. Typically, fragments are chosen in such a way that they are self-contained in the sense that the electron density is local and can be redistributed *within* a fragment due to polarization, but the density transfer *between* fragments should be kept small. Systematic tests evaluating various choices of fragmentation have been conducted maximizing the accuracy in the total properties, which often results in fragments being different from the well-established chemical moieties. For instance, for a protein fragments can differ from conventional amino acid residues.

This chapter is focused on the fragment molecular orbital (FMO) method [8] implemented [9, 10] in GAMESS [11, 12], which is available free of charge [13]. There are several reviews of FMO that can be consulted for details of the method [14–17].

Because of their lost cost, molecular mechanics (MM) simulations are often applied to study biochemical systems. QM calculations are also increasingly used for large systems [18, 19]. Traditional and most commonly used force fields are not polarizable, neglect charge transfer, and can fail to describe the directionality of bonding if the potentials depend only on distances between atoms, not on angles, for example, for *p*-orbitals in halogens [20].

There are novel force fields based on machine learning [21] and those including polarization [22, 23]. Alternatively, there are fast parametrized QM methods, fitted to experiment or higher-order QM methods [24, 25], and fragment-based methods have been hailed as a new generation of force fields [4].

Nevertheless, many MM applications are still conducted with nonpolarizable force fields. Although biased to reproducing properties for certain model systems, these methods are fine-tuned and fast, which are their two main advantages. The disadvantages include the issue of mixing parameters for different kinds of subsystems, e.g., water and protein, the problem of defining parameters for ligands, and treating the coordination of metal ions. Force fields are often parametrized for well-defined hybridizations and standard bond lengths and angles and are less accurate when real systems feature a departure from them, not to mention breaking and making chemical bonds.

Biochemical systems are typically not rigid, with a molecular structure fluctuating with time. This major complexity requires very expensive time evolution studies, usually accomplished with molecular dynamics (MD) simulations. In this aspect, MM approaches are still advantageous as they allow longer time simulations. However, QM methods catch up [26, 27] and can be used for MD simulations.

In general, molecular structures should be pre-optimized for QM calculations, because distorted bonds and steric clashes can cause convergence failure in QM calculations. Typically, experimental or otherwise constructed structures are first optimized with force fields. The structures can be refined using the frozen domain FMO [28, 29], FMO/MM [30, 31], and FMO-DFTB [32] methods. It is also possible to locate the minimum energy crossing of the energy surfaces of two different electronic states, for instance, to study reactivity of iron-containing proteins [33].

2 Binding Energy in Solution: Main Concepts

First, it should be made clear what energy is meant in this chapter. Although ideally one would like to look at the free energy (the Gibbs energy), it is rather hard to calculate it using quantum-mechanical approaches for large molecular systems. There are ways of evaluating entropy with force fields [34], an ensemble approach from statistical thermodynamics by treating isomers as states [35], or rotatable bond model [36]. Also, one can add the vibrational contribution using FMO to get the enthalpy at 0 K [37, 38].

In the context of this chapter, energy refers to the internal energy, i.e., QM energy. With an exception of the section below that deals with thermal fluctuations, it is the energy of a single structure, obtained either from a geometry optimization or MD.

Next, it is necessary to define interaction and binding energies, as there seem to be some confusion and ambiguity about them. The binding energy ΔE is the energy required to form a complex AB from two isolated (denoted by 0) molecules A and B, with A, B, and AB in their respective energy minima.

$$\Delta E = E^{\mathrm{AB}} - E_0^{\mathrm{A}} - E_0^{\mathrm{B}} \tag{1}$$

The interaction energy between A and B is the energy gained or lost due to the contact between two subsystems A and B in the complex AB:

$$\Delta E_{\mathrm{AB}} = E^{\mathrm{AB}} - E'^{\mathrm{A}} - E'^{\mathrm{B}} \tag{2}$$

Here, the energies of isolated systems do not appear; only the complex AB is computed. E'^{A} and E'^{B} are the internal energies of A and B in the complex AB.

In the subsystem analysis [39], the binding energy ΔE is decomposed as

$$\Delta E = \Delta E_{A}^{def} + \Delta E_{B}^{def} + \Delta E_{A}^{desolv} + \Delta E_{B}^{desolv} + \Delta E'^{pol}_{A} + \Delta E'^{pol}_{B} + \Delta E_{AB} \quad (3)$$

into the deformation (def) energies of A and B $\left(\Delta E_{A}^{def} + \Delta E_{B}^{def}\right)$, the desolvation (desolv) [40] energies of A and B $\left(\Delta E_{A}^{desolv} + \Delta E_{B}^{desolv}\right)$, the destabilization component of the polarization (pol) energies of A and B $\left(\Delta E'^{pol}_{A} + \Delta E'^{pol}_{B}\right)$, and the interaction energy ΔE_{AB}.

These concepts are visualized in Fig. 1. The binding pocket of protein A is enlarged for binding and the structure of ligand B also changes (deformation). The surfaces of A and B are partially desolvated (blue solvent is replaced by white vacuum). Then, in the complex, both A and B are polarized (turn pink) facing bare surfaces of each other. Finally, deformed, desolvated and polarized A and B interact.

The concept of polarization used in FMO [41] comes from the Kitaura-Morokuma energy decomposition analysis (KM-EDA) [42–44]. Isolated molecules A and B, forming a complex, are in their most stable state. The internal energy of A (or B) inside the complex is higher than in the isolated state. This difference is the destabilization contribution to the polarization energy. The stabilization component is a part of the interaction energy.

The polarization contribution to binding in solution is differential, not absolute. This is because in the isolated state, molecules A and B are already polarized by the solvent, and in the complex, a part of the polarization due to the solvent is replaced by the polarization by the partner molecule. The difference in these two types of polarization contributes to the binding energies. This differential effect may be expected to be relatively small for neutral molecules, which is why nonpolarizable force fields although missing it may still give reasonable results.

For charged systems the interaction energy is typically a manyfold overestimate of the binding energy. For example, for the binding of Trp-cage (1L2Y) to an anionic ligand [39], the total binding and interaction energies were computed to be −9.5 and −66.9 kcal/mol, different by a factor of 7. The difference between the binding interaction and binding energies is the contribution of fragments, i.e., polarization, deformation, and desolvation of monomers (Eq. 3). It is clear that interaction energies may be very misleading, and instead, it was suggested [39] to divide the

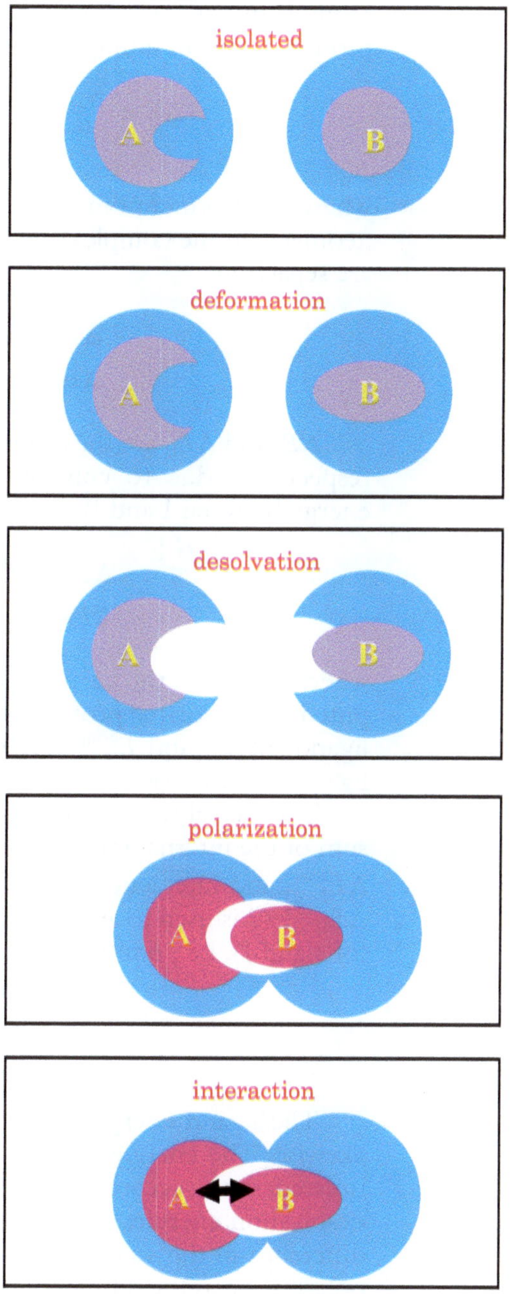

$$\Delta E = \Delta E_A^{\mathrm{def}} + \Delta E_B^{\mathrm{def}} + \Delta E_A^{\mathrm{desolv}} + \Delta E_B^{\mathrm{desolv}} + \Delta E'^{\mathrm{pol}}_A + \Delta E'^{\mathrm{pol}}_B + \Delta E_{AB}$$

Fig. 1 Concepts used to analyze binding between subsystems A and B (blue circles represent the solvent, purple and pink objects represent the solute)

binding energy into fragment binding energies. For a complex of a protein A and an unfragmented ligand B,

$$\Delta E = \sum_{\mathrm{I} \in \mathrm{A,B}} \Delta E_{\mathrm{I}}^{\mathrm{bind}} + \Delta E_{\mathrm{A}}^{\mathrm{def}} + \Delta E_{\mathrm{B}}^{\mathrm{def}} \quad (4)$$

where the fragment binding energies $\Delta E_{\mathrm{I}}^{\mathrm{bind}}$ are computed for the geometry in the complex and deformation energies $\Delta E_{\mathrm{A}}^{\mathrm{def}} + \Delta E_{\mathrm{B}}^{\mathrm{def}}$ are separated.

$$\begin{aligned} \Delta E_{\mathrm{I}}^{\mathrm{bind}} &= \Delta E_{\mathrm{I}} + \Delta E_{\mathrm{IB}} \\ \Delta E_{\mathrm{B}}^{\mathrm{bind}} &= \Delta E_{\mathrm{B}} \end{aligned} \quad (5)$$

ΔE_{I} and ΔE_{B} are the changes in the partial energies of I and B, respectively, due to complexation, and ΔE_{IB} is the interaction energy between I and B.

$$\begin{aligned} \Delta E_{\mathrm{I}} &= E_{\mathrm{I}}^{\mathrm{bound}} - E_{\mathrm{I}}^{\mathrm{isolated}} \\ \Delta E_{\mathrm{B}} &= E_{\mathrm{B}}^{\mathrm{bound}} - E_{\mathrm{B}}^{\mathrm{isolated}} \end{aligned} \quad (6)$$

where $E_{\mathrm{I}}^{\mathrm{isolated}}$ is the partial energy of fragment (≈residue) I computed in the isolated protein and $E_{\mathrm{B}}^{\mathrm{isolated}}$ is the energy of the isolated ligand. $E_{\mathrm{I}}^{\mathrm{bound}}$ and $E_{\mathrm{B}}^{\mathrm{bound}}$ are the partial energy of fragment I and ligand B in the complex, respectively.

The partial residue energy $E_{\mathrm{I}}^{\mathrm{bound}}$ (and $E_{\mathrm{I}}^{\mathrm{isolated}}$) in Eq. 6 is the sum of the internal fragment energy E'_{I}, fragment solvation energy $\Delta E_{\mathrm{I}}^{\mathrm{solv}}$, and interactions of I with other residues J in the protein, ΔE_{IJ} (1/2 is used to avoid double counting of the interaction both for I and J), i.e., omitting the subscript "bound" and "isolated":

$$E_{\mathrm{I}} = E'_{\mathrm{I}} + \Delta E_{\mathrm{I}}^{\mathrm{solv}} + \frac{1}{2} \sum_{\mathrm{J} \neq \mathrm{I} \in \mathrm{A}} \Delta E_{\mathrm{IJ}} \quad (7)$$

The ligand energies $E_{\mathrm{B}}^{\mathrm{bound}}$ and $E_{\mathrm{B}}^{\mathrm{isolated}}$ in Eq. 6 are defined likewise. The energies in Eq. 7 are subtracted for the bound and isolated states according to Eq. 6 yielding the polarization and desolvation energies. The desolvation arises as the bound-isolated difference in the solvation energies of fragment I, $\Delta E_{\mathrm{I}}^{\mathrm{solv}}$, and the polarization as the difference in both the internal E'_{I} and interaction energies ΔE_{IJ}. The presence of ligand B changes the residue-residue interactions ΔE_{IJ} in protein A (a three-body effect involving I, J, and B) via the embedding.

The total interaction energy in Eq. 3 is

$$\Delta E_{\mathrm{AB}} = \sum_{\mathrm{I} \in \mathrm{A}} \Delta E_{\mathrm{IB}} \quad (8)$$

and the connection of the other two terms in Eq. 3 to partial energies is

$$\sum_{\mathrm{I}\in\mathrm{A}} \Delta E_{\mathrm{I}} = \Delta E_{\mathrm{A}}^{\mathrm{desolv}} + \Delta E'^{\mathrm{pol}}_{\mathrm{A}}$$
$$\Delta E_{\mathrm{B}} = \Delta E_{\mathrm{B}}^{\mathrm{desolv}} + \Delta E'^{\mathrm{pol}}_{\mathrm{B}} \tag{9}$$

Equation 3 is the "macroscopic" way of analyzing the interactions treating the protein as a subsystem, whereas Eq. 4, equivalent to Eq. 3, is the "microscopic" way based on residues. Finally, if ligand B is split into several fragments, then the quantities related to it are split as for residues in the protein. Summarizing, fragment binding energies $\Delta E_{\mathrm{I}}^{\mathrm{bind}}$ including the polarization and desolvation are a better measure of binding than interaction energies ΔE_{IB}.

For protein–protein interactions (PPI), Eq. 5 (ligand B is replaced with protein B) is slightly modified by splitting equally the inter-protein interaction energy as

$$\Delta E_{\mathrm{I}}^{\mathrm{bind}} = \Delta E_{\mathrm{I}} + \frac{1}{2}\sum_{\mathrm{J}}^{\text{protein 2}} \Delta E_{\mathrm{IJ}} \quad \text{for} \quad \mathrm{I}\in\text{protein 1}$$
$$\Delta E_{\mathrm{J}}^{\mathrm{bind}} = \Delta E_{\mathrm{J}} + \frac{1}{2}\sum_{\mathrm{I}}^{\text{protein 1}} \Delta E_{\mathrm{IJ}} \quad \text{for} \quad \mathrm{J}\in\text{protein 2} \tag{10}$$

Also, Eq. 7 is applied to residues in each protein separately, where the intra-protein interaction energy is also split equally among residues. An example of an application of FMO to study PPI is given in Chapter 13 of this book.

3 Interaction Analyses in FMO

3.1 Overview

For a molecular system divided into N fragments, the two-body expansion [45] of the energy E in FMO (FMO2) is

$$E = \sum_{\mathrm{I}}^{\mathrm{N}} E'_{\mathrm{I}} + \sum_{\mathrm{I}>\mathrm{J}}^{\mathrm{N}} \Delta E_{\mathrm{IJ}} \tag{11}$$

where E'_{I} is the internal energy of fragment I and ΔE_{IJ} is the pair interaction energy (PIE) of I and J, sometimes called the interfragment interaction energy (IFIE).

In many FMO applications, the focus is on analyzing ΔE_{IJ} values [46], on the basis of which residues important for binding a ligand are identified. There are several extensions: using configuration interaction, it was proposed to define the polarization and charge transfer in the configuration analysis for fragment interaction (CAFI) [47] and using localized orbitals to define the dispersion in the fragment interaction analysis based on local MP2 (FILM) [48]. Both CAFI and FILM are based on localized orbitals and as such provide useful information at a subfragment level, e.g., at the level of functional groups. Later, it was suggested to augment IFIE with statistical corrections (SC) to account in a parametrized way the effect of temperature in SCIFIE [49].

Another group of analyses is based on the incorporation of KM-EDA into FMO in the PIE decomposition analysis (PIEDA) [50]. Alternatively, there is the interaction energy analysis (IEA) [51] of solute-solvent interactions based on the effective fragment potential (EFP) model of solvation combined with FMO [52].

PIEs can be corrected using averaged many-body contributions [53] and the counterpoise (CP) corrections for basis set superposition error (BSSE) [54, 55]. Several levels of QM and MM calculations can be mixed in our own N-layer integrated molecular orbital molecular mechanics (ONIOM) interfaced with FMO [56].

Data analysis tools can be applied to categorize interactions using clustering [57], topological [58], and other schemes [59]. Pair interactions are also used as descriptors in structure-activity relationships [36, 60, 61] or as parameters in macroscopic models, for instance, in dissipative particle dynamics [62].

An analysis described for binding can be applied to analyze chemical reactions (binding can be taken to be a special case of a chemical reaction, A + B → AB). Such analysis can be done for enzymes in order to identify residues affecting the transition state barrier [63–65]. One can also compare isomers (an isomerization A → A′) as related to protein folding [66, 67] and preferences of α-helices over β-turns and other secondary motifs [50].

3.2 Decomposition Analyses

PIEDA was originally formulated for closed shell systems in gas phase [50], later extended to take into account solvent effects [68], to treat open-shell (radical) systems [69] and to incorporate basis set corrections [70, 71]. Excluding density functional tight binding (DFTB) [72], the decomposition is

$$\Delta E_{IJ} = \Delta E_{IJ}^{ES} + \Delta E_{IJ}^{EX} + \Delta E_{IJ}^{CT+MIX} + \Delta E_{IJ}^{DI} + \Delta E_{IJ}^{RC} + \Delta E_{IJ}^{BS} + \Delta E_{IJ}^{SOLV} \quad (12)$$

The DFTB case is described separately below. For non-DFTB methods, the interaction energy is a sum of at most 7 contributions: the electrostatic (ES); exchange-repulsion (EX) term, charge-transfer (CT) and MIX (CT + MIX) terms, dispersion (DI), remainder correlation (RC), basis set correction (BS), and solvent screening (SOLV). A schematic picture of various terms is shown in Fig. 2.

The electrostatic (ES) term is the Coulomb interaction between polarized fragments in vacuum. The ES term is computed by adding electron-electron, electron-nucleus, and nucleus-nucleus

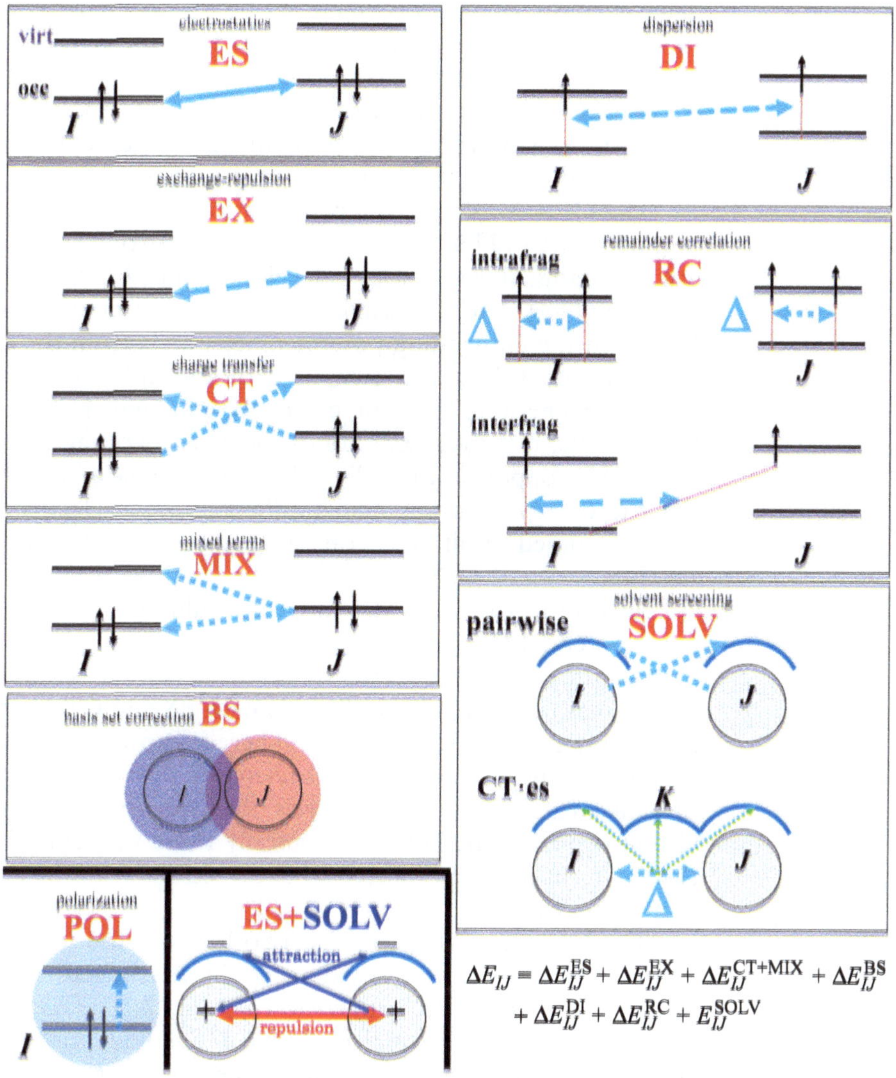

$$\Delta E_{IJ} = \Delta E_{IJ}^{\mathrm{ES}} + \Delta E_{IJ}^{\mathrm{EX}} + \Delta E_{IJ}^{\mathrm{CT+MIX}} + \Delta E_{IJ}^{\mathrm{BS}} + \Delta E_{IJ}^{\mathrm{DI}} + \Delta E_{IJ}^{\mathrm{RC}} + E_{IJ}^{\mathrm{SOLV}}$$

Fig. 2 Polarization and components of the pair interaction energy decomposition analysis. Δ indicates a differential effect. See main text for other notations

contributions; the former is calculated using the QM electron density distribution polarized by the environment. By separating the electron density into isolated and polarization contributions, one can define the stabilization component to the polarization, which is included in the ES term and can be separated [50]. Schematically, ES is the Coulomb interaction between charge distributions determined by the occupied orbitals (Fig. 2).

The exchange-repulsion (EX) term comes from the Pauli repulsion between fermions (electrons). This term can be decomposed into exchange and repulsion interaction, corresponding to the energy of the exchange operator and repulsion due to the delocalization of the occupied monomer orbitals, respectively [43]. Although in some EDA schemes [73] the two components of the EX term are separated, both in KM-EDA and PIEDA, they are considered together. The EX term is due to the short-ranged Pauli repulsion and the corresponding exchange between electron densities determined by occupied orbitals (Fig. 2).

The CT + MIX term is a combination of CT and MIX. In the original EDA [42] incorporated into PIEDA, this combination is not disentangled, but in a later development of EDA, the two were separated [43]. The CT term is the charge-transfer energy between I and J in the embedding potential. Separately from the energetic contribution, CT is also computed in the units of electrons. It happens sometimes that there are two charge-transfer channels, for example, two hydrogen bonds, between two fragments. If the direction is opposite, the CT in the units of charge may be small, but the energy is not. Schematically, CT is the electron transfer from occupied orbitals of one fragment to virtual orbitals of another.

According to the models used in building the effective fragment potential (EFP) [74], both EX and CT energies are dependent on the overlap between molecular orbitals of two fragments. A large (small) overlap leads to large (small) EX and CT terms, so the two terms are usually correlated. There is, however, a substantial difference (Fig. 2): the EX term is related to the overlap of occupied orbitals, whereas the CT terms is the overlap of occupied and virtual orbitals. The latter are typically more delocalized, so that the CT term may be expected to decay slower with the separation than the EX term.

The MIX term is the coupling of other contributions in PIEDA defined for monomers in the dimer [43]. A schematic representation of one kind of MIX term is shown in Fig. 2. The EX interaction (the lower arrow) is coupled to the CT interaction (the upper arrow), because within the dimer, the charge transfer between fragments affects the Pauli repulsion, so that there is an EX·CT coupling, which is a part of the MIX term. Normally, however, the MIX contribution is small (less than 1 kcal/mol) and the CT + MIX term is dominated by CT.

The dispersion (DI) and remainder correlation (RC) terms can be disentangled for density functional theory (DFT), because the DI contribution in DFT is computed explicitly using a parametrized model [75] and the RC term is also separately defined. For RHF, RC is zero, and DI is computed with the same parametrized model as in DFT. Second-order Møller-Plesset perturbation theory (MP2) and coupled cluster (CC) are examples of methods where the sum DI + RC is computed as one number. Although in literature one can find recipes [76, 77] for separating DI from RC, at present they are not implemented in PIEDA. A schematic representation of the DI term is shown in Fig. 2. The dispersion can be thought of as the interaction between two instantaneous dipoles, due to two electron excitations, one per fragment. The DI term is due to the correlation between the electron distributions.

The remainder correlation (RC) is any kind of electron correlation in dimer IJ which is not dispersion. The RC terms can be divided into intra- and interfragment contributions. The former arises because the dispersion interaction in fragment I (or J) within dimer IJ is different from the value in the polarized fragment by itself. This is because the electron density in the dimer is different from the sum of two monomers; in other words, the first type of the RC term is a mainly the coupling of DI and charge transfer. The other, interfragment type of RC is due to the correlation between two electron excitations, one of the intrafragment type ($I \leftrightarrow I$ in Fig. 2, of the polarization nature) and the other of the interfragment type ($I \leftrightarrow J$, of the CT nature). A similar coupling of $J \leftrightarrow J$ to $I \leftrightarrow J$ excitations is not shown in Fig. 2 for simplicity.

In case of DFT, RC is computed from the correlation energy as determined by the correlation functional. Most approximate functionals [78] in DFT are of a local, one-electron nature that do not describe long-ranged two-electron dispersion, but they can describe short-ranged local electron correlation.

The dispersion model used in RHF, DFT, and DFTB is independent of the electronic state, so that there is no coupling of DI to CT (intrafragment RC in Fig. 2). There is a concept of the charge conserving electron correlation, used to define the dispersion [79], where a few extra diagrams in addition to those listed for DI and RC in Fig. 2 are given.

The BS term describes a basis set correction. In QM, results depend on the basis set. There are various families of basis sets, for instance, cc-pV*n*Z [80], systematically improving the accuracy as a function of an integer n ($n \geq 2$). Ultimately, there is a limit known as the complete basis set limit, which can be extrapolated to [81]. A finite and thus incomplete set of atomic orbitals used in QM calculations gives rise to the basis set superposition error (BSSE). This error arises when two systems A and B form a complex AB. Then, a part of the AB–A–B energy difference attributed to BSSE is due to the lowering of the internal energy of A and B, not to the binding. The BS terms is shown schematically in Fig. 2: the black circles represent fragments, and basis sets are shown as shaded larger circles. A basis set on fragment I extends into its neighbor J, lowering the energy of J in the dimer because the basis set is incomplete, but this energy lowering is not due to binding but rather due to the lowering of the internal energy.

Although large basis sets exist, the cost of using them is very high. They have convergence problems, and in case of fragment-based methods, the accuracy of fragmentation decreases when the basis set size increases. So, in practice four choices are available: (a) use a medium basis set only (no BS correction); (b) use a small basis set plus a parametrized BS correction [82]; (c) use a medium basis set with an explicit BS correction obtained as a difference of two values, one with a large and the other with a medium basis set; and (d) use a CP correction. The third choice, an explicit BS correction is possible with the auxiliary polarization (AP) approach [70], in which the BS correction is the difference in the pair interaction energies for fragments calculated without electrostatic embedding:

$$\Delta E_{IJ}^{BS} = \Delta E_{IJ}^{\text{no embedding}}(\text{large basis set}) - \Delta E_{IJ}^{\text{no embedding}}(\text{medium basis set}) \quad (13)$$

In the normal case of using a medium (e.g., cc-pVDZ) and large (e.g., aug-cc-pVDZ) basis sets, the interaction energy $\Delta E_{IJ}^{\text{no embedding}}$ in Eq. 13 is overestimated for the medium basis set because of BSSE, and ΔE_{IJ}^{BS} is typically repulsive. However, if one subtracts values of a small, minimum basis set (e.g., STO-3G) from those of a medium basis set (e.g., 3-21G), then the BS term can be attractive and as large as −5 kcal/mol. This is because minimum basis sets tend to be very compact and have a small BSSE, and medium basis sets have a large BSSE, so this type of a calculation makes no sense and should be avoided in general. The parametrized BS term [82] is repulsive.

The solvent screening (SOLV) is implemented for polarizable continuum model (PCM) and solvent model density (SMD) methods (and a similar term has been implemented for the Poisson-Boltzmann model [83]). There are two contributions to the SOLV term. The first screening term (labeled pairwise in Fig. 2) describes how much of the solute electrostatic energy is screened by the solvent. To clarify it, the scheme on the bottom of Fig. 2 shows how the repulsive ES term is counteracted by the attractive pairwise SOLV term, for two solute cations that each induce an anionic charge on the solvent surface represented by arcs (the pairwise solute-solvent interactions are halved [68] so that there is no double counting). In addition to the electrostatic screening, there is a contribution of non-electrostatic (dispersion and repulsion) terms to the pairwise screening; this is a non-electrostatic interaction between solute atoms in fragment I and solvent charges around fragment J and the opposite combination. It complements the electrostatic screening.

Recently, the screening model for FMO has been improved [84] by decoupling the charge quenching effect; the revised pairwise SOLV term can be described by a scheme similar to that in Fig. 2 for the original term, with the difference that in the new scheme the charges induced by each fragment on the whole surface are used to get the screening, as opposed to local sets of total induced charges in the old model. In the applications of FMO in this chapter, the old scheme is used.

The second screening term to the solvent screening, denoted by CT·es, represents the coupling of the charge transfer (CT) between solute fragments I and J to the electrostatic (es) embedding due to the solvent (the convention is to use capital letters such as "CT" for solute and small letters such as "es" for solvent). This embedding is from the whole solvent cavity, including the surfaces of I, J, and all $K \neq I, J$ (Fig. 2, Δ for the SOLV term denotes CT). Of the two types of solvent screening terms, the pairwise one is the long-ranged solvent screening dominant for charged fragments. The CT·es term is short-ranged because CT is short-ranged.

The polarization (POL) is a term defined per monomer not per dimer. For completeness, it is also shown in Fig. 2. POL is a many-body effect that can be viewed as a mixing of occupied and virtual orbitals exerted by the embedding potential (shown by a mesh). This mixing changes the electronic state of fragments and hence the embedding potential and the calculations are iterated until self-consistency. The POL and CT terms are analogous: POL can be called intrafragment charge transfer, whereas CT is interfragment.

Now, for DFTB, the pair interaction is

$$\Delta E_{\mathrm{IJ}} = \Delta E_{\mathrm{IJ}}^{\mathrm{ES}} + \Delta E_{\mathrm{IJ}}^{0} + \Delta E_{\mathrm{IJ}}^{\mathrm{CT\cdot ES}} + \Delta E_{\mathrm{IJ}}^{\mathrm{DI}} + \Delta E_{\mathrm{IJ}}^{\mathrm{SOLV}} \tag{14}$$

DFTB is a parametrized QM method which has no explicit exchange (although implicitly it is present in the zeroth-order Hamiltonian). Therefore, the set of PIEDA terms in Eq. 14 is different from Eq. 12. The "0" term is the energy coming from the zeroth-order Hamiltonian in DFTB; it is the non-electrostatic, non-dispersion type of interaction. The CT·ES is the charge transfer coupled to the electrostatic (ES) embedding due to other fragments; it is a CT energy. The dispersion term is computed using a parametrized model, like in RHF and DFTB. Although most commonly used DFTB parameter sets are based on a minimum basis set, the BSSE is reduced by the fact that core electrons are not retained in DFTB, and BSSE is especially large for core electrons, as was also reported for FMO combined with core potentials [85].

In FMO, there is the problem of analyzing interactions for a pair of fragments that are connected by a covalent bond. In this case, there is a very strong interaction on the order of 14–15 hartree for C–C bonds, due to the nature of the fragment definition and the close proximity of two fragments [86]. There is a way [50] of correcting for the artefacts with the BDA correction scheme, used in the application of PIEDA to γ-peptides [87]. Another problematic issue is that fragmenting at a peptide bond results in poor accuracy, so polypeptides are usually fragmented at Cα atoms, i.e., fragments differ from residues by CO groups.

What use can one make of individual PIEDA components? One usage is to label interactions [50], because several components in PIEDA provide unique signatures for specific interactions, such as CH–π or hydrogen bonding. One can also define measures of binding based on a novel functional form taking the PIEDA components as parameters, different from the classical combination of the simple addition (Eq. 12), as in the attraction energy signature [46]. The EX component, if not accompanied by other attractive terms, points at a substantial steric repulsion and a need to optimize the structure (in some cases steric hindrance is inevitable to gain more energy for other contacts). The balance between electrostatic and non-electrostatic terms can be used to suggest modifications of ligands by changing polarity of substituents. To compare different ligands, the concept of fragment efficiency was applied to FMO [88].

3.3 Thermal Fluctuations

Typically, analyses in FMO are done for a single static structure. Although there were attempts to average the results for a small set of structures from MD [89], these results have the problem that a small set may not be representative. The robust way is to average results in FMO/MD simulations for all structures. The current time scale of FMO/MD is 30 and 1000 ps using RHF [90] and DFTB [91], respectively. In the fluctuation analysis [72] for FMO, one averages the energy in Eq. 11 during MD (as shown by brackets):

$$\langle E \rangle = \sum_{\mathrm{I}}^{\mathrm{N}} \langle E'_{\mathrm{I}} \rangle + \sum_{\mathrm{I}>\mathrm{J}}^{\mathrm{N}} \langle \Delta E_{\mathrm{IJ}} \rangle \tag{15}$$

The averaged quantities in Eq. 15 can be separated into the values for the minimum energy structure (the minimum found during MD) and the deviation from it, the fluctuation. Due to thermal movement, strong binding tends to be weakened relative to the minimum because the averaged distance between two interacting systems is longer in MD compared to the minimum [72]. However, some weak interactions in the global minimum may be strong in a local minimum, and such interactions may be enhanced by thermal fluctuations. Temperature thus tends to reduce large and enhance some small interactions.

4 Demonstrative Applications

4.1 Water Dimer

To get a feeling for various PIEDA terms, a calculation of a water dimer was conducted using GAMESS [11]. DFT was used in PIEDA so that all 7 components can be separated. The geometry was optimized in the gas phase at the MP2/6-311++G∗∗ level. A

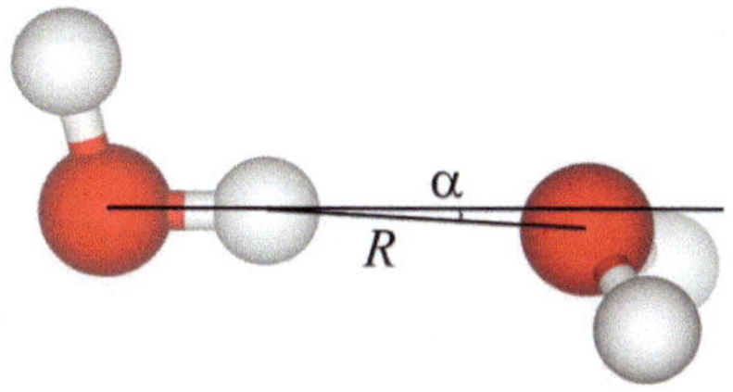

Fig. 3 Water dimer. *R* is the separation between H (left molecule) and O (right molecule). α is the angle showing the deviation of the O (right molecule) from the straight line formed by OH (left molecule)

B3LYP optimization resulted in minimum different from the high-level calculation [92], but the MP2 structure was very similar to it: the α angles (Fig. 3) in this and the high-level work [92] was 3.4 and 5.1°, respectively, whereas the O–O distance value at the minimum was 2.911 and 2.930 Å, respectively. After minimization, the right molecule in Fig. 3 was shifted along the line passing through the horizontal OH bond of the left molecule, so that the separation R was equal to some desired values (at the minimum, R is 1.944 Å). PIEDA/C-PCM<1> (IDISP = 0) calculations were conducted at the level of B3LYP/6-31G∗∗, and the BS term was estimated with the AP scheme as a difference between 6-311++G∗∗ and 6-31G∗∗ results without embedding.

The binding energy ΔE was estimated as the difference between the FMO energies of the dimer and two monomers, $\Delta E = E(\text{dimer}) - 2E(\text{monomer})$, taking the energy of a water molecule in its minimum geometry. For simplicity, Eq. 3 is rewritten as

$$\Delta E = \Delta E_{12} + \Delta E_1 + \Delta E_2 \tag{16}$$

That is, the binding energy is written as a sum of the interaction energy ΔE_{12} and two monomer contributions:

$$\Delta E_{\mathrm{I}} = \Delta E_{\mathrm{I}}^{\mathrm{def}} + \Delta E_{\mathrm{I}}^{\mathrm{desolv}} + \Delta E_{\mathrm{I}}'^{\mathrm{pol}} \tag{17}$$

The results are shown in Fig. 4a. A small hump near 3.5 A may be caused by the simplified way of generating the curve, i.e., the structure at each R was not reoptimized for other degrees of freedom (instead, two molecules were simply shifted).

The components of pair interactions are shown in Figs. 4b and 5. For this system consisting of two neutral water fragments, the interaction energy ΔE_{12} is only a ~50% overestimate of the binding energy ΔE. The total interaction energy is a uniformly decaying function of R, which becomes nearly zero (in this section, it means <0.1 kcal/mol) at about 4.5 Å. Curiously, the total PIE is almost identical to ES except at $R < 2$ Å.

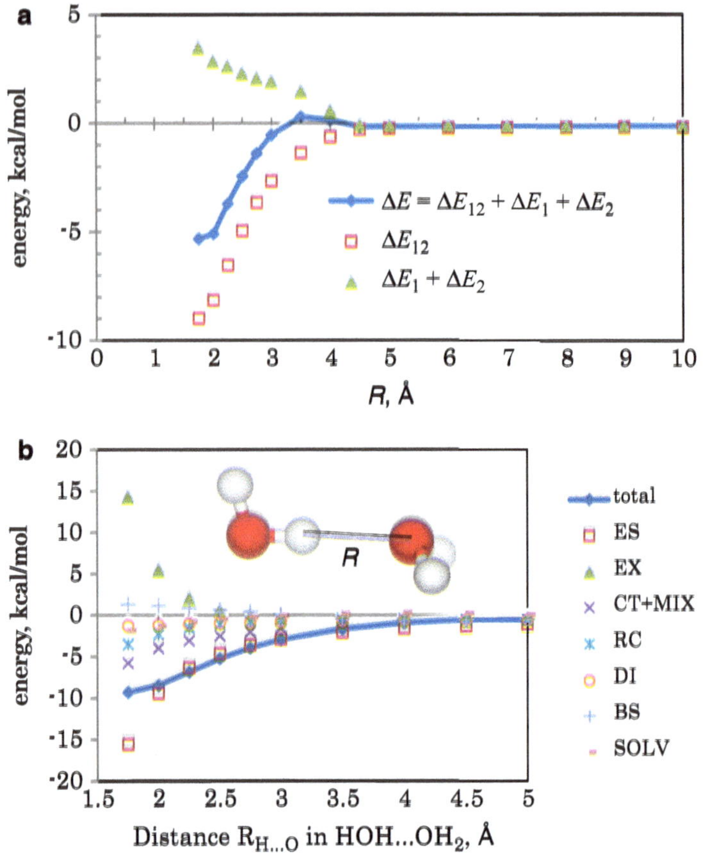

Fig. 4 (**a**) Binding energy ΔE in water dimer as a function of the separation R decomposed into pair interaction ΔE_{12} and monomer $\Delta E_1 + \Delta E_2$ contributions. (**b**) Total pair interaction energy between water molecules (numbered 1 and 2) and its components (B3LYP/AP:6-31G∗∗:6311++G∗∗)

The electrostatic component decays slowly as dictated by its $1/R$ Coulomb nature at long distances (zero at 10 Å). The EX term is repulsive and short-ranged (nearly zero at 3.5 Å). It decays quicker than ES, which may be related to the exponential decay of MOs and their overlap. The CT + MIX term is nearly zero at 4 Å, somewhat further than the EX term. This feature as described above is attributed to the involvement of virtual orbitals in CT.

The RC term is nearly zero at 3 Å. It is attractive, as it describes electron correlation that lowers the energy. Apparently, the charge transfer in the dimer leads to a gain in the electron correlation energy in each monomer (intrafragment RC) and in the interfragment RC term (Fig. 2). The DI term is small (about 0.7 kcal/mol at the shortest computed separation) because water fragments have a very small, one-atom type of contact (Fig. 3). The DI term becomes nearly zero at 3.5 Å. The BS term is short-ranged (nearly 0 at 3.5 Å) and repulsive due to BSSE.

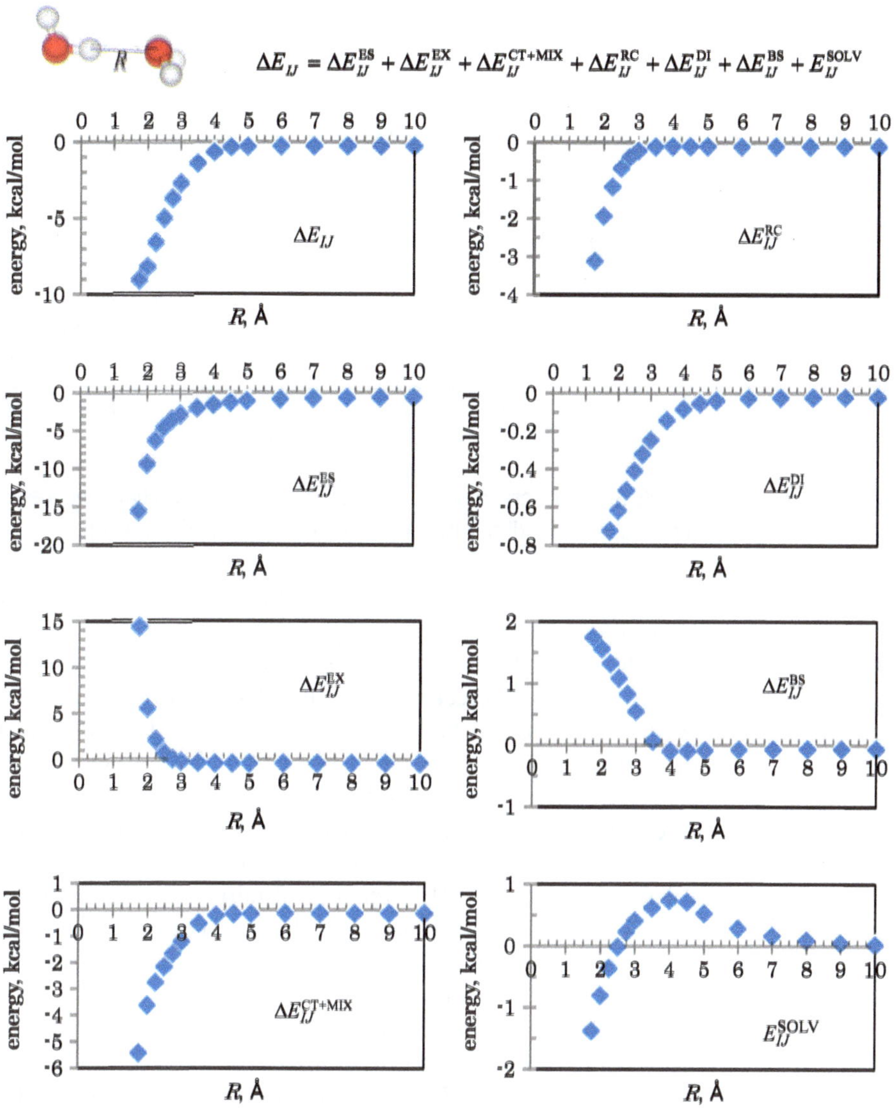

Fig. 5 Components of the pair interaction energy between two water molecules in the dimer (B3LYP/AP:6-31G**:6311++G**)

The most interesting dependence on R is observed for the SOLV term, nearly zero at 10 Å. At longer separations (about 4 Å and larger), the solvent screening is repulsive. It is the direct solvent screening, counteracting the electrostatic interaction between solute fragments, so that the long-range behavior of the solute-solute ES term is accompanied by a repulsive solute-solvent interaction similar in magnitude. However, at distances shorter than 4 Å, there is an opposite trend: the SOLV term is an increasing function of R. The solvent screening can be understood by looking at the two contributions separately, as shown in Fig. 6.

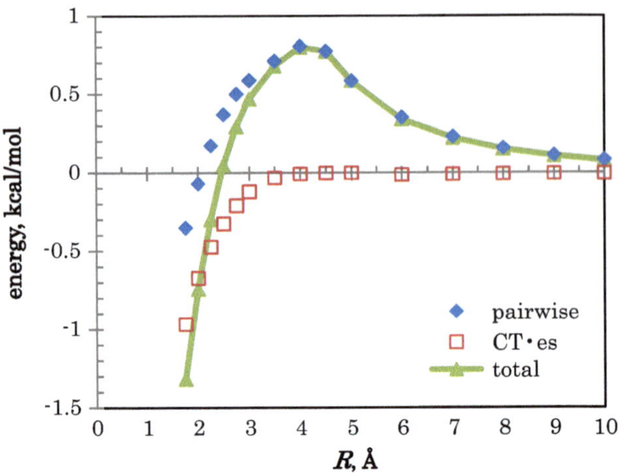

Fig. 6 Solvent screening contributions for the interaction between two water molecules in the dimer (B3LYP/AP:6-31G∗∗:6311++G∗∗)

The pairwise screening of the Coulomb interaction $\frac{Q_1 Q_2}{R}$ between two solute charges Q_1 and Q_2 at long distances R behaves [68] for $N = 2$ with the charge escape correction as $-\frac{\varepsilon-1}{\varepsilon R} Q_1 Q_2$, where ε is the dielectric constant of the solvent, so that the two terms when added yield the expected $\frac{Q_1 Q_2}{\varepsilon R}$ behavior of the total electrostatic interaction (for more than 2 fragments, $N > 2$, the screening is complicated by the many-body charge quenching effect [68]). The attractive CT·es term decays quickly with distance. When the CT·es and pairwise components are added, the observed form of the total SOLV term is obtained. In this case both fragments are neutral so that especially at short separation, both pairwise and CT·es are of a similar magnitude, but for ions the former term is typically dominant.

In this calculation, IDISP was set to 0, which neglects the non-electrostatic screening contributions. Normally, in applications IDISP is set to 1. There were two reasons for using IDISP = 0 here: the trends are easier to discuss for a purely electrostatic screening, and even more importantly, the IDISP = 1 model becomes invalid for multiple cavities at long separations. Namely, in this system, starting from 2.75 Å, the solute cavity splits into two distinct cavities, one for each water molecule. Schematically, it is shown in Fig. 7. Normally, if one studies only the complex, the problem of multiple cavities does not arise though.

4.2 Residue–Ligand Interactions

For a biochemical example, the interaction energies in the complex of prostaglandin H(2) synthase-1 with the reversible competitive inhibitor ibuprofen (PDB: 1EQG) containing 17,767 atoms were computed using PIEDA/C-PCM<1> and two QM methods, MP2/cc-pVDZ and DFT(CAM-B3LYP/D3(BJ))/AP:6-31G∗∗:6311++G∗∗ (in the latter, auxiliary polarization calculations were done in the gas phase to estimate the BS term). The

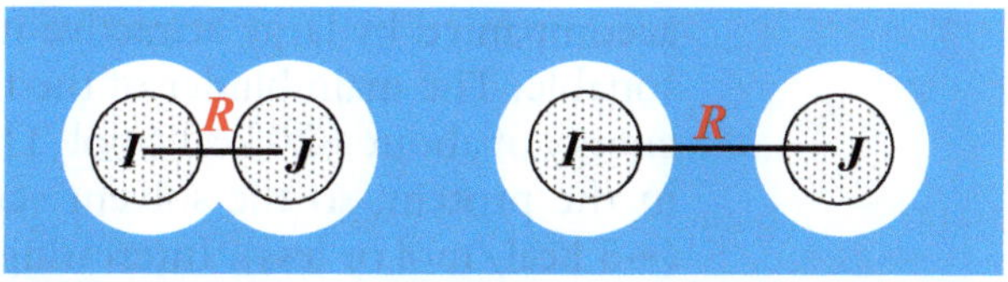

Fig. 7 Splitting of the solute cavity for dimer IJ into two separate monomer cavities at a longer separation *R*

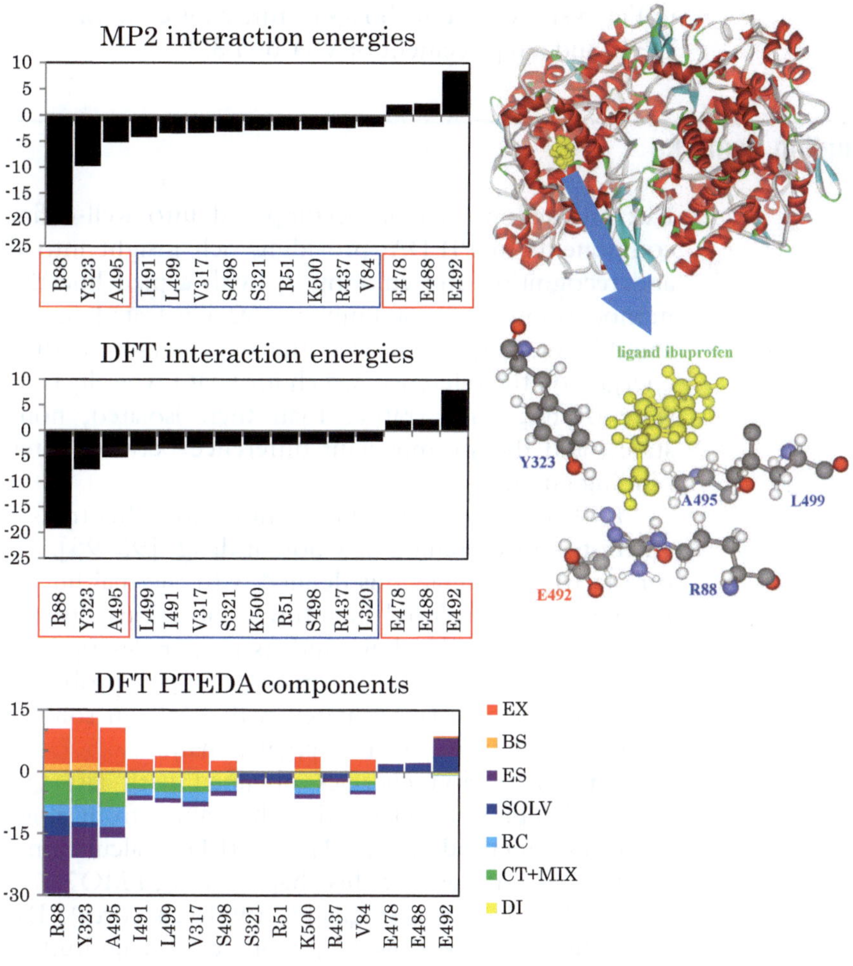

Fig. 8 Prostaglandin H(2) synthase-1 in complex (PDB: 1EQG) with the reversible competitive inhibitor ibuprofen (yellow), interaction energies of residue fragments with the ligand, and their pied components (all in kcal/mol). Conserved and different residues in the comparison of MP2/cc-pVDZ and CAM-B3LYP/AP:6-31G∗∗:6311++G∗∗ are shown as red and blue boxes, respectively

structure was optimized with the frozen domain FMO method earlier [28] using explicit water, replaced by continuum in this work.

All significant interactions $|\Delta E_{IJ}| > 2$ kcal/mol are plotted in Fig. 8: 12 attractive and 3 repulsive. Large EX components are accompanied by large attractive terms, so the structure seems reasonable. The main binder of the ligand is R88 with the interaction energy of about −20 kcal/mol. The ligand is neutral and it is buried in the protein, so the solvent screening SOLV is relatively small (−5 kcal/mol or less). Interestingly, it enhances the ES interaction for several residues. The results for DFT with dispersion correction and MP2 are very similar for the six main interactions (R88, Y323, A495, E478, E488, and E492 shown as red boxes), with a difference of 2.1 kcal/mol for Y323). However, the other nine residues in Fig. 8 shown as blue boxes feature a different order (MP2 versus DFT) and a replacement of V84 by L320.

5 Concluding Remarks

Pair interactions can be decomposed into well-defined physical components in PIEDA, providing rich insight into the binding and recognition and coloring up with a pied hue the otherwise monochromic rows of numbers (as illustrated in Fig. 8). One should not forget, however, the dependence of the interaction energies on the reference, which for FMO typically is the polarized state of fragments (rather than their isolated, non-interacting state), and the so important difference between interaction and binding energies.

FMO can be used in combination with other tools in medicinal chemistry for designing new potent drugs [93–96]. The temperature factor to interactions defined with thermal fluctuations [72] remains to be explored in future applications. Entropy plays an important role in binding, and its proper description and incorporation into the analysis remains an open problem. Molecular electrostatic potential evaluated with FMO in solution [97] can be used to supplement the energetic analysis.

The interaction energies obtained with fast QM methods [82, 32] highly correlate with the values from slower accurate traditional methods [98, 72]. A PIEDA calculation of crambin (1CRN, 642 atoms) for this chapter using FMO2-DFTB3/PCM took only 19 s on a single PC (with 24 cores). DFT or MP2 calculations may be slower by 3 orders of magnitude [98], which currently means hours on a small PC cluster, still doable. Quantum-mechanical fragmentation methods are an affordable and promising route to improving the efficiency of drug discovery.

Acknowledgments

The author thanks Prof. Kazuo Kitaura for innumerous fruitful discussions of molecular interactions. This work was supported by JSPS KAKENHI Grant 19H02682.

References

1. Kussmann J, Beer M, Ochsenfeld C (2013) Linear-scaling self-consistent field methods for large molecules. WIREs Comput Mol Sci 3:614–636
2. Akimov AV, Prezhdo OV (2015) Large-scale computations in chemistry: a bird's eye view of a vibrant field. Chem Rev 115:5797–5890
3. Gordon MS, Pruitt SR, Fedorov DG, Slipchenko LV (2012) Fragmentation methods: a route to accurate calculations on large systems. Chem Rev 112:632–672
4. Gao J, Truhlar DG, Wang Y, Mazack MJM, Löffler P, Provorse MR, Rehak P (2014) Explicit polarization: a quantum mechanical framework for developing next generation force fields. Acc Chem Res 47:2837–2845
5. Collins MA, Bettens RPA (2015) Energy-based molecular fragmentation methods. Chem Rev 115:5607–5642
6. Raghavachari K, Saha A (2015) Accurate composite and fragment-based quantum chemical models for large molecules. Chem Rev 115:5643–5677
7. Erlanson DA, Fesik SW, Hubbard RE, Jahnke W, Jhoti H (2016) Twenty years on: the impact of fragments on drug discovery. Nat Rev Drug Discov 15:605–619
8. Kitaura K, Ikeo E, Asada T, Nakano T, Uebayasi M (1999) Fragment molecular orbital method: an approximate computational method for large molecules. Chem Phys Lett 313:701–706
9. Fedorov DG, Kitaura K (2004) The importance of three-body terms in the fragment molecular orbital method. J Chem Phys 120:6832–6840
10. Alexeev Y, Mazanetz MP, Ichihara O, Fedorov DG (2012) GAMESS as a free quantum-mechanical platform for drug research. Curr Top Med Chem 12:2013–2033
11. Schmidt MW, Baldridge KK, Boatz JA, Elbert ST, Gordon MS, Jensen JH, Koseki S, Matsunaga N, Nguyen KA, Su S, Windus TL, Dupuis M, Montgomery JA (1993) General atomic and molecular electronic structure system. J Comput Chem 14:1347–1363
12. Gordon MS, Schmidt MW (2005) Advances in electronic structure theory: GAMESS a decade later. In: Dykstra CE, Frenking G, Kim KS, Scuseria GE (eds) Theory and applications of computational chemistry, the first forty years. Elsevier, Amsterdam, pp 1167–1189, Chapter 41
13. GAMESS Homepage. http://www.msg.ameslab.gov/gamess/index.html. Accessed May 17, 2019
14. Fedorov DG, Kitaura K (2007) Extending the power of quantum chemistry to large systems with the fragment molecular orbital method. J Phys Chem A 111:6904–6914
15. Fedorov DG, Nagata T, Kitaura K (2012) Exploring chemistry with the fragment molecular orbital method. Phys Chem Chem Phys 14:7562–7577
16. Tanaka S, Mochizuki Y, Komeiji Y, Okiyama Y, Fukuzawa K (2014) Electron- correlated fragment-molecular-orbital calculations for biomolecular and nano systems. Phys Chem Chem Phys 16:10310–10344
17. Fedorov DG (2017) The fragment molecular orbital method: theoretical development, implementation in GAMESS, and applications. WIREs Comput Mol Sci 7:e1322
18. Merz KM (2014) Using quantum mechanical approaches to study biological systems. Acc Chem Res 47:2804–2811
19. Ryde U, Söderhjelm P (2016) Ligand-binding affinity estimates supported by quantum-mechanical methods. Chem Rev 116:5520–5566
20. Kolář MH, Hobza P (2016) Computer modeling of halogen bonds and other σ-hole interactions. Chem Rev 116:5155–5187
21. Botu V, Batra R, Chapman J, Ramprasad R (2017) Machine learning force fields: construction, validation, and outlook. J Phys Chem C 121:511–522
22. Xu P, Guidez EB, Bertoni C, Gordon MS (2018) Ab initio force field methods derived from quantum mechanics. J Chem Phys 148:090901
23. Jing Z, Liu C, Cheng SY, Qi R, Walker BD, Piquemal JP, Ren P (2019) Polarizable force fields for biomolecular simulations: recent advances and applications. Annu Rev Biophys 48:371–394
24. Christensen AS, Kubar T, Cui Q, Elstner M (2016) Semiempirical quantum mechanical methods for noncovalent interactions for chemical and biochemical applications. Chem Rev 116:5301–5337
25. Grimme S (2003) Improved second-order Møller-Plesset perturbation theory by separate scaling of parallel- and antiparallel-spin pair correlation energies. J Chem Phys 118:9095–9102
26. Xie W, Orozco M, Truhlar DG, Gao J (2009) X-pol potential: an electronic structure-based force field for molecular dynamics simulation of a solvated protein in water. J Chem Theory Comput 5:459–467

27. Luehr N, Jin AGB, Martínez TJ (2015) Ab initio interactive molecular dynamics on graphical processing units (GPUs). J Chem Theory Comput 11:4536–4544
28. Fedorov DG, Alexeev Y, Kitaura K (2011) Geometry optimization of the active site of a large system with the fragment molecular orbital method. J Phys Chem Lett 2:282–288
29. Nakata H, Fedorov DG (2016) Efficient geometry optimization of large molecular systems in solution using the fragment molecular orbital method. J Phys Chem A 120:9794–9804
30. Fedorov DG, Ishida T, Uebayasi M, Kitaura K (2007) The fragment molecular orbital method for geometry optimizations of polypeptides and proteins. J Phys Chem A 111:2722–2732
31. Fedorov DG, Asada N, Nakanishi I, Kitaura K (2014) The use of many-body expansions and geometry optimizations in fragment-based methods. Acc Chem Res 47:2846–2856
32. Nishimoto Y, Fedorov DG (2016) The fragment molecular orbital method combined with density-functional tight-binding and the polarizable continuum model. Phys Chem Chem Phys 18:22047–22061
33. Kaliakin DS, Fedorov DG, Alexeev Y, Varganov SA (2019) Locating minimum energy crossings of different spin states using the fragment molecular orbital method. J Chem Theory Comput 15:6074–6084. https://doi.org/10.1021/acs.jctc.9b00641
34. Sawada T, Fedorov DG, Kitaura K (2010) Role of the key mutation in the selective binding of avian and human influenza hemagglutinin to sialosides revealed by quantum- mechanical calculations. J Am Chem Soc 132:16862–16872
35. Jensen JH (2015) Predicting accurate absolute binding energies in aqueous solution: thermodynamic considerations for electronic structure methods. Phys Chem Chem Phys 17:12441–12451
36. Mazanetz MP, Ichihara O, Law RJ, Whittaker M (2011) Prediction of cyclin- dependent kinase 2 inhibitor potency using the fragment molecular orbital method. J Cheminform 3:2
37. Green MC, Nakata H, Fedorov DG, Slipchenko LV (2016) Radical damage in lipids investigated with the fragment molecular orbital method. Chem Phys Lett 651:56–61
38. Nakata H, Fedorov DG (2019) Simulations of infrared and Raman spectra in solution using the fragment molecular orbital method. Phys Chem Chem Phys 21:13641–13652
39. Fedorov DG, Kitaura K (2016) Subsystem analysis for the fragment molecular orbital method and its application to protein-ligand binding in solution. J Phys Chem A 120:2218–2231
40. Murata K, Fedorov DG, Nakanishi I, Kitaura K (2009) Cluster hydration model for binding energy calculations of protein-ligand complexes. J Phys Chem B 113:809–817
41. Fedorov DG, Kitaura K (2006) Theoretical development of the fragment molecular orbital (FMO) method. In: Starikov EB, Lewis JP, Tanaka S (eds) Modern methods for theoretical physical chemistry of biopolymers. Elsevier, Amsterdam, pp 3–38
42. Morokuma K (1971) Molecular orbital studies of hydrogen bonds. III. C=O...H-O hydrogen bond in $H_2CO...H_2O$ and $H_2CO...2H_2O$. J Chem Phys 55:1236–1244
43. Kitaura K, Morokuma K (1976) A new energy decomposition scheme for molecular interactions within the Hartree-Fock approximation. Int J Quantum Chem 10:325–340
44. Chen W, Gordon MS (1996) Energy decomposition analyses for many-body interaction and applications to water complexes. J Phys Chem 100:14316–14328
45. Nakano T, Kaminuma T, Sato T, Fukuzawa K, Akiyama Y, Uebayasi M, Kitaura K (2002) Fragment molecular orbital method: use of approximate electrostatic potential. Chem Phys Lett 351:475–480
46. Heifetz A, Chudyk EI, Gleave L, Aldeghi M, Cherezov V, Fedorov DG, Biggin PC, Bodkin MJ (2016) The fragment molecular orbital method reveals new insight into the chemical nature of GPCR-ligand interactions. J Chem Inf Model 56:159–172
47. Mochizuki Y, Fukuzawa K, Kato A, Tanaka S, Kitaura K, Nakano T (2005) A configuration analysis for fragment interaction. Chem Phys Lett 410:247–253
48. Ishikawa T, Mochizuki Y, Amari S, Nakano T, Tokiwa H, Tanaka S, Tanaka K (2007) Fragment interaction analysis based on local MP2. Theor Chem Accounts 118:937–945
49. Tanaka S, Watanabe C, Okiyama Y (2013) Statistical correction to effective interactions in the fragment molecular orbital method. Chem Phys Lett 556:272–277
50. Fedorov DG, Kitaura K (2007) Pair interaction energy decomposition analysis. J Comput Chem 28:222–237
51. Nagata T, Fedorov DG, Sawada T, Kitaura K (2012) Analysis of solute-solvent interactions

in the fragment molecular orbital method interfaced with effective fragment potentials: theory and application to a solvated griffithsin-carbohydrate complex. J Phys Chem A 116:9088–9099

52. Nagata T, Fedorov DG, Kitaura K (2012) Analytic gradient and molecular dynamics simulations using the fragment molecular orbital method combined with effective potentials. Theor Chem Accounts 131:1136
53. Nakano T, Mochizuki Y, Yamashita K, Watanabe C, Fukuzawa K, Segawa K, Okiyama Y, Tsukamoto T, Tanaka S (2012) Development of the four-body corrected fragment molecular orbital (FMO4) method. Chem Phys Lett 523:128–133
54. Kamiya M, Hirata S, Valiev M (2008) Fast electron correlation methods for molecular clusters without basis set superposition errors. J Chem Phys 128:074103
55. Ishikawa T, Ishikura T, Kuwata K (2009) Theoretical study of the prion protein based on the fragment molecular orbital method. J Comput Chem 30:2594–2601
56. Asada N, Fedorov DG, Kitaura K, Nakanishi I, Merz KM (2012) An efficient method to evaluate intermolecular interaction energies in large systems using overlapping multicenter ONIOM and the fragment molecular orbital method. J Phys Chem Lett 3:2604–2610
57. Amari S, Aizawa M, Zhang J, Fukuzawa K, Mochizuki Y, Iwasawa Y, Nakata K, Chuman H, Nakano T (2006) VISCANA: visualized cluster analysis of protein-ligand interaction based on the ab initio fragment molecular orbital method for virtual ligand screening. J Chem Inf Model 46:221–230
58. Sladek V, Tokiwa H, Shimano H, Shigeta Y (2018) Protein residue networks from energetic and geometric data: are they identical? J Chem Theory Comput 14:6623–6631
59. Maruyama K, Sheng Y, Watanabe H, Fukuzawa K, Tanaka S (2018) Application of singular value decomposition to the inter-fragment interaction energy analysis for ligand screening. Comput Theor Chem 1132:23–34
60. Yoshida T, Yamagishi K, Chuman H (2008) QSAR study of cyclic urea type HIV-1PR47 inhibitors using ab initio MO calculation of their complex structures with HIV-1PR. QSAR Comb Sci 27:694–703
61. Zhang Q, Yu C, Min J, Wang Y, He J, Yu Z (2011) Rational questing for potential novel inhibitors of FabK from *Streptococcus pneumoniae* by combining FMO calculation, CoMFA 3D-QSAR modeling and virtual screening. J Mol Model 17:1483–1492
62. Doi H, Okuwaki K, Mochizuki Y, Ozawa T, Yasuoka K (2017) Dissipative particle dynamics (DPD) simulations with fragment molecular orbital (FMO) based effective parameters for 1-Palmitoyl-2-oleoyl phosphatidyl choline (POPC) membrane. Chem Phys Lett 684:427–432
63. Nakata H, Fedorov DG, Nagata T, Kitaura K, Nakamura S (2015) Simulations of chemical reactions with the frozen domain formulation of the fragment molecular orbital method. J Chem Theory Comput 11:3053–3064
64. Pruitt SR, Steinmann C (2017) Mapping interaction energies in chorismate mutase with the fragment molecular orbital method. J Phys Chem A 121:1798–1808
65. Abe Y, Shoji M, Nishiya Y, Aiba H, Kishimoto T, Kitaura K (2017) The reaction mechanism of sarcosine oxidase elucidated using FMO and QM/MM methods. Phys Chem Chem Phys 19:9811–9822
66. He X, Fusti-Molnar L, Cui G, Merz KM Jr (2009) Importance of dispersion and electron correlation in ab initio protein folding. J Phys Chem B 113:5290–5300
67. Simoncini D, Nakata H, Ogata K, Nakamura S, Zhang KYJ (2015) Quality assessment of predicted protein models using energies calculated by the fragment molecular orbital method. Mol Inform 34:97–104
68. Fedorov DG, Kitaura K (2012) Energy decomposition analysis in solution based on the fragment molecular orbital method. J Phys Chem A 116:704–719
69. Green MC, Fedorov DG, Kitaura K, Francisco JS, Slipchenko LV (2013) Open-shell pair interaction energy decomposition analysis (PIEDA): formulation and application to the hydrogen abstraction in tripeptides. J Chem Phys 138:074111
70. Fedorov DG, Kitaura K (2014) Use of an auxiliary basis set to describe the polarization in the fragment molecular orbital method. Chem Phys Lett 597:99–105
71. Fedorov DG (2018) Analysis of solute-solvent interactions using the solvation model density combined with the fragment molecular orbital method. Chem Phys Lett 702:111–116
72. Fedorov DG, Kitaura K (2018) Pair interaction energy decomposition analysis for density functional theory and density-functional tight-binding with an evaluation of energy fluctuations in molecular dynamics. J Phys Chem A 122:1781–1795

73. Su P, Li H (2009) Energy decomposition analysis of covalent bonds and intermolecular interactions. J Chem Phys 131:014102
74. Jensen JH (2001) Intermolecular exchange-induction and charge transfer: derivation of approximate formulas using nonorthogonal localized molecular orbitals. J Chem Phys 114:8775–8783
75. Grimme S, Ehrlich S, Goerigk L (2011) Effect of the damping function in dispersion corrected density functional theory. J Comput Chem 32:1456–1465
76. Cybulski SM, Chałasiński G, Moszyński R (1990) On decomposition of second- order Møller-Plesset supermolecular interaction energy and basis set effects. J Chem Phys 92:4357–4363
77. Iwata S (2011) Dispersion energy evaluated by using locally projected occupied and excited molecular orbitals for molecular interaction. J Chem Phys 135:094101
78. Civalleri B, Presti D, Dovesi R, Savin A (2012) On choosing the best density functional approximation. In: Chemical modelling: applications and theory, vol 9. Royal Society of Chemistry, London, pp 168–185
79. Thirman J, Head-Gordon M (2015) An energy decomposition analysis for second- order Møller-Plesset perturbation theory based on absolutely localized molecular orbitals. J Chem Phys 143:084124
80. Kendall RA, Dunning TH Jr, Harrison RJ (1992) Electron affinities of the first-row atoms revisited. Systematic basis sets and wave functions. J Chem Phys 96:6796–6806
81. Truhlar DG (1998) Basis-set extrapolation. Chem Phys Lett 294:45–48
82. Fedorov DG, Kromann JC, Jensen JH (2018) Empirical corrections and pair interaction energies in the fragment molecular orbital method. Chem Phys Lett 702:111–116
83. Okiyama Y, Nakano T, Watanabe C, Fukuzawa K, Mochizuki Y, Tanaka S (2018) Fragment molecular orbital calculations with implicit solvent based on the Poisson-Boltzmann equation: implementation and DNA study. J Phys Chem B 122:4457–4471
84. Fedorov DG (2019) Solvent screening in zwitterions analyzed with the fragment molecular orbital method. J Chem Theory Comput 15 (10):5404–5416. https://doi.org/10.1021/acs.jctc.9b00715
85. Ishikawa T, Mochizuki Y, Nakano T, Amari S, Mori H, Honda H, Fujita T, Tokiwa H, Tanaka S, Komeiji Y, Fukuzawa K, Tanaka K, Miyoshi E (2006) Fragment molecular orbital calculations on large scale systems containing heavy metal atom. Chem Phys Lett 427:159–165
86. Fedorov DG, Slipchenko LV, Kitaura K (2010) Systematic study of the embedding potential description in the fragment molecular orbital method. J Phys Chem A 114:8742–8753
87. James WHIII, Buchanan EG, Mueller CW, Dean JC, Kosenkov D, Slipchenko LV, Guo L, Reidenbach AG, Gellman SH, Zwier TS (2011) Evolution of amide stacking in larger g-peptides: Triamide H-bonded cycles. J Phys Chem A 115:13783–13798
88. Mazanetz MP, Chudyk E, Fedorov DG, Alexeev Y (2016) Applications of the fragment molecular orbital method to drug research. In: Zhang W (ed) Computer aided drug discovery. Springer, New York, pp 217–255
89. Ishikawa T, Kuwata K (2010) Interaction analysis of the native structure of prion protein with quantum chemical calculations. J Chem Theory Comput 6:538–547
90. Pruitt SR, Brorsen KR, Gordon MS (2015) Ab initio investigation of the aqueous solvation of the nitrate ion. Phys Chem Chem Phys 17:27027–27034
91. Vuong VQ, Nishimoto Y, Fedorov DG, Sumpter BG, Niehaus TA, Irle S (2019) The fragment molecular orbital method based on long-range corrected density-functional tight-binding. J Chem Theory Comput 15:3008–3020
92. Lane JR (2013) CCSDTQ optimized geometry of water dimer. J Chem Theory Comput 9:316–323
93. Heifetz A, Trani G, Aldeghi M, MacKinnon CH, McEwan PA, Brookfield FA, Chudyk EI, Bodkin M, Pei Z, Burch JD, Ortwine DF (2016) Fragment molecular orbital method applied to lead optimization of novel interleukin-2 inducible T-cell kinase (ITK) inhibitors. J Med Chem 59:4352–4363
94. Quintero-Ronderos P, Mercier E, Fukuda M, Gonzalez R, Suarez CF, Patarroyo MA, Vaiman D, Gris JC, Laissue P (2017) Novel genes and mutations in patients affected by recurrent pregnancy loss. PLoS One 12: e0186149
95. Turk S, Merget B, Eid S, Fulle S (2018) From cancer to pain target by automated selectivity inversion of a clinical candidate. J Med Chem 61:4851–4859
96. Li S, Qin C, Cui S, Xu H, Wu F, Wang J, Su M, Fang X, Li D, Jiao Q, Zhang M, Xia C, Zhu L, Wang R, Li J, Jiang H, Zhao Z, Li J, Li H

(2019) Discovery of a natural- product-derived preclinical candidate for once-weekly treatment of type 2 diabetes. J Med Chem 62:2348–2361

97. Fedorov DG, Brekhov A, Mironov V, Alexeev Y (2019) Molecular electrostatic potential and electron density of large systems in solution computed with the fragment molecular orbital method. J Phys Chem A 123:6281–6290

98. Morao I, Fedorov DG, Robinson R, Southey M, Townsend-Nicholson A, Bodkin MJ, Heifetz A (2017) Rapid and accurate assessment of GPCR-ligand interactions using the fragment molecular orbital-based density-functional tight-binding method. J Comput Chem 38:1987–1990

Chapter 5

Underappreciated Chemical Interactions in Protein–Ligand Complexes

Andrew Anighoro

Abstract

Non-covalent interactions lie at the bases of the molecular recognition process. In medicinal chemistry, understanding how bioactive molecules interact with their target can help to explain structure–activity relationships (SAR) and improve potency of lead compounds. In particular, computational analysis of protein–ligand complexes can help to unravel key interactions and guide structure-based drug design.

The literature describing protein-ligand complexes is typically focused on few types of non-covalent interactions (e.g., hydrophobic contacts, hydrogen bonds, and salt bridges). Stacking interactions involving aromatic rings are also relatively well known to medicinal chemistry practitioners. Potency optimization efforts are often focused on targeting these interactions. However, a variety of underappreciated interactions were shown to have a relevant effect on the stabilization of protein–ligand complexes. This chapter aims at listing selected non-covalent interactions and discuss some examples on how they can impact drug design.

Key words Non-covalent interactions, Drug design, Medicinal chemistry, Quantum mechanics, Non-classical hydrogen bond, π–π stacking, Halogen–π, Halogen bond, Cation–π, Hydrophobic interactions, Orthogonal multipolar interactions

1 Introduction

Drug discovery is a multi-objective optimization problem where a promising candidate is supposed to modulate the target(s) of interest with good potency but also have acceptable safety and pharmacokinetic profiles. In order to address potency, when the three-dimensional (3D) structure of the biological target is available, it is possible to apply structure-based drug design techniques [1]. Some drugs act as covalent inhibitors [2]; however the majority of drug discovery programs are focused on the design of non-covalent ligands. The ability of a small molecule to form a reversible complex with a protein or a nucleic acid and thus exert a biological activity depends on its capability to form productive non-covalent interaction with the macromolecule binding site

Alexander Heifetz (ed.), *Quantum Mechanics in Drug Discovery*, Methods in Molecular Biology, vol. 2114, https://doi.org/10.1007/978-1-0716-0282-9_5, © Springer Science+Business Media, LLC, part of Springer Nature 2020

[3]. Careful analysis of experimentally elucidated complexes [4] as well as ab initio calculations [5, 6] can help to unravel key protein–ligand interactions. The insight gained from such analysis can be used to guide the rational design of more potent compounds by optimizing existing interactions or by targeting new ones. Notably, other factors affect the overall binding affinity of small molecules to their target such as desolvation and the change in entropy of the system [7]. While these factors may be just as important, this chapter will focus only on the discussion of non-covalent interactions.

Protein–ligand complexes are usually described in literature by means of few well-known interactions such as classical hydrogen bonds, salt bridges, hydrophobic contacts, and π–π stacking. Nevertheless, statistical analysis of small molecule crystal structures in the Cambridge Structural Database (CSD) [8] as well as protein–ligand complexes deposited in the Protein Data Bank (PDB) [9] has highlighted the existence of several less known weak interactions [3, 4]. Notable examples are weak hydrogen bonds, a variety of interactions involving π systems, and contacts established by halogen atoms. Such underappreciated interactions are not usually encoded in the functional form of molecular mechanics (MM) force fields and docking scoring functions which may only hope to implicitly grasp some of their effects. Ab initio methods based on quantum mechanics (QM), on the other hand, offer a more rigorous way to calculate the contribution of such interactions [5].

A very useful compendium of non-covalent interactions for medicinal chemists which includes strong as well as a number of weak contacts has been published by Bissantz et al. [3]. A wide body of literature is available on specific weak interactions. For example, the first accounts on CH···O contacts appeared in the 1930s [10]. Nevertheless, it is easy to forget the importance or simply overlook such interactions which remain in some respect underappreciated. The flip side of the coin is the risk of overinterpretation, e.g., draw conclusion based on limited data collected from low-resolution structures. It may be hard to draw a clear-cut conclusion on when drug design may benefit by targeting weak interactions. However, there are such examples in literature; therefore it is good to be aware of this possibility. The scope of this chapter is to list representative non-covalent interactions together with some examples relevant to medicinal chemistry.

2 Non-classical Hydrogen Bond

The hydrogen bond is defined on the *IUPAC Gold Book* as:

> *a particular type of multicenter (three center—four electron) X–H···Y in which the central hydrogen atom covalently linked to an electronegative atom*

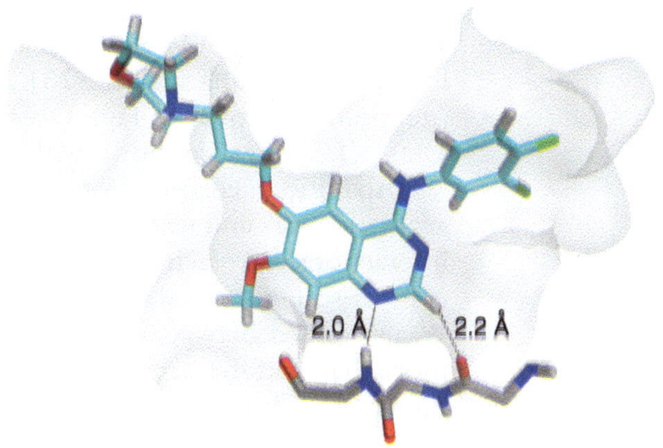

Fig. 1 Non-classical hydrogen bond. Non-classical hydrogen bond formed between gefitinib (cyan) and the kinase domain of the epidermal growth factor receptor (EGFR) (PDB code 4WKQ). Side chains of displayed residues are omitted for clarity

> *X (C, N, O, S. . .) forms an additional weaker bond with atom Y (N, O, S. . .) in the direction of its lone electron pair orbital. The energy of hydrogen bonds, which is usually in the range of 3–15 kcal/mol (12–65 kJ/mol), results from the electrostatic interaction and also from the orbital interaction of the antibonding σ* (XH)MO of the molecule acting as the hydrogen donor and the non-bonding lone electron pair MOn_Y of the hydrogen acceptor molecule* [11]

This definition builds on classical works such as Linus Pauling's *The Nature of the Chemical Bond* [12] as well as numerous studies that since then have been conducted to further characterize the nature of this ubiquitous attractive interaction [13–15]. Classical hydrogen bonds where a hydrogen atom acts as bridge between two electronegative atoms such as oxygen and nitrogen are arguably better known and more often the subject of optimization efforts. Nevertheless, the importance of "non-classical" interactions where a CH can act as a hydrogen bond donor has been extensively studied and recognized [14]. The most relevant interaction of this class is probably represented by the CH· · ·O interaction (Fig. 1). This type of interaction was first proposed by Glasstone in 1937 [10], while the first crystallographic evidence was provided by June Sutor in 1962 [16] followed by another paper in 1963 [17]. For several years, the nature and the very existence of this attractive interaction remained controversial [18]. Following a rigorous statistical analysis conducted on crystallographic databases [19], the significance of this interaction became more widely accepted and investigated [20, 21].

A good example of the importance of this interaction in the context of drug binding is represented by ATP-competitive protein kinase inhibitors. Typically, these compounds establish a network of hydrogen bonds with the backbone of the hinge, a stretch of residues that connect the N- and C-lobe of the kinase and provide an anchoring point for the adenine moiety of ATP. It has been shown that CH···O contacts may play an important role in stabilizing these complexes [22, 23]. A representative kinase inhibitor (gefitinib) is shown in Fig. 1, and a classical hydrogen bond and CH···O contact are shown with dashed lines.

Ab initio calculations for the interaction between representative aromatic rings and water have shown that the strength of CH···O interactions can reach −3.48 kcal/mol in the case of isoxazole. For comparison, this is about four times stronger than the van der Waals contact between methane and water (−0.87 kcal/mol) and more than half the strength (63%) of the hydrogen bond in a water dimer (−5.51 kcal/mol) [22].

3 π–π Stacking

π-interactions can be generally defined as attractive interactions between a π system (a molecule containing π electrons) and another molecule. Among the best known examples of this class of interactions is the π–π (stacking) interaction (Fig. 2) [24]. A typical example of a stacking interaction is represented by the benzene dimer which may exist in a sandwich, T-shaped, or parallel-displaced configurations [25]. Interaction energies for this system range between 1.8 and 2.8 kcal/mol [25]. Benzene rings as well as different heterocycles are often found in drug molecules [26]; it is well known that these moieties can contribute to the binding affinity of a small molecule to its target by forming stacking interactions. What may be underappreciated in this case is the extent to

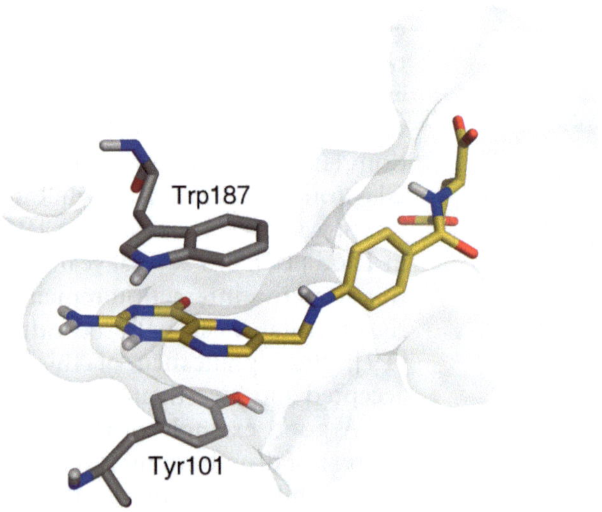

Fig. 2 π–π stacking. Interaction between folate (gold) and two aromatic residues in the binding site of the folate receptor beta (PDB code 4KMZ)

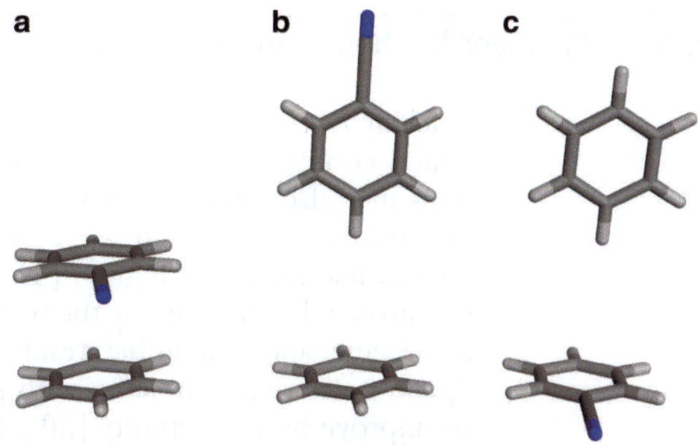

Fig. 3 Representative π–π stacking configurations used in a theoretical study addressing the effect of substituents on these interactions. The benzene–benzonitrile system is shown here in a sandwich (**a**) and two alternative T-shaped configurations that differ for the position of the relative orientation of the benzene and the substituted ring (**b** and **c**)

which individual properties of different aromatic rings can affect these interactions. In particular, the electrostatic properties of different heterocycles can affect the energy and geometry of stacking interactions [27]. The effect of different substituents on the aromatic rings can also be prominent. A theoretical study compared π–π interactions in the following systems: benzene dimer, benzene-phenol, benzene-toluene, benzene-fluorobenzene, and benzene-benzonitrile [28]. Three different configurations were considered, i.e., sandwich (Fig. 3a) and two alternative T-shaped configurations (Fig. 3b and c). In a sandwich configuration, all systems including a substituted ring lead to an improved interaction with respect to the benzene dimer. Phenol, toluene, and fluorobenzene improved the calculated π–π interaction from −1.7 (calculated for the benzene dimer) to about −2.3 kcal/mol, while benzonitrile had the most marked effect resulting in −3.3 kcal/mol. In the T-shaped configuration shown in Fig. 3b, phenol and toluene resulted in a less energetic interaction with benzene with respect to the benzene dimer, while fluorobenzene and benzonitrile had the opposite effect. This trend was reversed in the configuration shown in Fig. 3c where fluorobenzene and benzonitrile resulted in an improved interaction with benzene while phenol and toluene where found to be detrimental.

Among aromatic amino acids, the role of histidine is particularly interesting as its imidazole ring besides participating in classical hydrogen bonds, and coordinating metallic cations can also be involved in π–π stacking, cation–π (as either the cation, when protonated, or π system), and hydrogen–π interactions [29]. Interestingly, these effects can have potential applications in scaffold replacement strategies by accounting for effects that may not be obvious by visual inspection.

4 π–π Stacking Involving Non-aromatic Groups

Stacking interactions are often formed by aromatic rings in protein–ligand complexes (see above). Nevertheless, non-aromatic π systems may also establish such interactions. Amide stacking can also be observed fairly often in protein–ligand contacts, especially in trypsin-like proteins (Fig. 4) [4]. Amide groups may be exposed in a protein binding site by the backbone as well as the side chain of asparagine and glutamine residues. Amide stacking formed by ligand heteroarene moieties with protein amides can be exploited to improve binding affinity [30]. In particular, the energy of such stacking interactions can be improved by (1) orienting the fragment dipole vector such that it is aligned in an antiparallel fashion with the dipole of the interacting protein amide group, (2) increasing the heteroarene dipole moment, and (3) decreasing its π-electron density [30].

The guanidinium group of arginine can also form π–π stacking interactions. This has been observed in protein–protein contacts where arginine can form stacked dimers despite the repulsive charge or interact with other π systems [31]. Furthermore, π–π stacking involving arginine has been observed also in protein–ligand complexes [32]. While lysine–arene interaction is dominated by electrostatics, arginine–arene interaction involves a nearly equal mix of electrostatics and dispersion forces [32]; this has been also referred to as π^{+}–π interaction [33].

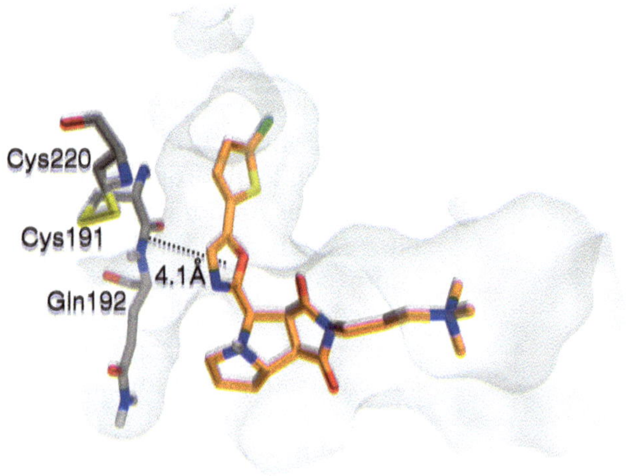

Fig. 4 π–π stacking involving non-aromatic groups. Amide stacking between a backbone amide of FXa and the oxazole moiety of a co-crystallized inhibitor (PDB code 2Y5G)

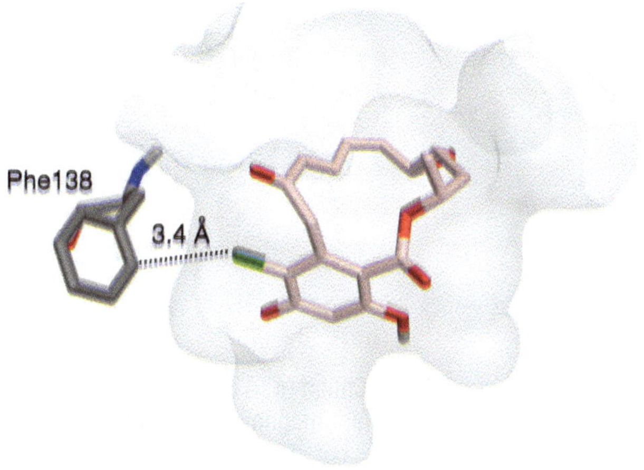

Fig. 5 Halogen–π. Cl–π interaction between radicicol (pink) and Phe138 of Hsp90 (PDB code 4EGK)

5 Halogen–π

Contacts between chlorine and fluorine atoms of the ligand and aromatic side chains are often observed in experimentally determined complexes [4]. Cl–π interaction (Fig. 5) accounts for −2.01 kcal/mol in a model system (benzene-chlorohydrocarbon); this is mainly attributed to dispersion forces [34]. F–π attractive forces, on the other hand, are mainly of electrostatic nature [35]. An analysis of protein–ligand complexes extracted from the PDB showed that the distance between Cl and the nearest aromatic atom was between 3 and 4 Å with an average of 3.6 Å. Furthermore, edge-on geometry (chlorine pointing toward an aromatic atom or bond between them) is generally predominant in crystallographic complexes; however π density plays a role as face-on geometry (chlorine pointing toward the center of the aromatic ring) is favored in π electron-rich rings [34]. Fluorine–π interactions are typically stronger than chlorine–π interactions and have an average distance of about 3 Å. This trend is opposite with respect to halogen bonds where less electronegative and more polarizable halogens tend to form stronger interactions [35].

6 Cation–π

Cation–π interactions are formed between a cation and the π-electron density of a conjugated system, typically an aromatic ring (Fig. 6) [36]. This interaction contributes to stabilize the protein fold and protein–ligand complexes [37]. A cation–π interaction can improve binding affinity of ligands of 2–5 kcal/mol, which is competitive with the contribution of hydrogen bonds

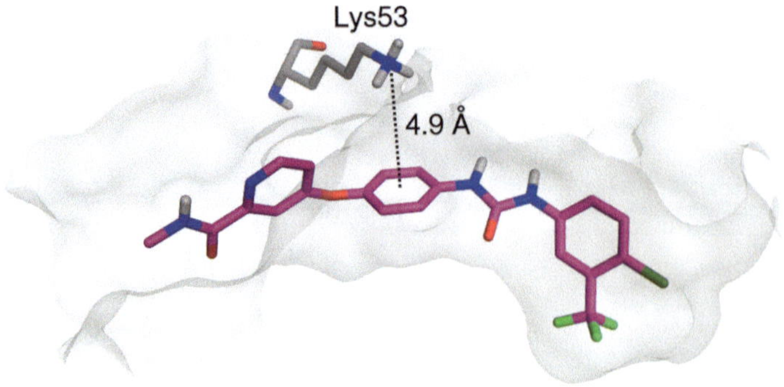

Fig. 6 Cation–π. A cation–π interaction between Lys53 of MAP kinase p38α and sorafenib (magenta) (PDB code 3GCS)

and ion pairs [36]. This interaction typically occurs within a distance of 6 Å [37]. In principle, either the ligand or the protein may provide a protonated nitrogen to form this type of interaction. However, as aromatic rings are ubiquitous in drug-like molecules, while positively charged groups are rarer, it is more often observed that a basic residue (lysine or arginine) interacts with a ligand's aromatic moiety [4]. As mentioned in the paragraph dedicated to π–π stacking involving non-aromatic groups, lysine interaction with aromatic rings is largely electrostatic in nature, while in the case of arginine, dispersive forces are also playing a relevant role [32]. A relevant example of cation–π interaction where the positive charge is on the ligand is represented by certain neurotransmitters (e.g., acetylcholine) binding to the aromatic box motifs in G protein-coupled receptors (GPCRs) [38, 39].

7 Halogen Bond

A halogen bond is a non-covalent interaction that can be formed between the positively charged sigma hole of a halogen and a Lewis base acting as an electron donor moiety (Fig. 7) [40]. Halogens involved in this type of interaction usually include chlorine, iodine, and bromine, while, in the context of protein binding sites, the electron donating moiety is often represented by the ubiquitous backbone carbonyl or by various side chain groups (e.g., hydroxyl, carboxyl, sulfur atoms, nitrogen on a histidine) [40]. Interactions between halogens and π systems, which sometimes are also referred to as halogen bonds, have been discussed in the previous paragraph. Calculations done on model systems including a backbone carbonyl as a Lewis base showed for CX···O contacts a distance between 3 and 3.5 Å, with angles close to 180° [40], and energy roughly comparable to a hydrogen bond (5.8 kcal/mol) [41] in the following order of strength I > Br > Cl. Halogen bonds have been often found by chance in protein–ligand complexes; however

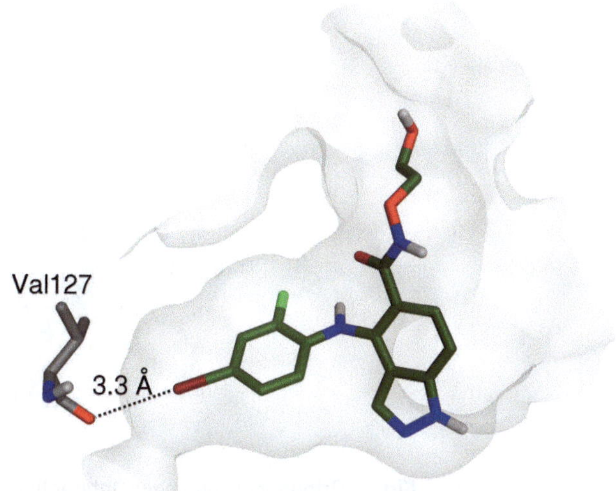

Fig. 7 Halogen bond. Interaction between iodine of an allosteric inhibitor of dual specificity mitogen-activated protein kinase kinase 1 (MEK) and a backbone carbonyl (PDB code 3V04)

computational methods have been increasingly applied to try to account for these interactions in rational drug design [42]. For example, up to 35-fold increase in binding affinity was obtained for serotonin receptor 5-HT7R ligands by following a protocol aimed at identifying binding site regions prone to interact via halogen bonds (hot spots) [43].

8 Orthogonal Multipolar Interactions

Orthogonal multipolar interactions are dipolar electrostatic interactions that are observed in crystal structures of small molecules as well as protein–ligand complexes [44]. Such interactions can be formed by two carbonyl groups orthogonally oriented between each other or also by CF···C=O contacts (Fig. 8). CF···C=O interactions have proven particularly interesting in the context of drug design. For example, the systematic replacement of one or more hydrogens in thrombin inhibitors resulted in up to fivefold gain in potency [45]. These observations have driven the development of computational approaches to map protein binding sites trying to uncover fluorine-favorable regions [46].

9 Hydrophobic Interactions

The role of hydrophobic interactions is vital for biomolecular recognition [3]. Historically, hydrophobic interactions have been accounted for with terms that are delocalized, typically in the

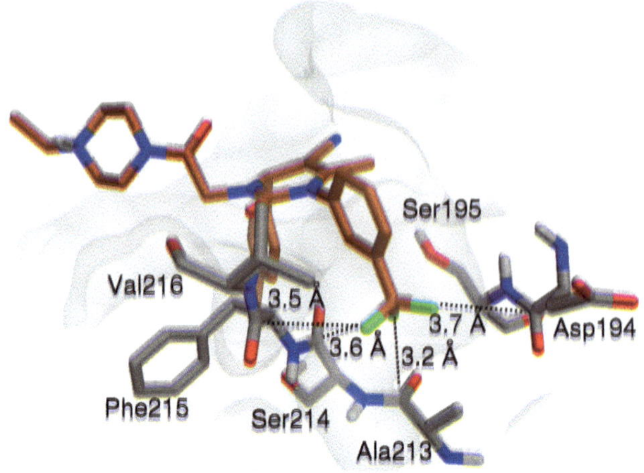

Fig. 8 Orthogonal multipolar interaction. Interaction between the CF_3 group, a dihydropyrimidinone inhibitor of neutrophil elastase, and several backbone carbonyls in the binding site (PDB code 5A0B)

form of either a shape complementarity term or solvation/entropy penalty. Such approaches are useful in providing an overall estimation of the contribution to the affinity but do not readily provide spatial decomposition that can be readily incorporated into a design strategy for future compounds [47]. Visual inspection of protein–ligand complexes operated by experts is also carried out to try to understand hydrophobic as well as other types of interactions. However, this approach is subjective and nonquantitative in nature. The strength of hydrophobic interactions can be evaluated with QM methods, which are considered to be a reliable approach for the exploration of receptor–ligand interactions and an improvement over dispersion potentials in force fields [47].

10 Final Remarks

In this chapter, selected weak interactions found in protein–ligand complexes were presented. This list is not intended to be exhaustive; instead it serves as a reminder that a variety of potentially underappreciated interactions may affect binding of small molecules to a protein binding site and that their knowledge can be useful to medicinal chemistry applications. The mere visual inspection of crystallographic complexes—even when performed by a trained eye—may result in overlooking some relevant contacts. In this context, QM methods can be applied to unravel weak interactions [6, 48, 49]. A potential limitation is represented by the accuracy of crystal structures. Fuzzy electron density maps as well as imperfect model refinement may result in overlooking a considerable fraction of non-covalent interactions [50]. This has been shown to be the case even for relatively strong contacts such as classical hydrogen bonds [50]; therefore it is likely that weaker contacts may suffer from this effect at least as much. QM/MM optimization can help recover overlooked interactions [50]. Within these limits, a careful analysis of intermolecular contacts has the potential to guide rational drug design toward more potent compounds.

References

1. Anderson AC (2003) The process of structure-based drug design. Chem Biol 10:787–797
2. Singh J, Petter RC, Baillie TA et al (2011) The resurgence of covalent drugs. Nat Rev Drug Discov 10:307–317
3. Bissantz C, Kuhn B, Stahl M (2010) A medicinal Chemist's guide to molecular interactions. J Med Chem 53:5061–5084
4. Ferreira de Freitas R, Schapira M (2017) A systematic analysis of atomic protein–ligand interactions in the PDB. Med Chem Commun 8:1970–1981
5. Raha K, Peters MB, Wang B et al (2007) The role of quantum mechanics in structure-based drug design. Drug Discov Today 12:725–731
6. Kitaura K, Ikeo E, Asada T et al (1999) Fragment molecular orbital method: an approximate computational method for large molecules. Chem Phys Lett 313:701–706
7. Fox JM, Zhao M, Fink MJ et al (2018) The molecular origin of enthalpy/entropy compensation in biomolecular recognition. Annu Rev Biophys 47:223–250
8. Groom CR, Bruno IJ, Lightfoot MP et al (2016) The Cambridge structural database. Acta Crystallogr B 72:171–179
9. Berman HM, Westbrook J, Feng Z et al (2000) The Protein Data Bank. Nucleic Acids Res 28:235–242
10. Glasstone S (1937) The structure of some molecular complexes in the liquid phase. Trans Faraday Soc 33:200–206
11. IUPAC—hydrogen bond (HT07050). https://goldbook.iupac.org/terms/view/HT07050
12. Linus P (1931) The nature of the chemical bond. Application of results obtained from the quantum mechanics and from a theory of paramagnetic susceptibility to the structure of molecules. J Am Chem Soc 53:1367–1400
13. Lippincott ER, Schroeder R (1955) One-dimensional model of the hydrogen bond. J Chem Phys 23:1099–1106
14. Desiraju GR, Steiner T (1999) The weak hydrogen bond in structural chemistry and biology. Oxford University Press/International Union of Crystallography, Oxford
15. Espinosa E, Molins E, Lecomte C (1998) Hydrogen bond strengths revealed by topological analyses of experimentally observed electron densities. Chem Phys Lett 285:170–173
16. Sutor DJ (1962) The C–H···O hydrogen bond in crystals. Nature 195:68–69
17. Sutor DJ (1963) 204. Evidence for the existence of C–H···O hydrogen bonds in crystals. J Chem Soc:1105–1110
18. Donohue J (1968) Selected topics in hydrogen bonding. In: Rich A, Davidson NR, Pauling L (eds) Structural chemistry and molecular biology. W. H. Freeman, San Francisco
19. Taylor R, Kennard O (1982) Crystallographic evidence for the existence of CH···O, CH···N and CH···Cl hydrogen bonds. J Am Chem Soc 104:5063–5070
20. Desiraju GR (1991) The C-H···O hydrogen bond in crystals: what is it? Acc Chem Res 24:290–296
21. Steiner T (1997) Unrolling the hydrogen bond properties of C–H···O interactions. Chem Commun 8:727–734
22. Pierce AC, Sandretto KL, Bemis GW (2002) Kinase inhibitors and the case for CH···O hydrogen bonds in protein–ligand binding. Proteins 49:567–576
23. Pierce AC, ter Haar E, Binch HM et al (2005) CH···O and CH···N hydrogen bonds in ligand design: a novel Quinazolin-4-ylthiazol-2-ylamine protein kinase inhibitor. J Med Chem 48:1278–1281
24. Hunter CA, Sanders JKM (1990) The nature of π–π Interactions. J Am Chem Soc 112:5525–5534
25. Sinnokrot MO, Valeev EF, Sherrill CD (2002) Estimates of the ab initio limit for π–π interactions: the benzene dimer. J Am Chem Soc 124:10887–10893

26. Taylor RD, MacCoss M, Lawson ADG (2014) Rings in Drugs. J Med Chem 57:5845–5859
27. Huber RG, Margreiter MA, Fuchs JE et al (2014) Heteroaromatic π-stacking energy landscapes. J Chem Inf Model 54:1371–1379
28. Sinnokrot MO, Sherrill CD (2004) Substituent effects in π–π interactions: Sandwich and T-shaped configurations. J Am Chem Soc 126:7690–7697
29. Liao S-M, Du Q-S, Meng J-Z et al (2013) The multiple roles of histidine in protein interactions. Chem Cent J 7:44
30. Harder M, Kuhn B, Diederich F (2013) Efficient stacking on protein amide fragments. ChemMedChem 8:397–404
31. Vernon RM, Chong PA, Tsang B et al (2018) Pi-pi contacts are an overlooked protein feature relevant to phase separation. elife 7:e31486
32. Kumar K, Woo SM, Siu T et al (2018) Cation–π interactions in protein–ligand binding: theory and data-mining reveal different roles for lysine and arginine. Chem Sci 9:2655–2665
33. Singh NJ, Min SK, Kim DY et al (2009) Comprehensive energy analysis for various types of π-interaction. J Chem Theory Comput 5:515–529
34. Imai YN, Inoue Y, Nakanishi I et al (2008) Cl–π interactions in protein–ligand complexes. Protein Sci 17:1129–1137
35. Li P, Maier JM, Vik EC et al (2017) Stabilizing fluorine–π interactions. Angew Chem Int Ed Engl 56:7209–7212
36. Dougherty DA (2013) The Cation–π interaction. Acc Chem Res 46:885–893
37. Gallivan JP, Dougherty DA (1999) Cation-π interactions in structural biology. Proc Natl Acad Sci U S A 96:9459–9464
38. Dougherty DA (2007) Cation-π interactions involving aromatic amino acids. J Nutr 137:1504S–1508S
39. Olsen JA, Balle T, Gajhede M et al (2014) Molecular recognition of the neurotransmitter acetylcholine by an acetylcholine binding protein reveals determinants of binding to nicotinic acetylcholine receptors. PLoS One 9: e91232
40. Wilcken R, Zimmermann MO, Lange A et al (2013) Principles and applications of halogen bonding in medicinal chemistry and chemical biology. J Med Chem 56:1363–1388
41. Shinada NK, de Brevern AG, Schmidtke P (2019) Halogens in protein–ligand binding mechanism: a structural perspective. J Med Chem. In Press
42. Coates M, Ho PS (2016) Computational tools to model halogen bonds in medicinal chemistry. J Med Chem 59:1655–1670
43. Kurczab R, Canale V, Satała G et al (2018) Amino acid hot spots of halogen bonding: a combined theoretical and experimental case study of the 5-HT7 receptor. J Med Chem 61:8717–8733
44. Paulini R, Müller K, Diederich F (2005) Orthogonal multipolar interactions in structural chemistry and biology. Angew Chem Int Ed Engl 44:1788–1805
45. Olsen JA, Banner DW, Seiler P et al (2003) A fluorine scan of thrombin inhibitors to map the fluorophilicity/fluorophobicity of an enzyme active site: evidence for C-F···C=O interactions. Angew Chem Int Ed Engl 42:2507–2511
46. Pollock J, Borkin D, Lund G et al (2015) Rational design of orthogonal multipolar interactions with fluorine in protein–ligand complexes. J Med Chem 58:7465–7474
47. Heifetz A, James T, Southey M et al (2019) Characterising GPCR–ligand interactions using a fragment molecular orbital-based approach. Curr Opin Struct Biol 55:85–92
48. Heifetz A, Aldeghi M, Chudyk EI et al (2016) Using the fragment molecular orbital method to investigate agonist–orexin-2 receptor interactions. Biochem Soc Trans 44:574–581
49. Zhang J-X, Sheong FK, Lin Z (2018) Unravelling chemical interactions with principal interacting orbital analysis. Chemistry 24:9639–9650
50. Xu Z, Zhang Q, Shi J et al (2019) Underestimated noncovalent interactions in protein data Bank. J Chem Inf Model. In Press

Chapter 6

Geometry Optimization, Transition State Search, and Reaction Path Mapping Accomplished with the Fragment Molecular Orbital Method

Hiroya Nakata and Dmitri G. Fedorov

Abstract

Recent development of the fragment molecular orbital (FMO) method related to energy gradients, geometry optimization, transition state search, and chemical reaction mapping is summarized. The frozen domain formulation of FMO is introduced in detail, and the structure of related GAMESS input files for FMO is described.

Key words FMO, Fragment molecular orbital method, Geometry optimization, Transition state search, GAMESS, DFT, DFTB, FMO-DFTB, OPTFMO, SADPOINT, OPTIMIZE, FMO/MM, FMO/MD, FMO/FD

1 Introduction

The molecular structure determines chemical properties, being the key to understanding many phenomena. Although there is a remarkable progress in the field of experimental determination of structures, by far not every structure has been or can be easily determined experimentally. In practice, most computational studies rely on structures that are refined using some computational technique (often guided by available experimental data). For biochemical systems, there is the major complexity that structures are very flexible, and multiple local minima may be important, in addition to the global minimum.

A majority of methods to determine the structure are based on energy gradients (derivatives of the energy with respect to nuclear coordinates) and the energy is minimized, yielding a minimum energy structure. Traditionally, molecular mechanics (MM) methods have been used for structure minimization, and later hybrids of MM with quantum mechanics (QM) have become popular [1–3], in particular, for protein–ligand complexes because ligands may not

Alexander Heifetz (ed.), *Quantum Mechanics in Drug Discovery*, Methods in Molecular Biology, vol. 2114, https://doi.org/10.1007/978-1-0716-0282-9_6,

have predefined parameters, or, some ligands including transition metals may not be well treated with force fields.

To accelerate QM calculations, fragmentation is very helpful and many methods have been developed [4–13] including the fragment molecular orbital (FMO) method [14–18]. FMO has been applied to a variety of biochemical systems [19–22].

In FMO one divides a molecular system into fragments and performs QM calculations on each fragment (monomer) and pair of fragments (dimers). From the properties of monomers and dimers, the total properties are calculated using the many-body expansion [23]. Details can be found elsewhere [18, 24]. This chapter focuses on the development of FMO in GAMESS [25, 26] and other programs can also be used for FMO calculations [27, 28].

2 Methodology

2.1 Analytic Gradient

The energy of a molecular system divided into N fragments in FMO is calculated as

$$E = \sum_{I}^{N} E'_I + \sum_{I>J}^{N} \left[E'_{IJ} - E'_I - E'_J + \mathrm{Tr}(\Delta\mathbf{D}^{IJ}\mathbf{V}^{IJ}) \right] \tag{1}$$

where $\Delta\mathbf{D}^{IJ}$ is the electron density transfer matrix of dimer IJ, $\mathbf{V}^{IJ}$ is the embedding potential matrix, and E'_X is the internal energy of monomers ($X = I$) or dimers ($X = IJ$).

The main difficulty in deriving the FMO gradient is in the electrostatic embedding $\mathbf{V}^X$, which couples the electronic states of all fragments, implicitly via polarization in E'_X and explicitly in $\Delta\mathbf{D}^{IJ}\mathbf{V}^{IJ}$. Because the embedding is determined self-consistently for monomers, there is no self-consistency between the electronic states of monomers and dimers, which complicates the gradient. One has to solve coupled-perturbed (CP) equations to get any FMO gradient in the fully analytic way. This is accomplished in the SCZV (self-consistent Z-vector) method [29], and the derivations have to be done for each QM method separately.

It is possible to neglect the response terms arising in CP equations, and obtain an approximate FMO gradient. Numerous tests show that the errors for this gradient are on the order of 10^{-3} hartree/bohr. Practically, this means that geometry optimizations because of the residual error cannot use a tight convergence threshold (roughly, below 10^{-3} hartree/bohr), and if it is specified, the geometry optimization would typically feature an oscillation in the energy and the gradient would never fall below the desired threshold. This approximate gradient is a big problem for numerical second derivatives and molecular dynamics (MD) simulations.

The cost of doing SCZV calculations is not big; typically, it is only a fraction (below 50%) of the total timing, so if the analytic gradient with SCZV is available for a QM method one is interested in, it is definitely to be used.

Table 1
Summary of approximate (appr) and fully analytic (exact) gradients in the FMO method[a]

Method	Closed shell		Unrestricted		Restricted open shell	
	appr	exact	appr	exact	appr	exact
MP2	[35]	[36]				
HF	[37]	[29]	[38]	[39]	[40]	[41]
DFT	[42]	[43]	[44]	[45]		
DFTB	[46]	[47]				
PCM	[48]	[49]	[50]	[50]	[50]	[50]

[a] Also, approximate MCSCF [51] and TDDFT [52] gradients are available

There is another problem: for the adaptive frozen orbital (AFO) boundary treatment [30], SCZV has been done only for DFTB [31] and all other methods can only use an approximate gradient [32]. AFO is typically used for inorganic systems, whereas hybrid orbital projection operator (HOP) [33] is used for bio systems. For HOP, analytic gradients are available [34] for many QM methods. The current state of FMO/HOP gradient development is described in Table 1.

Most geometry optimizations are done at the two-body level of FMO2. However, in some cases, one can consider using the more accurate FMO3 method [53]. As an example, it was shown [54] that FMO2 overestimates the coordination number of solvated sodium cation, driven by the insufficiently accurate description of the solute-solvent charge transfer. In FMO3, in addition to monomers and dimers, trimers are also computed.

Flexible systems take many steps to converge to a minimum. For smaller systems (containing up to a few hundreds of atoms), it is feasible to do an FMO optimization with a variety of methods, such as RHF or DFT. For larger systems, the practical choice is limited to four types of approaches, listed in Table 2. HF-3c has been validated for protein structure optimization [55]. Solvent effects can be described with a continuum [48, 56] or explicit solvent. For the latter, one can either define water molecules as FMO or EFP [57] fragments or use a force field.

To save the time, one can optimize only a subset of atoms. This can be done in two ways: in regular FMO calculations (recomputing the whole system at each step), and in frozen domain calculations (recomputing only the active site). The latter is more efficient. There are several kinds of geometry optimizations (Table 3), a regular optimization, and three specially constrained optimizations.

Table 2
Summary of FMO methods suitable for geometry optimizations of large systems

Method	References	Optimized area	Details
FMO/MM	[23, 58]	Full system	Uses GAMESS/Tinker
FMO/FD	[59, 60]	Active site	Other coordinates are frozen
FMO-DFTB	[46, 47]	Full system	Needs DFTB parameters
FMO-HF3c	[61]	Full system	Active site may also be optimized

Table 3
Types of FMO geometry optimizations

Method	References	Details
Minimum location	[58]	Energy minimization
Saddle point location	[62]	Transition state search
Intrinsic reaction coordinate	[60]	Reaction path mapping
Minimum energy crossing point	[63]	Crossing of surfaces

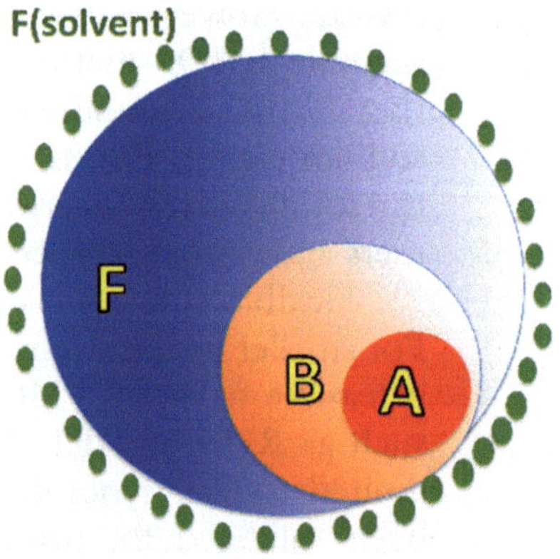

Fig. 1 Schematic illustration of the domain definition in FMO/FD

2.2 FMO with a Frozen Domain

In the frozen domain approach, the molecular system is divided into three domains: active **A**, polarizable buffer **B**, and frozen **F** (Fig. 1). A domain division is done fragment-wise, each fragment as a whole is assigned to a domain. Only atoms in domain **A** are optimized; it is possible to optimize only some and not all atoms in **A**; for example, only H atoms.

FMO/FD uses the multilayer formulation of FMO [64]. For the initial structure, the whole system is recomputed at the level of FMO1, i.e., neglecting the dimer terms in Eq. 1. This is done at a lower level of theory and basis set. Then, for each structure (initial and consequent), the polarizable buffer domain **B** is recomputed at a higher level of theory in the frozen embedding of domain **F**. Domain **F** contributes in two ways: to the embedding and by adding the interaction energy between fragments in domains **F** and **B** to the total energy. The reason why one computes a polarizable domain **B** rather than **A** is to increase the accuracy by considering polarization and other effects in an extra area adjacent to **A**. In other words, when some atoms in **A** move, that changes the electronic state of fragments outside of **A**; limited to fragments in domain **B**. Domain **A** is always included in domain **B** by definition (that is, fragments in **A** are also polarized). Domains **F** and **B** form layers 1 and 2, respectively, in the multilayer FMO setup.

There are two kinds of FMO/FD calculations, full and approximate FMO/FD. The latter is denoted as FMO/FDD, and it is much faster and about as accurate as the full FD method, so typically, FMO/FDD is used. Specifically, pair interactions in FDD are limited to dimers where at least one of the two fragments is in domain **A**. This greatly reduces the number of dimers to be computed.

2.3 Engines for Geometry Optimizations

There are several drivers for FMO optimizations in GAMESS, chosen with RUNTYP. Their summary is given in Table 4.

FMO/FD(D) and FMO/MM cannot use OPTFMO, and for them, the number of atoms is limited only for the polarizable domain **B** and QM layer in FMO/FD(D) and FMO/MM, respectively. In OPTIMIZE, the approximately updated Hessian matrix is always diagonalized at each step (a large cost). For a few hundreds of atoms, OPTIMIZE is a good choice despite the diagonalization cost, because the normal modes guide a geometry optimization well. When more atoms are present, then often the approximate

Table 4
Drivers for regular and constrained FMO geometry optimizations in GAMESS

Driver	Coordinates	Max atoms	Features
OPTIMIZE	Cart,int[a]	2000[b]	Regular energy minimization
OPTFMO	Cart	Unlimited	Regular energy minimization
SADPOINT	Cart	2000	Transition state search
IRC	Cart	2000	Mapping a reaction path
MEX	Cart	2000	Finding a minimum energy crossing

[a] Cartesian (Cart) or internal (int)

[b] The limit can be raised by running a script that alters the source code increasing the limit (the script is included in GAMESS), and recompiling GAMESS

update scheme results in imaginary modes that cause a termination of the optimization. For large system optimizations (a thousand of atoms or more), OPTFMO is recommended. It has no limit to the system size and requires moderate memory.

Although OPTIMIZE has an extra choice of using internal coordinates, there is a substantial cost of the internal conversion of data (the gradient is always computed in Cartesian coordinates). A potential use of internal coordinates is to freeze some of them, which may be more convenient than freezing Cartesian coordinates.

SADPOINT and IRC runs can be done with and without the frozen domain treatment; for FD, the limit of the number of atoms applies to the polarizable domain **B**. A saddle point runs requires a guess of the transition state structure, and a Hessian calculation for it. One imaginary mode should describe the reaction. An IRC run is performed starting from the saddle point, with a precomputed Hessian; an IRC calculation is done twice, in the forward and backward direction (i.e., towards products and reactants).

It was observed [65] that for large and complex systems the constrained search of the reaction path gets "lost" and does not actually reach reactants and products. Supplementing the partial path mapped automatically, one can add the separately optimized values of the energies of the reactants and products, and connect the two minima to the path around the transition state. On the other hand, with a manual definition of the reaction coordinate, the whole path was mapped for an enzyme [66].

Of course, there is no guarantee that a geometry optimization can find the global minimum. A geometry optimization can overcome small barriers connecting multiple minima, but eventually, it converges to a minimum, which may not be global. A practical solution to get the global minimum may be to run MD, possibly with FMO [67]. Modern methods to find the global minimum combine MD with geometry optimizations [68].

There is the problem of preparing an initial structure for a geometry optimization. QM methods are very sensitive to elongated bonds or sterical clashes. A bad structure can result in a divergence of the energy calculation at the first step and such structure cannot be optimized. One can preoptimize the structure with a force field. Also, QM methods with a minimum basis set, DFTB and HF-3c, are slightly better for poor initial structures, so one could run a refinement with them first.

To accelerate optimizations, a Hessian may be provided; otherwise, if it is needed (in OPTIMIZE), it can be calculated approximately. A transition state search and IRC, on the other hand, always require a Hessian for the initial structure.

So far, FMO has been used mainly to optimize the structure of ground states. However, because approximate MCSCF and TDDFT gradients are available, it is possible to optimize the structure of excited states. An excited state may be chosen by changing the multiplicity; for example, if the ground state is a singlet, then by choosing an open-shell method (such as UHF), one can optimize the geometry of the lowest triplet state, which relative to the singlet is an excited state. Some examples of excited state optimizations have been published [52]. Here, an extra concern is that during a geometry optimization the order of states may change. Excited state geometries are relevant to emission spectra, and may be useful to biochemical studies of photoactive proteins.

2.4 GAMESS Input Files for Geometry Optimizations

Input files for running FMO calculations can be generated using several available graphical user interfaces, for example, Facio [69]. It is useful, however, to understand how an input file is organized so that one can manually modify it.

A GAMESS input file (Table 5) is divided into groups, beginning with $X and ending with $END, where *X* is a group name. Within each group, several keywords are specified separated by

Table 5
Main groups of options in GAMESS input files

Option	Value	Comments
$CONTRL		
RUNTYP	optimize	optfmo, sadpoint, irc
NPRINT	−5	Minimum output
SCFTYP	rhf	Other choices: uhf, rohf, mcscf
DFTTYP	b3lyp[a]	B3LYP, LCBLYP, etc.
MPLEVL	2[a]	For MP2
$BASIS[a]		
GBASIS	n31	6-31G(d) basis set
NGAUSS	6	Also denoted as 6-31G(*)
NDFUNC	1	Polarization function (*)
$SYSTEM		
MWORDS	400	Replicated memory allocation (1 Mword = 8 MB)
MEMDDI	50	Memory shared in parallel (in Mwords)
$GDDI		
NGROUP	8	Division of nodes into groups
$SCF		
DIRSCF	.t.	Direct SCF
DIIS	.t.	DIIS convergence method

[a] The basis set, DFT and MP2 are selected as shown only for one layer runs. For two layers (FD) a different way is used, not described here

Table 6
Main input group for FMO, $FMO

$FMO[a]		
NFRAG	5	Number of fragments
NLAYER	1	Number of layers
MODGRD	42	Analytic gradient with SCZV
	10	Approximate gradient without SCZV
ICHARG(1)	0,0,1,...	Charge for each fragment
LAYER(1)	1,1,2,...	Layer for each fragment
INDAT(1)	0 1 − 6 0	List of atoms for each fragment
		"1 − 6" means from atom 1 to atom 6 (0 ends a fragment)

[a] Options defining arrays, contain the index of the first element modified, usually 1

spaces. For example, $CONTRL includes keywords to specify the type of calculation with RUNTYP (geometry optimization, normal mode calculation, etc.) and the SCF type (RHF, UHF, or ROHF) in SCFTYP. Other general groups are: $BASIS (basis set), $SYSTEM (memory allocation), $GDDI (parallelization options), and $SCF (SCF converger). The DFT functional is usually specified in the $CONTRL group (although it is considered obsolete in GAMESS), for example, DFTTYP = B3LYP.

There are several FMO specific groups, and $FMO is the most important one (Table 6). The approximate and analytic gradient can be specified as MODGRD set to 10 and 42, respectively (42 is obtained by adding 32 to 10, and 32 means turn on SCZV). In most FMO calculations, a single layer is used; but in FMO/FD (D) always two layers are specified (NLAYER = 2). For example, the active center in an enzyme (in the polarizable buffer **B**) can be treated with a higher level of B3LYP/6-31G(d,p), while the rest (the frozen domain **F**) is computed at a lower level theory of RHF/STO-3G. The number of fragments N is specified as NFRAG; ICHARG and FRGNAM are the arrays of charges and names of each fragment, respectively; INDAT lists atomic numbers for each fragment. Occasionally, one may like to manually merge fragments, which can be done by manipulating the list of atoms in INDAT, and ICHARG, LAYER, etc. have to be adjusted.

$FMOHYB defines hybrid orbitals (such as 1s and 4 sp^3 orbitals for C, the total of $N_{MO} = 5$) used in HOP, precomputed for CH_4 (Table 7). They are basis set specific. $FMOBND defines pairs of atoms for each bond detached between fragments.

To add solvent in PCM, the $PCM group is used (Table 8). SOLVENT = water chooses water and IEF = −10 denotes C-PCM

Table 7
Input groups for fragment boundary definition

$FMOHYB[a]		
6-31G*	15 5	Label, N_{AO}, N_{MO}
1 0	−0.06 ...	N_{MO} Hybridized orbitals expanded
		in the basis of N_{AO} AOs
$FMOBND[b]		
− 18 19	6-31G*	Two atoms per bond: the second
− 38 39	6-31G*	Atom in the pair retains two
− 58 59	6-31G*	Electrons from the bond

[a] This group is used for HOP only and is omitted for AFO. 6-31G* here is a text label (name). The label itself does not define anything. It is only used as an index in $FMOBND. − 0.06, etc. are expansion coefficients defining hybridized orbitals (here, for C atom)

[b] For AFO, the third column (6-31G* above) is omitted

Table 8
Other input options[a]

$DFT		
DC	.t.	Dispersion correction
IDCVER	3	Version three (D3)
SG1	.t.	Fast grid for DFT
$PCM		
SOLVNT	Water	Solvent from a list
ICOMP	2	Induced charge renormalization scheme 2
ICAV	1	Add cavitation free energy
IDISP	1	Add solute-solvent repulsion and dispersion
IEF	−10	Iterative solver for C-PCM
IFMO	−1	Level of PCM in FMO
$STATPT[b]		
OPTTOL	0.0001	Optimization threshold (hartree/bohr)
IACTAT(1)	6327 −6334	List of active atoms in domain **A**
$IRC		
OPTTOL	0.0001	Optimization threshold (hartree/bohr)
FORWRD	.t.	Mapping direction in IRC
NPOINT	16	Number of points in the reaction path
IACTAT(1)	6327 −6334	List of active atoms in domain **A**

[a].t. is a short notation for TRUE

[b] This group is used for stationary point search (minima and saddle points)

(conductor-like PCM). ICOMP denotes the scheme for renormalizing induced solvent charges. ICAV and IDISP are set up to calculate cavitation energy and dispersion energy in PCM, respectively. IFMO = − 1 denotes PCM⟨1⟩ [49].

Finally, a complete example of an input file is given in Table 9. It is a geometry optimization using HF-3c, executed on four computer nodes (set in ngroup).

Table 9
GAMESS/FMO input file for a small peptide (coordinates are truncated)

```
 $contrl nprint=-5 ispher=1 runtyp=optimize $end
 $system memory=10000000 $end
 $gddi ngroup=4 $end
 $scf dirscf=.t. NPUNCH=0 soscf=.t. $end
 $fmo nfrag=4 nlayer=1 layer(1)=1,1,1,1 icharg(1)=0,0,0,0
      indat(1)=0 1 -11 0 12 -18 0 19 -25 0 26 -33 0
      modgrd=42 $end
 $fmoprp nprint=9 $end
 $basis gbasis=hf-3c $end
 $pcm solvnt=water ief=-10 icav=1 idisp=1 ifmo=-1 $end
 $statpt nstep=200 opttol=1e-4 nprt=-2 npun=-2 nprths=0 $end
 $fmohyb
 hf-3c 5 5
 1 0 -0.109774 0.515059  0.000000  0.000000  0.864504
 0 1 -0.109774 0.515059  0.815063  0.000000 -0.288167
 0 1 -0.109774 0.515059 -0.407531 -0.705865 -0.288168
 0 1 -0.109774 0.515059 -0.407531  0.705865 -0.288168
 0 1  0.996474 0.015610  0.000000  0.000000  0.000000
 $end
 $fmobnd
  -9 12 hf-3c
 -16 19 hf-3c
 -23 26 hf-3c
 $end
 $data
Basis set library
c1
h 1
c 6
n 7
o 8
 $end
 $fmoxyz
 H 1  0.00000  0.00000  0.00000
 C 6  0.00000  1.09000  0.00000
 H 1  1.02800  1.45300  0.00000
 H 1 -0.51400  1.45300 -0.89000
 C 6 -0.72100  1.60000  1.24900
 O 8 -0.83900  2.80600  1.45300
 N 7 -1.22700  0.72800  2.12500
 H 1 -1.12400 -0.26100  1.94700
 C 6 -1.91800  1.15900  3.32300
...
 $end
```

3 Applications

3.1 Geometry Optimizations

The accuracy of geometries optimized with FMO/FD was established by comparison to full FMO calculations for Trp-cage (1L2Y) and crambin (1CRN) proteins. Domain **B** was chosen to include fragments for which at least some atoms are within a certain distance from A; this distance is called the size of **B**. The domain **A** was a single residue: Trp-6 and Tyr-44 for Trp-cage and crambin, respectively. The size of **B** varied from 3.9 to 6.5 Å, and it was found [65] that RMSD values from full FMO do not exceed 0.03 Å.

The accuracy of the binding energies for Trp-cage was evaluated as a function of the size of **B** for a neutral and anionic ligands [65]. The neutral ligand had a good accuracy for all tested sizes (3.9–6.5 Å), but for the anionic ligand the use of a small polarizable buffer B (3.9 Å) resulted in a large error of 21.2 kcal/mol in the binding energy. The main reason for the large error is the fact that several residues assigned to the frozen domain **F**, most importantly, negatively charged Asp-9, are very strongly polarized by the ligand, which is properly accounted for when Asp-9 is placed in **B** (e.g., for the size of **B** of ≥5.2 Å). FMO/FDD was thus found to be prone to large errors in the binding energies for charged ligands when **B** is too small.

It is feasible to optimize the structure with FMO/FDD and then refine the energy with full FMO at the obtained minimum. By doing this [65], the error in the binding energy for the anionic ligand was drastically reduced, to 5.7 and 0.3 kcal/mol for FDD and FD, respectively (both at 3.9 Å). For 5.2 Å, the FDD error was 0.1 kcal/mol. Summarizing, charged ligands in general require larger **B** domains (5.9 Å), whereas neutral ligands can be used with medium size of 3.9 Å [59, 65].

FMO/FDD was applied to optimize the structure of a complex of the K-Ras protein with its ligand (4Q03). Two domain **A** definitions were used: only ligand and ligand plus 4 residues in the binding pocket. The RMSD between the optimized and X-ray structures were 0.199 and 0.241 Å for the smaller and larger setups, respectively (calculated for heavy atoms) [65] (*see* Fig. 2).

A summary of various geometry optimizations performed with FMO is given in Table 10 (all atoms were optimized). For traditional methods, only very small proteins or rigid inorganic systems can be fully optimized, but FMO-DFTB is fast enough to optimize full-size proteins.

3.2 Transition States and Reaction Paths

In the validation tests of FMO/FDD [60], transition states and reaction paths were compared to full FMO and QM without fragmentation, for the S_N2 reaction $XCH_3 + OH^- \rightarrow X^- + CH_3OH$ in explicit 41 water molecules, X = F, Cl and Br [65]. The active domain **A** was chosen to be the reaction fragment, containing the

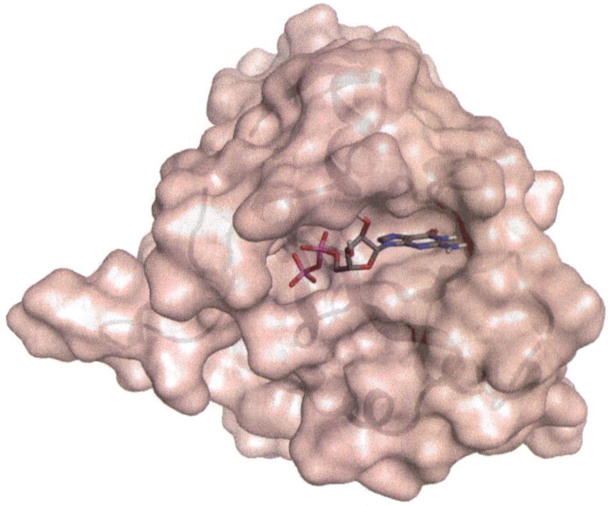

Fig. 2 The optimized structure of a complex of the K-Ras protein with its ligand

Table 10
Summary of systems optimized with FMO (all atoms optimized)

System	Atoms	Method	Validation and reference
Chignolin (1UAO)	138	FMO-DFT	RMSD to experiment 0.453 Å [70]
Chignolin (1UAO)	138	FMO-DFTB	RMSD to experiment 0.414 Å [71]
Trp-cage (1L2Y)	304	FMO-DFTB	RMSD to experiment 0.412 Å [71]
Crambin (1CRN)	642	FMO-RHF	RMSD to experiment 0.525 Å [72]
Crambin (1CRN)	642	FMO-DFTB	RMSD to experiment 0.435 Å [71]
Lysozyme (1IO5)	1961	FMO-DFTB	RMSD to experiment 0.913 Å [71]
Chymotripsinogen A (2CGA)	3578	FMO-DFTB	RMSD to experiment 0.720 Å [71]
Iβ cellulose	10,944	FMO-DFTB	RMSD to experiment 0.467 Å [73]
Mn complex	43	FMO-UHF	RMSD to experiment 0.291 Å [39]
BN nano ring	7878	FMO-DFT	no comparison to experiment [74]
Fullerite surface	1,030,440	FMO-DFTB	experimental cell length within 0.02 Å [46]

solute as well as six surrounding water molecules; the size of **B** was 5.2 Å, which for this small system meant that domain **F** was empty. It was found that FMO/FDD accurately predicted the reaction barriers (the largest error in the Gibbs free energy was 1.6 kcal/mol), and the mapped reaction paths were also nearly identical to FMO without domains and full calculations without fragmentation [60].

The Gibbs free energy for the S_N2 reaction was evaluated using statistical thermodynamics via partition functions based on harmonic frequencies from the Hessian, rotational contribution from the rigid rotor model and translational contribution [60].

FMO/FDD was applied to find the transition state and map the reaction path for enzymatic reactions of chorismate mutase (2CHT) [66] and TIM (7TIM) [60]. In the former, the reaction coordinate was manually chosen, in the latter, IRC was used to automatically determine it.

4 Conclusions

The frozen domain formulation of the fragment molecular orbital method has been described in detail. The recent progress in developing analytic gradient for FMO has been summarized. A few representative cases of geometry optimizations, transition state search, and reaction mapping have been introduced.

Until recently, the cost of geometry optimizations using FMO was very high precluding its usage in practical applications. Now one has the option to refine structures with FMO: using DFT with the frozen domain approach to reoptimize an active site and DFTB to optimize the whole structure. It is essential to add solvent in geometry optimizations, which can be conveniently done using continuum models. If a few water molecules mediate protein–ligand binding, they can be added explicitly as FMO fragments.

Once a structure is optimized, FMO can be used for performing an analysis of protein–ligand interactions, which is most conveniently done with the subsystem analysis [75] and pair interaction energy decomposition analysis [76, 77]. Also, one can do Hessian calculations to simulate infrared and Raman spectra and obtain zero point energy correction to binding [78].

Also, FMO/MD is becoming accessible with 1 ns MD simulations of small proteins already reported [31]. Alternatively to using a single structure, one can average properties in MD in the fluctuation analysis for FMO [79].

There is an ongoing work [80, 81] to develop massively parallel code in GAMESS, taking advantage of the recent progress in computational hardware. All of these make FMO applications to biochemical problems [82] possible today.

References

1. Warshel A, Karplus M (1972) Calculation of ground and excited state potential surfaces of conjugated molecules. I. Formulation and parametrization. J Am Chem Soc 94 (16):5612–5625
2. Hai L, Truhlar DG (2007) QM/MM: what have we learned, where are we, and where do we go from here? Theor Chem Acc 117 (2):185–199
3. Senn HM, Thiel W (2009) QM/MM methods for biomolecular systems. Angew Chem Int Ed 48(7):1198–1229
4. Gordon MS, Fedorov DG, Pruitt SR, Slipchenko LV (2012) Fragmentation methods: a

route to accurate calculations on large systems. Chem Rev 112:632–672

5. Otto P, Ladik J (1975) Investigation of the interaction between molecules at medium distances: I. SCF LCAO MO supermolecule, perturbational and mutually consistent calculations for two interacting HF and CH_2O molecules. Chem Phys 8:192–200
6. Gao JL (1997) Toward a molecular orbital derived empirical potential for liquid simulations. J Phys Chem B 101:657–663
7. Kurbanov EK, Leverentz HR, Truhlar DG, Amin EA (2012) Electrostatically embedded many-body expansion for neutral and charged metalloenzyme model systems. J Chem Theory Comput 8:1–5
8. Thapa B, Beckett D, Erickson J, Raghavachari K (2018) Theoretical study of protein-ligand interactions using the molecules-in-molecules fragmentation-based method. J Chem Theory Comput 14:5143–5155
9. Kobayashi M, Fujimori T, Taketsugu T (2018) Automated error control in divide and conquer self consistent field calculations. J Comput Chem 39:909–916
10. Liu J, Zhang JZH, He X (2016) Fragment quantum chemical approach to geometry optimization and vibrational spectrum calculation of proteins. Phys Chem Chem Phys 18:1864–1875
11. Sahu N, Gadre SR (2016) Vibrational infrared and Raman spectra of polypeptides: fragments-in-fragments within molecular tailoring approach. J Chem Phys 144:114113
12. Liu J, Herbert JM (2016) Pair-pair approximation to the generalized many-body expansion: an alternative to the four-body expansion for ab initio prediction of protein energetics via molecular fragmentation. J Chem Theory Comput 12:572–584
13. Fang T, Li Y, Li S (2017) Generalized energy-based fragmentation approach for modeling condensed phase systems. WIREs: Comput Mol Sci 7:e1297
14. Kitaura K, Ikeo E, Asada T, Nakano T, Uebayasi M (1999) Fragment molecular orbital method: an approximate computational method for large molecules. Chem Phys Lett 313:701–706
15. Fedorov DG, Kitaura K (2007) Extending the power of quantum chemistry to large systems with the fragment molecular orbital method. J Phys Chem A 111:6904–6914
16. Fedorov DG, Nagata T, Kitaura K (2012) Exploring chemistry with the fragment molecular orbital method. Phys Chem Chem Phys 14:7562–7577
17. Tanaka S, Mochizuki Y, Komeiji Y, Okiyama Y, Fukuzawa K (2014) Electron-correlated fragment-molecular-orbital calculations for biomolecular and nano systems. Phys Chem Chem Phys 16:10310–10344
18. Fedorov DG (2017) The fragment molecular orbital method: theoretical development, implementation in GAMESS, and applications. WIREs: Comput Mol Sci 7:e1322
19. Sawada T, Fedorov DG, Kitaura K (2010) Role of the key mutation in the selective binding of avian and human influenza hemagglutinin to sialosides revealed by quantum-mechanical calculations. J Am Chem Soc 132:16862–16872
20. Okiyama Y, Watanabe C, Fukuzawa K, Mochizuki Y, Nakano T, Tanaka S (2019) Fragment molecular orbital calculations with implicit solvent based on the Poisson-Boltzmann equation: II. Protein and its ligand-binding system studies. J Phys Chem B 123:957–973
21. Heifetz A, James T, Southey M, Morao I, Aldeghi M, Sarrat L, Fedorov DG, Bodkin JM, Townsend-Nicholson A (2019) Characterising GPCR-ligand interactions using a fragment molecular orbital-based approach. Curr Opin Struct Biol 55:85–92
22. Mazanetz MP, Chudyk E, Fedorov DG, Alexeev Y (2016) Applications of the fragment molecular orbital method to drug research. In: Zhang W (ed) Computer aided drug discovery. Springer, New York, pp 217–255
23. Fedorov DG, Asada N, Nakanishi I, Kitaura K (2014) The use of many-body expansions and geometry optimizations in fragment-based methods. Acc Chem Res 47:2846–2856
24. Fedorov DG, Kitaura K (2009) The fragment molecular orbital method: practical applications to large molecular systems. CRC Press, Boca Raton, FL
25. Schmidt NW, Baldridge KK, Baldridge JA, Boatz JA, Elbert ST, Gordon MS, Jensen JJ, Koseki S, Matsunaga N, Nguyen KA, Su S, Windus TL, Dupuis M, Montgomery JA (1993) General atomic and molecular electronic structure system. J Comput Chem 14:1347–1363
26. Gordon MS, Schmidt MW (2005) Advances in electronic structure theory: GAMESS a decade later. In: Dykstra CE, Frenking G, Kim KS, Scuseria GE (eds) Theory and applications of computational chemistry, the first forty years. Elsevier, Amsterdam, pp 1167–1189
27. T Ishikawa KK N Yamamoto (2010) Partial energy gradient based on the fragment molecular orbital method: application to geometry optimization. Chem Phys Lett 500:149–154

28. Tsukamoto T, Mochizuki Y, Watanabe N, Fukuzawa K, Nakano T (2012) Partial geometry optimization with FMO-MP2 gradient: application to Trpcage. Chem Phys Lett 535:157–162
29. Nagata T, Brorsen K, Fedorov DG, Kitaura K, Gordon MS (2011) Fully analytic energy gradient in the fragment molecular orbital method. J Chem Phys 134:124115
30. Fedorov DG, Jensen JH, Deka RC, Kitaura K (2008) Covalent bond fragmentation suitable to describe solids in the fragment molecular orbital method. J Phys Chem A 112 (46):11808–11816
31. Vuong VQ, Nishimoto Y, Fedorov DG, Sumpter BG, Niehaus TA, Irle S (2019) The fragment molecular orbital method based on long-range corrected density-functional tight-binding. J Chem Theory Comput 15:3008–3020
32. Fedorov DG, Avramov PV, Jensen JH, Kitaura K (2009) Analytic gradient for the adaptive frozen orbital bond detachment in the fragment molecular orbital method. Chem Phys Lett 477:169–175
33. Nakano T, Kaminuma T, Sato T, Akiyama Y, Uebayasi M, Kitaura K (2000) Fragment molecular orbital method: application to polypeptides. Chem Phys Lett 318(6):614–618
34. Nagata T, Fedorov DG, Kitaura K (2010) Importance of the hybrid orbital operator derivative term for the energy gradient in the fragment molecular orbital method. Chem Phys Lett 492:302–308
35. Fedorov DG, Kitaura K (2004) Second order Møller-Plesset perturbation theory based upon the fragment molecular orbital method. J Chem Phys 121:2483–2490
36. Nagata T, Fedorov DG, Ishimura K, Kitaura K (2011) Analytic energy gradient for second-order Møller-Plesset perturbation theory based on the fragment molecular orbital method. J Chem Phys 135(4):044110
37. Kitaura K, Sugiki SI, Nakano T, Komeiji Y, Uebayasi M (2001) Fragment molecular orbital method: analytical energy gradients. Chem Phys Lett 336(1,2):163–170
38. Komeiji Y, Mochizuki Y, Nakano T, Mori H (2012) Recent advances in fragment molecular orbital-based molecular dynamics(FMO-MD) simulations. In: Wang L (ed) Molecular dynamics - theoretical developments and applications in nanotechnology and energy. InTech, Rijeka, pp 3–24
39. Nakata H, Fedorov DG, Nagata T, Yokojima S, Ogata K, Kitaura K, Nakamura S (2012) Unrestricted Hartree-Fock based on the fragment molecular orbital method: energy and its analytic gradient. J Chem Phys 137:044110
40. Pruitt SR, Fedorov DG, Gordon MS (2012) Geometry optimizations of open-shell systems with the fragment molecular orbital method. J Phys Chem A 116:4965–4974
41. Nakata H, Schmidt MW, Fedorov DG, Kitaura K, Nakamura S, Gordon MS (2014) Efficient molecular dynamics simulations of multiple radical center systems based on the fragment molecular orbital method. J Phys Chem A 118(41):9762–9771
42. Fedorov DG, Kitaura K (2004) On the accuracy of the 3-body fragment molecular orbital method (FMO) applied to density functional theory. Chem Phys Lett 389:129–134
43. Brorsen KR, Zahariev F, Nakata H, Fedorov DG, Gordon MS (2014) Analytic gradient for density functional theory based on the fragment molecular orbital method. J Chem Theory Comput 10(12):5297–5307
44. Nakata H, Fedorov DG, Yokojima S, Kitaura K, Sakurai M, Nakamura S (2014) Unrestricted density functional theory based on the fragment molecular orbital method for the ground and excited state calculations of large systems. J Chem Phys 140(14):144101
45. Nakata H, Fedorov DG (2020) Development of the analytic second derivatives for the fragment molecular orbital method. In: Tanaka S (ed), Springer Japan, Tokyo, in press
46. Nishimoto Y, Fedorov DG, Irle S (2014) Density-functional tight-binding combined with the fragment molecular orbital method. J Chem Theory Comput 10(11):4801–4812
47. Nishimoto Y, Nakata H, Fedorov DG, Irle S (2015) Large-scale quantum-mechanical molecular dynamics simulations using density-functional tight-binding combined with the fragment molecular orbital method. J Phys Chem Lett 6(24):5034–5039
48. Li H, Fedorov DG, Nagata T, Kitaura K, Jensen JH, Gordon MS (2010) Energy gradients in combined fragment molecular orbital and polarizable continuum model (FMO/PCM) calculation. J Comput Chem 31(4):778–790
49. Nagata T, Fedorov DG, Li H, Kitaura K (2012) Analytic gradient for second order Møller-Plesset perturbation theory with the polarizable continuum model based on the fragment molecular orbital method. J Chem Phys 136 (20):204112
50. Nakata H, Fedorov DG, Kitaura K, Nakamura S (2015) Extension of the fragment molecular orbital method to treat large open-shell systems in solution. Chem Phys Lett 635:86–92

51. Fedorov DG, Kitaura K (2005) Multiconfiguration self-consistent-field theory based upon the fragment molecular orbital method. J Chem Phys 122:0541081
52. Chiba M, Fedorov DG, Nagata T, Kitaura K (2009) Excited state geometry optimizations by time-dependent density functional theory based on the fragment molecular orbital method. Chem Phys Lett 474(13):227–232
53. Fedorov DG, Kitaura K (2004) The importance of three-body terms in the fragment molecular orbital method. J Chem Phys 120 (15):6832–6840
54. Nishimoto Y, Fedorov DG (2017) Three-body expansion of the fragment molecular orbital method combined with density-functional tight-binding. J Comput Chem 38 (7):406–418
55. Goerigk L, Collyer CA, Reimers JR (2014) Recommending Hartree-Fock theory with London-dispersion and basis-set-superposition corrections for the optimization or quantum refinement of protein structures. J Phys Chem B 118:14612–14626
56. Fedorov DG (2018) Analysis of solute-solvent interactions using the solvation model density combined with the fragment molecular orbital method. Chem Phys Lett 702:111–116
57. Nagata T, Fedorov DG, Sawada T, Kitaura K, Gordon MS (2011) A combined effective fragment potential - fragment molecular orbital method. II. Analytic gradient and application to the geometry optimization of solvated tetraglycine and chignolin. J Chem Phys 134:034110
58. Fedorov DG, Ishida T, Uebayasi M, Kitaura K (2007) The fragment molecular orbital method for geometry optimizations of polypeptides and proteins. J Phys Chem A 111 (14):2722–2732
59. Fedorov DG, Alexeev Y, Kitaura K (2011) Geometry optimization of the active site of a large system with the fragment molecular orbital method. J Phys Chem Lett 2 (4):282–288
60. Nakata H, Fedorov DG, Nagata T, Kitaura K, Nakamura S (2015) Simulations of chemical reactions with the frozen domain formulation of the fragment molecular orbital method. J Chem Theory Comput 11(7):3053–3064
61. Fedorov DG, Kromann JC, Jensen JH (2018) Empirical corrections and pair interaction energies in the fragment molecular orbital method. Chem Phys Lett 706:328–333
62. Nakata H, Nagata T, Fedorov DG, Yokojima S, Kitaura K, Nakamura S (2013) Analytic second derivatives of the energy in the fragment molecular orbital method. J Chem Phys 138:164103
63. Kaliakin DS, Fedorov DG, Alexeev Y, Varganov SA (2019) Locating minimum energy crossings of different spin states using the fragment molecular orbital method. J Chem Theory Comput 15:6074–6084
64. Fedorov DG, Ishida T, Kitaura K (2005) Multilayer formulation of the fragment molecular orbital method (FMO). J Phys Chem A 109:2638–2646
65. Nakata H, Fedorov DG (2016) Efficient geometry optimization of large molecular systems in solution using the fragment molecular orbital method. J Phys Chem A 120:9794–9804
66. Steinmann C, Fedorov DG, Jensen JH (2013) Mapping enzymatic catalysis using the effective fragment molecular orbital method: towards all ab initio biochemistry. PLOS one 8:e60602
67. Komeiji Y, Nakano T, Fukuzawa K, Ueno Y, Inadomi Y, Nemoto T, Uebayasi M, Fedorov DG, Kitaura K (2003) Fragment molecular orbital method: application to molecular dynamics simulation, ab initio FMO-MD. Chem Phys Lett 372:342–347
68. Mironov V, Alexeev Y, Mulligan VK, Fedorov DG (2019) A systematic study of minima in alanine dipeptide. J Comput Chem 40:297–309
69. Suenaga M (2008) Development of GUI for GAMESS / FMO calculation. J Comput Chem Jpn 7:33–53
70. Fedorov DG, Kitaura K (2014) Use of an auxiliary basis set to describe the polarization in the fragment molecular orbital method. Chem Phys Lett 597:99–105
71. Nishimoto Y, Fedorov DG (2016) The fragment molecular orbital method combined with density-functional tight-binding and the polarizable continuum model. Phys Chem Chem Phys 18(32):22047–22061
72. Nakata H, Fedorov DG, Yokojima S, Kitaura K, Nakamura S (2014) Simulations of Raman spectra using the fragment molecular orbital method. J Chem Theory Comput 10:3689–3698
73. Nishimoto Y, Fedorov DG, Irle S (2015) Third-order density-functional tight-binding combined with the fragment molecular orbital method. Chem Phys Lett 636(11):90–96
74. Avramov PV, Fedorov DG, Sorokin PB, Sakai S, Entani S, Ohtomo M, Y Matsumoto HN (2012) Intrinsic edge asymmetry in narrow zigzag hexagonal heteroatomic nanoribbons causes their subtle uniform curvature. J Phys Chem Lett 3:2003–2008

75. Fedorov DG, Kitaura K (2016) Subsystem analysis for the fragment molecular orbital method and its application to protein-ligand binding in solution. J Phys Chem A 120:2218–2231
76. Fedorov DG, Kitaura K (2007) Pair interaction energy decomposition analysis. J Comput Chem 28:222–237
77. Fedorov DG, Kitaura K (2012) Energy decomposition analysis in solution based on the fragment molecular orbital method. J Phys Chem A 116:704–719
78. Nakata H, Fedorov DG (2019) Simulations of infrared and Raman spectra in solution using the fragment molecular orbital method. Phys Chem Chem Phys 21:13641–13652
79. Fedorov DG, Kitaura K (2018) Pair interaction energy decomposition analysis for density functional theory and density-functional tight-binding with an evaluation of energy fluctuations in molecular dynamics. J Phys Chem A 122:1781–1795
80. Fedorov DG, Olson RM, Kitaura K, Gordon MS, Koseki S (2004) A new hierarchical parallelization scheme: generalized distributed data interface (GDDI), and an application to the fragment molecular orbital method (FMO). J Comput Chem 25:872–880
81. Mironov V, Alexeev Y, Fedorov DG (2019) Multithreaded parallelization of the energy and analytic gradient in the fragment molecular orbital method. Int J Quant Chem 119: e25937
82. Alexeev Y, Mazanetz MP, Ichihara O, Fedorov DG (2012) GAMESS as a free quantum-mechanical platform for drug research. Curr Top Med Chem 12:2013–2033

Chapter 7

Taking Water into Account with the Fragment Molecular Orbital Method

Yoshio Okiyama, Kaori Fukuzawa, Yuto Komeiji, and Shigenori Tanaka

Abstract

This chapter describes the current status of development of the fragment molecular orbital (FMO) method for analyzing the electronic state and intermolecular interactions of biomolecular systems *in solvent*. The orbital energies and the inter-fragment interaction energies (IFIEs) for a specific molecular structure can be obtained directly by performing FMO calculations by exposing water molecules and counterions around biomolecular systems. Then, it is necessary to pay attention to the thickness of the water shell surrounding the biomolecules. The single-point calculation for snapshots from MD trajectory does not incorporate the effects of temperature and configurational fluctuation, but the SCIFIE (statistically corrected IFIE) method is proposed as a many-body correlated method that partially compensates for this deficiency. Furthermore, implicit continuous dielectric models have been developed as effective approaches to incorporating the screening effect of the solvent in thermal equilibrium, and we illustrate their usefulness for theoretical evaluation of IFIEs and ligand-binding free energy on the basis of the FMO-PBSA (Poisson–Boltzmann surface area) method and other computational methods.

Key words Fragment molecular orbital (FMO) method, Solvent effect, Water molecule, Inter-fragment interaction energy (IFIE), Statistically corrected IFIE (SCIFIE), Ligand binding, Hydration shell, Dielectric continuum, Poisson–Boltzmann equation

1 Introduction

Fragment molecular orbital (FMO) calculations in vacuum have shown their effectiveness in various fields such as ligand–protein interaction analysis [1]. However, it has also been recognized that incorporation of solvent effect, together with thermal fluctuation and entropy effects, can further improve the agreement between calculation and experimentation of ligand binding affinity, for instance. In addition, consideration of solvent effects is also important when we study the stable structure of biomolecular complex systems composed of proteins, nucleic acids, small ligand molecules, surrounding ions, and so on. There are two major ways to introduce the solvent effects into FMO calculations, that is, explicit approach that takes water molecules into account and implicit

Alexander Heifetz (ed.), *Quantum Mechanics in Drug Discovery*, Methods in Molecular Biology, vol. 2114, https://doi.org/10.1007/978-1-0716-0282-9_7,

approach that describes solvent in terms of continuum model. This article illustrates the current status of these approaches and some future issues.

2 Explicit Water

Treatment of explicit solvent water seems to be essential when considering the role of molecular water surrounding biomolecules. Biomolecules form hydrogen bonds with water at their interface, which involves charge transfer. The electrostatic potential of the surface of charged solute molecules changes dramatically depending on the presence or absence of molecular water [2]. Interfacial water gradually changes to bulk water as the distance from the solute increases. The presence of solvent water molecules also affects intramolecular interactions of solute biomolecules. Here, the role of molecular water is described from both viewpoints of shell waters as a solvent environment and crystal waters.

2.1 Shell Solvent

Effects of the explicit water solvent on FMO calculation for biological molecules have been investigated systematically with various solute molecules including a protein [3], double-stranded (ds) DNA [4, 5], single-stranded (ss) DNA, and its complex with an ssDNA binding protein (SSB) [6, 7]. As reviewed below, the overall tendency of the solvent effects was similar among the solute molecules as follows: the first solvent shell within 4 Å from the solute surface governs the solute internal energy and charge, while the second shell, up to ca. 8 Å, is also important.

The sequence of papers used the following protocol to investigate the electronic properties of the solute molecules as functions of the solvent thickness. The molecular configurations for FMO were prepared by classical molecular mechanics (MM) –molecular dynamics (MD) (MM–MD) at room temperature. Several configurations were selected from the generated trajectories, were annealed to 0 K, and were finally energy-minimized. From the structures thus optimized, the solutes (protein and/or DNA) and solvent shell (water and counterions) were excised to construct a systematic series of configurations with varying shell thicknesses. The thickness of the solvent shell was determined by measuring the distance between non-hydrogen atoms of the solute and solvent. As an example, *see* Fig. 1 for constructed series of molecular configurations of dsDNA [4, 5]. The configurations were subjected to FMO calculation.

The protein net charge, internal energy, and solvent–protein interaction were calculated as functions of the solvent thickness in the study of ubiquitin, the first comprehensive FMO study of the solvation on a protein [3]. The net charge converged at 4 Å, and the internal energy at 8 Å. The protein–solvent interaction showed a marginal convergence around 8 Å. Another finding was that water

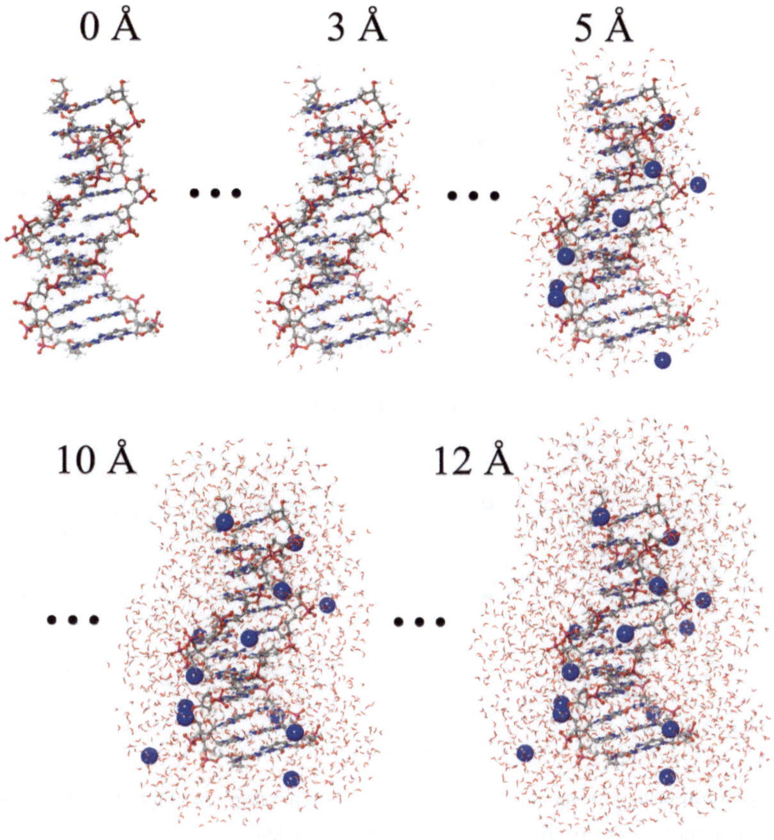

Fig. 1 Thickness of water solvent for DNA. Reproduced from Ref. 5 by permission of the publisher

molecules within 4 Å of the outer surface were significantly less polarized than the bulk water.

Similar results were obtained for dsDNA [4, 5], ssDNA, and ssDNA:SSB complex [6, 7]. *See* Fig. 2 for the graphs of solute charge, internal energy, and the solute–solvent interaction of the ssDNA:SSB. Frontier molecular orbital (MO) energies of fragments within dsDNA showed a different behavior [4, 5]; the highest occupied MOs (HOMOs) and lowest unoccupied MOs (LUMOs) converged only marginally at 5 Å, and they did not converge completely even at 12 Å. The quantity of solvent needed for conversion thus depends on the kind of physical property of interest.

The effect of counterions on the solute properties may be another concern. In the protein solvation study [3], the ubiquitin protein was chosen because it bears no net charge at the neutral pH, and hence no counterion was necessary to neutralize the simulated system. In the dsDNA study [4, 5], enough counterions were added to neutralize the molecular system in the preparative MM–MD calculation, but only those counterions present within the solvent shell were preserved for the subsequent FMO calculation,

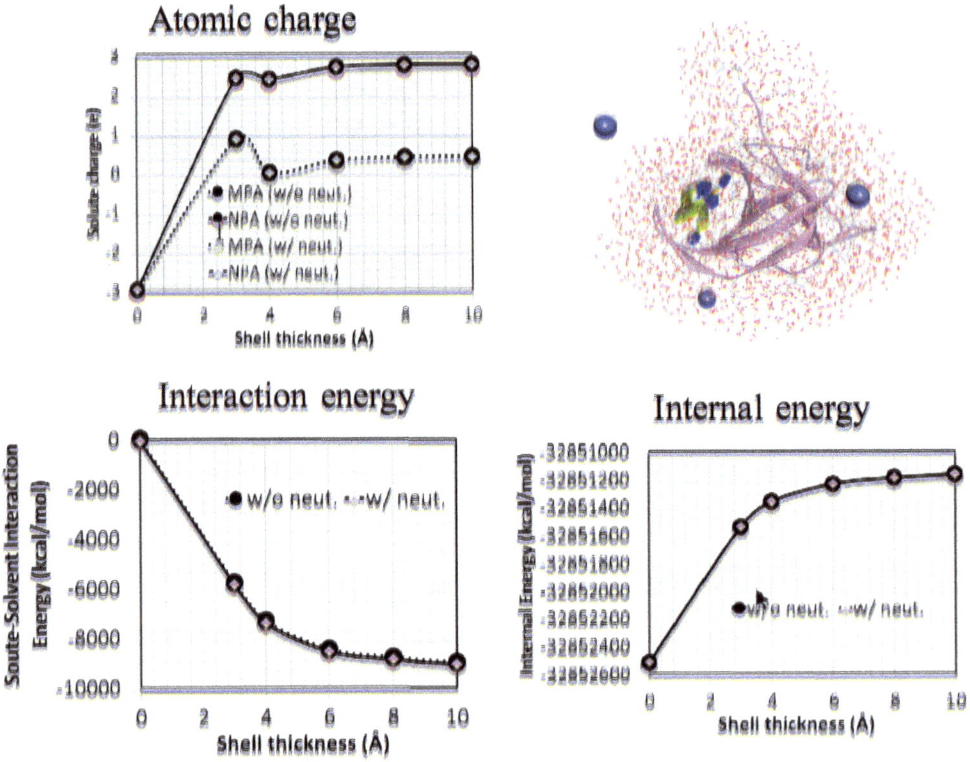

Fig. 2 Properties of the ssDNA/SSB complex as functions of solvent thickness with and without neutralization. Reproduced from Ref. 6 by permission of the publisher

resulting in FMO calculations of negatively charged molecular systems. The effect of counterions was minutely investigated later in the studies of the ssDNA:SSB complex [6]. As is clearly seen in Fig. 2, the solute charge and internal energy were virtually uninfluenced by the presence of enough counterions to neutralize the system. Even the solute–solvent interaction was only slightly influenced. Thus, neutralization by counterions did not affect much the internal properties of the solutes investigated.

This fact does not necessarily mean that counterions have no effect on the solutes. We should keep in mind that in Refs. 6, 7, FMO calculations were performed for only molecular configurations in which counterions surrounded the solutes at a distance, not in direct contact. In fact, in an earlier test FMO calculation of DNA fragments in vacuo [4, 5], neutralization of a fragment by a directly bound counterion affected greatly the orbital energies.

Taken together, the series of FMO calculations of protein and DNA have revealed that the solute net charge and internal energy are affected dominantly by the first water solvent shell within ca. 4 Å of the solutes and auxiliarily by the second shell up to ca. 8 Å and that the counterions do not influence much the solutes unless directly bound to the solutes.

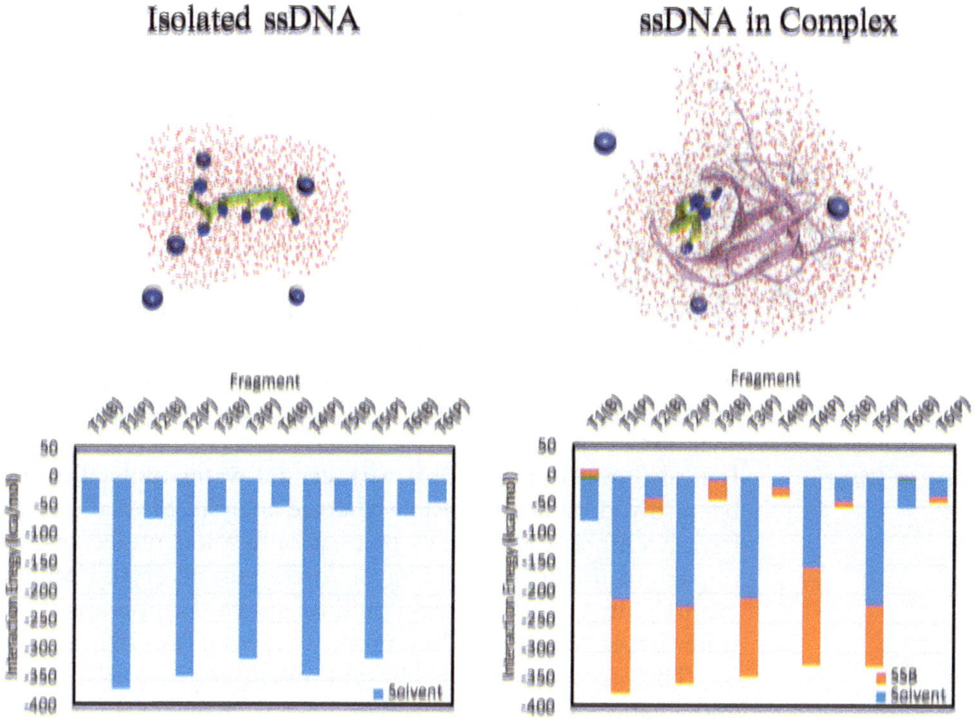

Fig. 3 Interaction of ssDNA with the solvent and SSB. Reproduced from Ref. 6 by permission of the publisher

Based upon these findings, minute solvation effects on the solutes were further analyzed. An important finding was modulation of interactions within dsDNA [4, 5]. Upon solvation, there is a minus charge transfer from the DNA to the solvent, which elicits charge rearrangement within the DNA molecules and eventually results in enhancement of the base–base H-bond and in weakening of the stacking. In the SSB recognition by ssDNA, desolvation was clearly seen by inter-fragment interaction energy (IFIE) analysis [7]. The nucleotide units of ssDNA interact with the solvent in the isolated state, but upon SSB binding, many of them change their partners to specific amino acid residues of SSB. This desolvation is clearly illustrated in the IFIE (Fig. 3) in which the interactions with the solvent are replaced with that with SSB.

While systematic evaluation of size of shell water has been discussed as above, there have been some studies for protein–ligand interaction analysis using a specific size of shell water. Kurita et al. suggested that the protonation state of histidine residue under water solvent environment plays an important role in the interaction between a chiral ligand and vitamin D receptor (VDR) that is covered with 6 Å shell water [8]. They have also shown that in VDR–ligand interaction surrounded by 6 Å shell water, the EC_{50} activity value of the ligand with tetrazole ring was dramatically changed by the difference of nitrogen position, and such "activity cliff" was caused by the hydrogen bond which induced CH/π interaction [9].

As another aspect of the role of molecular water, electrostatic potentials of the surface of charged solute biomolecules change dramatically due to charge transfer with surrounding water molecules. In the case of the estrogen receptor alpha (ERα) ligand binding domain, the total charge of the solute protein was originally at $-6.0e$, but changed to about $-2.3e$ by charge transfer with the surrounding 4 Å water shell. As a result, the electrostatic potential on the protein surface approached to neutral, and the characteristics of the charged amino acid residues exposed on the surface became clearer, which helps to understand molecular recognition [2].

2.2 Crystal Water

It is also important to deal with crystal water molecules that has been solved by X-ray crystal structure analysis. Several calculations have been made specifically to treat only water molecules that are important for ligand binding or to contain all crystal waters. For example, in a protein–ligand binding study, the presence of crystal water improved the binding property prediction [10].

The presence of special bound water, which cross-links protein and ligand, is also known. The crystal structure of ERα–ligand complex includes one bound water molecule, which is conserved in most crystal structures. FMO calculations revealed that the water molecule plays a key role in ligand binding as a member of the hydrogen bonding network among ERα and ligand (Fig. 4) [11].

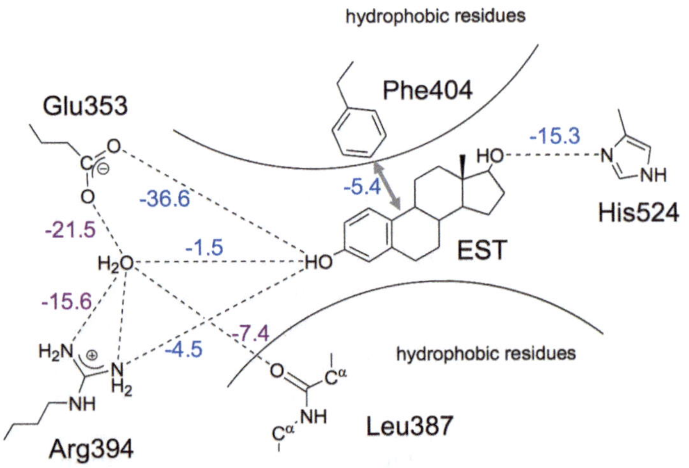

Fig. 4 Hydrogen bond network among estrogen receptor alpha, ligand, and water molecule (dotted line). The double arrow represents the CH/π interaction. The numbers shown along with the dotted line and double arrow indicate inter-fragment interaction energies (IFIEs) in kcal/mol between amino acid residue and ligand (blue) or water (purple) calculated at the MP2/6-31G(d)//HF/6-31G (d) level

3 SCIFIE: Effective Incorporation of Many-Body Screening

FMO calculations, in the ground state, give electronically polarized molecular state with minimal energy for a fixed molecular configuration. In this sense, the electrostatic screening effect has already been included in the (structurally) static and electronic ways. Thus, through the FMO calculations with explicit water molecules as discussed in Subheading 2, the electrostatic screening effect can be described in terms of the electron polarization with fixed molecular configuration. However, it is well recognized [1] that the IFIEs thus calculated often fail to account for the electrostatic screening effect due to surrounding water or distant ions even qualitatively, while it is empirically known to be significant. This defect can then be explained physicochemically by the fact that the calculations do not take into account the effects of configurational polarization of charged or polar molecules or functional groups at finite temperature. This important contribution to dielectric screening is predominantly attributed to polar water molecules surrounding biomolecules, whose measure (e.g., dielectric constant of about 80 in the case of water) is associated with the configurational dynamics of water molecules; its effective incorporation into FMO calculations by means of continuous dielectric model is discussed in Subheading 4. In this section, on the other hand, we employ a viewpoint of what is (mainly electrostatic) screening effect due to many-body correlations among (charged or polarized) classical "particles" and address an alternative scheme called statistically corrected inter-fragment interaction energy (SCIFIE) approach [12] in which a partial and effective correction for many-body screening is performed on the IFIEs obtained for a snapshot molecular configuration. This SCIFIE scheme is practically useful because FMO calculations are often carried out for a snapshot configuration due to their high computational cost, while the ensemble average over multiple configurations at finite temperature would also be possible at later stage.

Let us consider a molecular configuration for which an FMO-IFIE calculation has been performed. There are a variety of pairwise IFIEs with positive (repulsive) or negative (attractive) sign in that system due to mixed ionic, polar, dispersion, exchange-repulsion, charge-transfer, and other molecular interactions. Then, the basic idea of SCIFIE is as follows [12]. As for molecular interactions, in general, the contribution of electrostatic (Coulombic) interaction is predominantly large, and retaining the charge neutrality imposes a significant constraint on system energetics to avoid substantial increase in total energy without the charge neutrality. For example, Coulombic systems always have negative charges in the near vicinity of positive charges (and vice versa) to keep the charge neutrality locally; then considering an electric charge distantly interacting with these positive and negative charges, it interacts strongly with each individual positive or negative charge, but effectively feels a much milder interaction with their mixture due to the cancellation among positive and negative contributions. This picture thus provides an underlying basis for those physical accounts such as the Ewald summation and the electrostatic screening effect. Hence, instead of considering the strong interactions with high magnitudes between "bare" electric charges, it would be relevant to account for effective, screened interactions among "dressed" particles that are alternatively placed with opposite charges. The difficulties found in "bare" IFIEs are analogous to these observations, and we may (partly) overcome them by considering the effective interactions associated with or represented by interparticle correlations. Such a description can be given in the context of classical many-body problem [13] through the concepts of screened interaction and direct correlation function.

In the SCIFIE scheme, each fragment i in a molecular system is regarded as a classical particle embedded in the network of mutual interactions whose "bare" energy values are given as u_{ij} between the fragments i and j. The effective interaction or the potential of mean force w_{ij} is then related to the pair correlation function h_{ij} as.

$$h_{ij} = e^{-\beta w_{ij}} - 1.$$

The parameter β is usually related to the absolute temperature T and the Boltzmann constant k_B via $\beta = 1/k_B T$, but, in the present formalism, may be regarded as an optimization parameter to control the degree of screening, which will be specified afterward. The pair correlation function h_{ij} is related to the direct correlation function c_{ij} in terms of the Ornstein–Zernike (OZ) relation [13]:

$$h_{ij} = c_{ij} + \sum_{k \neq i,j} c_{ik} h_{kj}.$$

The direct correlation function is conceptually introduced as a difference between the total and indirect parts of interparticle correlation. We may thus express it in the Percus–Yevick (PY) approximation [13] for classical many-body problem as:

$$c_{ij} = e^{-\beta w_{ij}} - e^{-\beta\left(w_{ij} - u_{ij}\right)},$$

providing a closure equation to determine w_{ij} for a given set of u_{ij}. Here, one may also employ other closure approximations such as the hypernetted chain (HNC) approximation [13].

The procedure above on the basis of the PY approximation may be formulated as follows. By introducing the Mayer function,

$$f_{ij} = e^{-\beta u_{ij}} - 1,$$

we can rewrite the PY equation as:

$$c_{ij} = \frac{f_{ij}(h_{ij} + 1)}{f_{ij} + 1}.$$

Substituting this equation into the OZ relation, we find

$$h_{ij} = \frac{f_{ij}(h_{ij} + 1)}{f_{ij} + 1} + \sum_k \frac{f_{ik}(h_{ik} + 1)}{f_{ik} + 1} h_{kj},$$

or

$$h_{ij} = f_{ij} + \left(f_{ij} + 1\right) \sum_k \frac{f_{ik}(h_{ik} + 1)}{f_{ik} + 1} h_{kj}.$$

For given u_{ij} and f_{ij}, we can thus obtain h_{ij} through solution to this equation and also find the effective interaction as

$$w_{ij} = -\frac{1}{\beta} \ln \left(h_{ij} + 1\right).$$

In the formulation above, the effective, screened interactions are obtained through the inter-fragment correlations in a system with given set of bare interactions. The (physical) origin of the screening may partially be attributed to the conformational fluctuations and associated entropic effects in the system, but the parameter β should not be chosen according to room temperature, but rather as a measure of randomness and collectivity. The energy scale of IFIEs obtained by single-point FMO calculation is of electronic origin and hence much higher than that of thermal energy at room temperature. While the electronic state of the pertinent system is very sensitive to the conformational fluctuations due to thermal energy, the relevant value of β in the present screening model at a fixed conformation should be independent of the thermal energy associated with the molecular system whose conformations could fluctuate at finite temperature. The parameter β is thus regarded as not "thermodynamic" one but "information theoretic" one, and its optimal value can be determined by a "maximum screening" ansatz under the energy constraint condition so that a quantity characterizing the degree of screening.

$$\sigma = \sum_{i<j} \left| w_{ij} \right|$$

is minimized. It is here remarked that the pair correlation function h_{ij} can be related to the mutual information in information theory [14]. In addition, in order to find a more realistic evaluation for the effective interaction between fragments, the statistical averaging over the thermal motion at room temperature may be taken later via generating many conformations for FMO calculations with, e.g., MD simulations.

The SCIFIE scheme, as formulated above, has been successfully applied to some model and more realistic (biomolecular, organic, and inorganic) systems [2, 15–17], in which one observes the effective screening of "bare" IFIEs.

4 Implicit Water

4.1 Overview

An implicit solvent treatment is a practical way in quantum chemistry and a first choice to include solvent effects into the description of electronic state of a solute molecule. All biochemical processes take place in the environment of water solution that can strongly affect the reaction energies [18]. Therefore, such effects are conveniently introduced by treating surroundings as a dielectric continuum also in FMO. There are several FMO applications with such continuum solvent methods as polarizable continuous model (PCM) [19], solvation model density (SMD) [20], and Poisson–Boltzmann surface area (PBSA) model [21, 22]. An implicit treatment has the advantage of avoiding costly statistical sampling of the explicit solvent configurations while giving some sort of averaged results consistent with an explicit treatment. On the other hand, there is a discrepancy with an explicit treatment in the description of solute–solvent interactions including electron's outflow. Therefore, it would be indispensable to use the implicit or explicit treatment properly depending on the situation.

One of the essential features of the FMO application with the implicit solvent model is the ability to describe the reasonable MO on a local fragment, which is the most symbolic and critical characters defining the electronic properties of the system. HOMO is especially crucial in analyses because of its relation to the stability and reactivity, while its energy inside biomolecules often seems unstable in the gas phase. By imposing induced charges as an implicit solvent, the energy is successfully lowered to guarantee the stable electronic state for DNA [22] and proteins [23].

Another is the ability to give the reasonable solvation free energy of whole system. For example, the solvation free energy of dsDNA in an implicit solvent is asymptotically close to that in an explicit solvent sufficiently including water and ions in the previous section (Fig. 5). Based on such a solvated FMO ability, the qualification of modeled structures was performed by Simoncini et al. [24]. For eight protein targets, they assessed the quality of their

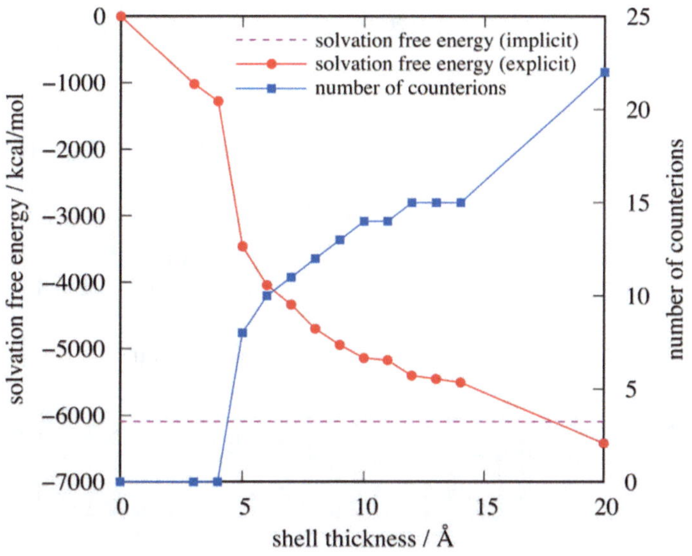

Fig. 5 Solvation free energy of dsDNA in an explicit solvent as a function of the shell thickness by comparison with that in an implicit solvent. The number of counterions contained in the explicit solvent is also plotted. (Reprinted with permission from Ref. 22. Copyright 2018 American Chemical Society)

predicted models by using FMO energy to extract the best model from each ensemble. The FMO energy was derived from FMO2-HF/6-31G(d) with Grimme's dispersion correction [25, 26] (HF-D3) in PCM solvent. They found FMO-based model ranking was superior to empirically derived one.

IFIE analysis and binding affinity prediction will be of special interest in drug discovery. The following sections present their significant analysis and application examples. Note that all FMO calculations combined with an implicit solvent model below use the FMO2-MP2/6-31G(d) level with some exceptions.

4.2 Fragment Interaction Energy

IFIE and its decomposition (PIEDA) analyses can reveal the electronic mechanisms leading to local stability inside biomolecules. How is then IFIE affected when a biomolecule is immersed in an implicit solvent? Actually, an internal IFIE in solution, that is, an IFIE purely considered between solute fragments, hardly changes compared to that in vacuo, even if one uses an implicit or explicit solvent model. Accordingly, Fedorov et al. explicitly included the solute–solvent interaction into IFIE as a natural division of total FMO energy in solution and treated its contribution as the solvation term in PIEDA [27]. The solvation term arises counteractively to the electrostatic (ES) term, and they eventually cancel out each other (*see* Fig. 6). This is called the solvent screening effect, and *the perfect screening is observed for the interaction of two-point charges* under ideal conditions [28].

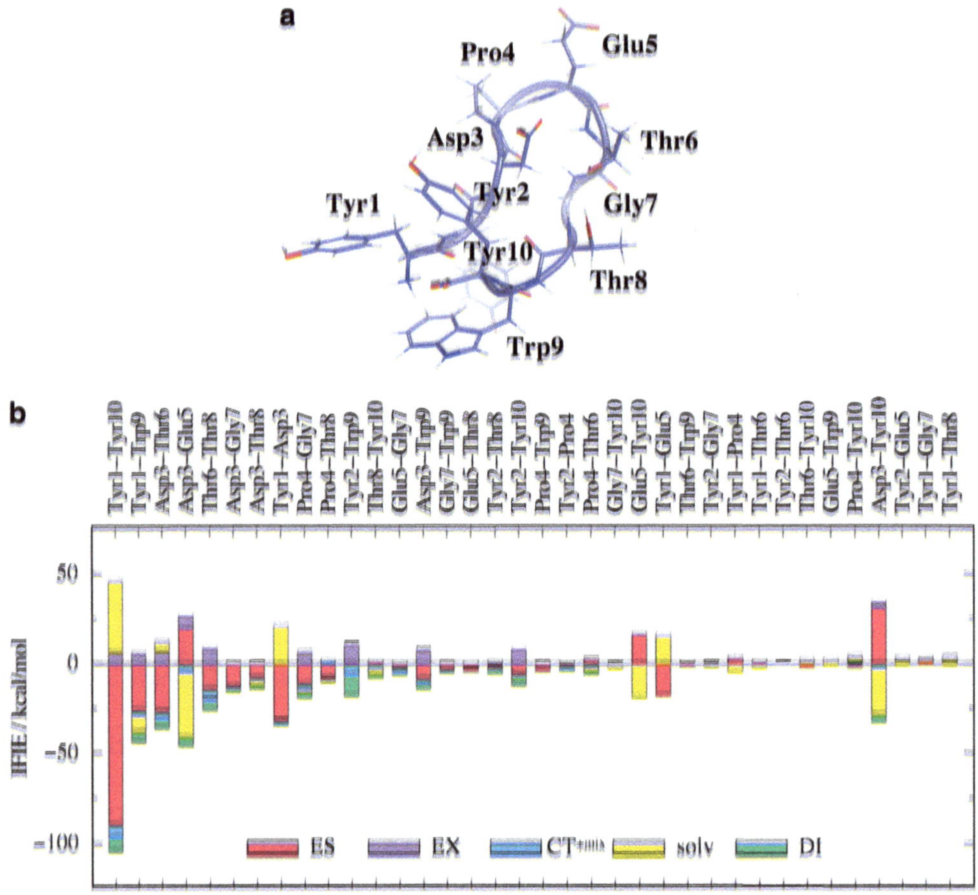

Fig. 6 (**a**) Crystal structure of superchignolin (PDB ID: 5AWL) with modeled hydrogen atoms. Both termini of the structure are in a zwitterionic state. (**b**) IFIEs of the structure in solution. The FMO-PBSA calculation at the MP2/6-31G(d) level yields the PIEDA components including the solvation term

4.2.1 Screening Effect

Okiyama et al. practically and systematically studied solvent screening effects on fragment interactions for DNA and bioactive proteins, ubiquitin and ERα, using FMO-PBSA approach [22, 29]. Solvent screening effectively reduces the excessive electrostatic interactions between charged fragments also inside such biomolecules while hardly affecting the electron-correlated ones. For example, IFIEs between backbone fragments including charged phosphate groups for DNA or between polar-residue fragments on the molecular surface for proteins are demonstratively damped according to the distance. Screening gives a clear view for significant interactions including secondary structure patterns.

On the other hand, if two fragments are both insufficiently exposed to the solvent or neutral, there is little or no influence on IFIE. This is exactly the case for proteins, whose fragments are never uniformly exposed to the solvent. It should be noted whether focused interactions are subject to screening effects, because especially a ligand complexed with a protein is more or less buried inside the receptor.

Here, we update the story as ***the screening between two fragments depends on their chargeability, solvent exposure, and distance*** in the case of biomolecular systems. When focused on the interactions with a ligand, application of subsystem analysis [30] in consideration of desolvation is also promising in the future.

4.2.2 Energetic Map

The energetic map analysis is an application of screening-included IFIE in the context of drug discovery. Śliwa et al. proposed a methodology to determine significant interactions in each protein–ligand complex by using a PIEDA-based energetic map of the orthosteric binding site [31]. They demonstrated the applicability to serotonin receptors, 5-HT_{1A}, 5-HT_{2A}, and 5-HT_7 complexed with selected long-chain arylpiperazines by using FMO-PCM approach. This methodology identifies significant attractive/repulsive areas in the binding sites that would provide hints for designing new binders to target proteins.

4.3 Binding Free Energy

Accurate prediction of binding free energy is the central theme in drug discovery. In the FMO-based approach, the binding energy can be estimated either by the supermolecular approach or by the summed energy of IFIEs (IFIE-sum) or total interaction energy (TIE). Likewise, the solvent effect can also be incorporated either by directly coupling an FMO polarizable medium with a continuum solvent or by adding molecular mechanically (MM) derived solvation free energy to an FMO energy obtained in vacuo. In short, there are four ways as the combination to estimate binding free energy depends on the computational cost or accuracy (*see* Table 1).

Table 1
Variation of binding free energy estimation based on FMO

	Solvent effect	
Binding energy	**MM energy additive**	**Directly coupled**
Interaction-based approach	Solute unpolarized Desolvation included Low cost Ref. 34	Solute polarized No desolvation Slightly expensive Ref. 36
Supermolecular approach	Solute unpolarized Desolvation included Moderate cost Ref. 35	Solute polarized Desolvation included Expensive Refs. 29, 33, 35

4.3.1 Protein-Mutative Prediction

Jensen et al. proposed a methodology to predict the cleavability of target peptides for the human immunodeficiency virus (HIV)-1 protease based on Chaudhuri and Gray's algorithm [32] applying FMO-derived binding energies [33]. They estimated the binding energies by using FMO-PCM approach. The energy-only methodology enables us to discriminate the cleavability of peptides for wild-type/mutant HIV-1 proteases and also to predict the capability of mutant HIV-1 protease to cleave a new target peptide as the substrate.

4.3.2 Ligand-Binding Prediction

Minor structural or chemical changes in ligands sometimes lead to dramatic changes in their activity, which we call the activity cliff. Because the cause of this cliff is critical for controlling the activity, its understanding is a big pharmaceutical issue. Watanabe et al. struggled with this theme on the serine/threonine kinase Pim1 and its inhibitors by using FMO [34]. The six benzofuranone class inhibitors have only difference in the nitrogen positions of indole ring. The binding free energies were estimated by combining IFIE-sum on ligand in vacuo and solvation free energy obtained from MM-PBSA. While comparing structural preparation between crystal geometry itself and MM- or QM-optimized one, this approach gives the good correlation with pIC50 ($R^2 = 0.85$ as the best) and suggestions on the cliff's mechanism with PIEDA.

Okimoto et al. considered different ways of deriving binding free energy to predict affinity between tankyrase 2 and its inhibitors [35]. They concluded the ability of FMO combined with MM-PBSA approach is the best to rank the binding affinity compared to those of other MM-based approaches. The correlation ($R = 0.856$) was derived for 23 compounds including two classes of common features with different flexibilities.

Morao et al. demonstrated the calculated results for several G protein-coupled receptors (GPCRs) [36]. The TIE values calculated by FMO-DFTB3 in PCM solvent showed good correlations with experimental values in pKi of ten compounds for human β2-adrenoceptor ($R^2 = 0.7833$), in pEC50 of ten compounds for κ-opioid receptor ($R^2 = 0.662$), and in pIC50 of seven compounds for human P2Y12 receptor ($R^2 = 0.8121$). The authors realized tremendous speedup for these calculations by using semi-empirical approach correlated well with FMO2-MP2/6-31G(d).

Okiyama et al. recently applied FMO-PBSA with fully polarizable media to predict ligand binding affinities of ERα [29]. The binding free energies of five bioactive compounds correlated well with their in vitro activities ($R = 0.990$). This study has a great impact in treating agonist/antagonist compounds with different charges.

4.3.3 Miscellaneous

On the structure. In the initial structure where one tries to perform FMO, its artificial strains would be often introduced when the coordinates are determined through experimental analyses or molecular simulations. Many of these causes are attributed to the accuracy of the MM force field. Such strains can be removed by QM relaxation to modify the structure into QM-preferred one. Actually, the ONIOM optimization for the pharmacophores of Pim1 [34] and tankyrase 2 [35] improved their correlation factors. Therefore, QM optimization including heavy atoms using QM/MM or ONIOM will lead to significant refinement of analysis results by mitigating strains, even if one employs an X-ray crystal structure with high resolution. In the ERα study, the positions of hydrogens making critical hydrogen bonds are optimized by taking the pharmacophore out once. If the strain energy remains in the structure, it appears as an increment of the deformation energy of a ligand.

On the dielectricity. When coupling an implicit solvent directly to FMO, there is no need to be aware of the inner (solute) dielectric constant because the solute is a polarizable medium. On the other hand, when combining the solvation free energy of MM-PBSA with FMO, one must arbitrarily set some value of the constant for the MM-PBSA calculation. One never knows the best value in advance; however, it would be recommended to employ unity from the results of several studies [34, 35].

On the nonpolar term and other contributions. While the electrostatic contribution is mainly the subject of debate in implicit solution, the nonpolar one is another key for accurate prediction of the binding free energy. The empirical SA-based approach in MM-PBSA is one of the effective and proven ways for estimating the nonpolar energy of biomolecular systems. The PCM nonpolar energy is, however, known to have a large disagreement with such an empirically derived one when applied to large biomolecules [35, 37]. It would be also future issue to incorporate the entropic and thermal corrections by using vibration analysis or estimation from the number of rotatable bonds.

5 Summary

This review article has comprehensively addressed the description of solvation (mainly hydration) effects in the FMO method. When we employ the explicit solvation model by water molecules, the electronic states and associated inter-/intra-molecular interactions are significantly affected by water shell with counterions, whose magnitudes are varied depending on the modeling of shell thickness. Whereas one may perform full ab initio MD simulations based on the FMO method in the future, we are currently limited to collecting finite numbers of snapshot FMO calculations for the interaction analysis. The SCIFIE analysis then works for incorporating the many-body screening effects into the FMO-IFIE evaluation. The implicit solvation model such as the FMO-PBSA, on the other hand, provides a good compromise between cost and accuracy to obtain the screened IFIEs and the binding free energy for the solvated system in thermodynamic equilibrium, which are compared well with corresponding experimental results.

Acknowledgments

This work was supported by JSPS KAKENHI Grant Numbers 17H06353 and 18K03825.

References

1. Tanaka S, Mochizuki Y, Komeiji Y, Okiyama Y, Fukuzawa K (2014) Electron-correlated fragment-molecular-orbital calculations for biomolecular and nano systems. Phys Chem Chem Phys 16:10310–10344. https://doi.org/10.1039/c4cp00316k
2. Watanabe C, Fukuzawa K, Tanaka S, Aida-Hyugaji S (2014) Charge clamps of lysines and hydrogen bonds play key roles in the mechanism to fix helix 12 in the agonist and antagonist positions of estrogen receptor α: intramolecular interactions studied by the ab initio fragment molecular orbital method. J Phys Chem B 118:4993–5008. https://doi.org/10.1021/jp411627y
3. Komeiji Y, Ishida T, Fedorov DG, Kitaura K (2007) Change in a protein's electronic structure induced by an explicit solvent: an ab initio fragment molecular orbital study of ubiquitin. J Comput Chem 28:1750–1762. https://doi.org/10.1002/jcc.20686
4. Fukuzawa K, Watanabe C, Kurisaki I, Taguchi N, Mochizuki Y, Nakano T, Tanaka S, Komeiji Y (2014) Accuracy of the fragment molecular orbital (FMO) calculations for DNA: total energy, molecular orbital, and inter-fragment interaction energy. Comput Theor Chem 1034:7–16. https://doi.org/10.1016/j.comptc.2014.02.002
5. Fukuzawa K, Kurisaki I, Watanabe C, Okiyama Y, Mochizuki Y, Tanaka S, Komeiji Y (2015) Explicit solvation modulates intra- and inter-molecular interactions within DNA: electronic aspects revealed by the ab initio fragment molecular orbital (FMO) method. Comput Theor Chem 1054:29–37. https://doi.org/10.1016/j.comptc.2014.11.020
6. Komeiji Y, Okiyama Y, Mochizuki Y, Fukuzawa K (2017) Explicit solvation of a single-stranded DNA, a binding protein, and their complex: a suitable protocol for fragment molecular orbital calculation. Chem-Bio Informatics J 17:72–84. https://doi.org/10.1273/cbij.17.72
7. Komeiji Y, Okiyama Y, Mochizuki Y, Fukuzawa K (2018) Interaction between a single-stranded DNA and a binding protein viewed by the fragment molecular orbital method. Bull Chem Soc Jpn 91:1596–1605. https://doi.org/10.1246/bcsj.20180150
8. Terauchi Y, Suzuki R, Takeda R, Kobayashi I, Kittaka A, Takimoto-Kamimura M, Kurita N (2019) Ligand chirality can affect histidine protonation of vitamin-D receptor: ab initio molecular orbital calculations in water. J Steroid Biochem Mol Biol 186:89–95. https://doi.org/10.1016/j.jsbmb.2018.09.020
9. Takeda R, Kobayashi I, Shimamura K, Ishimura H, Kadoya R, Kawai K, Kittaka A, Takimoto-Kamimura M, Kurita N (2017) Specific interactions between vitamin-D receptor and its ligands: Ab initio molecular orbital calculations in water. J Steroid Biochem Mol Biol 171:75–79. https://doi.org/10.1016/j.jsbmb.2017.02.018
10. Takeda R, Kobayashi I, Suzuki R, Kawai K, Kittaka A, Takimoto-Kamimura M, Kurita N (2018) Proposal of potent inhibitors for vitamin-D receptor based on ab initio fragment molecular orbital calculations. J Mol Graph Model 80:320–326. https://doi.org/10.1016/j.jmgm.2018.01.014
11. Fukuzawa K, Mochizuki Y, Tanaka S, Kitaura K, Nakano T (2006) Molecular interactions between estrogen receptor and its ligand studied by the ab initio fragment molecular orbital method. J Phys Chem B 110:24276–24276. https://doi.org/10.1021/jp065705n

12. Tanaka S, Watanabe C, Okiyama Y (2013) Statistical correction to effective interactions in the fragment molecular orbital method. Chem Phys Lett 556:272–277. https://doi.org/10.1016/j.cplett.2012.11.085
13. Hansen JP, McDonald IR (2006) Theory of simple liquids, 3rd edn. Academic, London
14. Jaynes ET (2003) Probability theory: the logic of science. Cambridge University Press, Cambridge
15. Okiyama Y, Tsukamoto T, Watanabe C, Fukuzawa K, Tanaka S, Mochizuki Y (2013) Modeling of peptide–silica interaction based on four-body corrected fragment molecular orbital (FMO4) calculations. Chem Phys Lett 566:25–31. https://doi.org/10.1016/j.cplett.2013.02.020
16. Kato K, Fukuzawa K, Mochizuki Y (2015) Modeling of hydroxyapatite–peptide interaction based on fragment molecular orbital method. Chem Phys Lett 629:58–64. https://doi.org/10.1016/j.cplett.2015.03.057
17. Ando H, Shigeta Y, Baba T, Watanabe C, Okiyama Y, Mochizuki Y, Nakano M (2015) Hydration effects on enzyme–substrate complex of nylon oligomer hydrolase: inter-fragment interaction energy study by the fragment molecular orbital method. Mol Phys 113:319–326. https://doi.org/10.1080/00268976.2014.941311
18. Ryde U, Söderhjelm P (2016) Ligand-binding affinity estimates supported by quantum-mechanical methods. Chem Rev 116:5520–5566. https://doi.org/10.1021/acs.chemrev.5b00630
19. Fedorov DG, Kitaura K, Li H, Jensen JH, Gordon MS (2006) The polarizable continuum model (PCM) interfaced with the fragment molecular orbital method (FMO). J Comput Chem 27:976–985. https://doi.org/10.1002/jcc.20406
20. Fedorov DG (2018) Analysis of solute–solvent interactions using the solvation model density combined with the fragment molecular orbital method. Chem Phys Lett 702:111–116. https://doi.org/10.1016/j.cplett.2018.05.002
21. Watanabe H, Okiyama Y, Nakano T, Tanaka S (2010) Incorporation of solvation effects into the fragment molecular orbital calculations with the Poisson–Boltzmann equation. Chem Phys Lett 500:116–119. https://doi.org/10.1016/j.cplett.2010.10.017
22. Okiyama Y, Nakano T, Watanabe C, Fukuzawa K, Mochizuki Y, Tanaka S (2018) Fragment molecular orbital calculations with implicit solvent based on the Poisson–Boltzmann equation: implementation and DNA study. J Phys Chem B 122:4457–4471. https://doi.org/10.1021/acs.jpcb.8b01172
23. Nishimoto Y, Fedorov DG (2016) The fragment molecular orbital method combined with density-functional tight-binding and the polarizable continuum model. Phys Chem Chem Phys 18:22047–22061. https://doi.org/10.1039/C6CP02186G
24. Simoncini D, Nakata H, Ogata K, Nakamura S, Zhang KYJ (2015) Quality assessment of predicted protein models using energies calculated by the fragment molecular orbital method. Mol Inform 34:97–104. https://doi.org/10.1002/minf.201400108
25. Grimme S (2006) Semiempirical GGA-type density functional constructed with a long-range dispersion correction. J Comput Chem 27:1787–1799. https://doi.org/10.1002/jcc.20495
26. Grimme S, Antony J, Ehrlich S, Krieg H (2010) A consistent and accurate ab initio parametrization of density functional dispersion correction (DFT-D) for the 94 elements H-Pu. J Chem Phys 132:154104. https://doi.org/10.1063/1.3382344
27. Fedorov DG, Kitaura K (2012) Energy decomposition analysis in solution based on the fragment molecular orbital method. J Phys Chem A 116:704–719. https://doi.org/10.1021/jp209579w
28. Mazanetz MP, Chudyk EI, Fedorov DG, Alexeev Y (2016) Applications of the fragment molecular orbital method to drug research. In: Zhang W (ed) Computer-aided drug discovery. methods in pharmacology and toxicology. Humana Press, New York, NY, pp 217–255
29. Okiyama Y, Watanabe C, Fukuzawa K, Mochizuki Y, Nakano T, Tanaka S (2019) Fragment molecular orbital calculations with implicit solvent based on the Poisson–Boltzmann equation: II. Protein and its ligand-binding system studies. J Phys Chem B 123:957–973. https://doi.org/10.1021/acs.jpcb.8b09326
30. Fedorov DG, Kitaura K (2016) Subsystem analysis for the fragment molecular orbital method and its application to protein–ligand binding in solution. J Phys Chem A 120:2218–2231. https://doi.org/10.1021/acs.jpca.6b00163
31. Śliwa P, Kurczab R, Kafel R, Drabczyk A, Jaśkowska J (2019) Recognition of repulsive and attractive regions of selected serotonin receptor binding site using FMO-EDA approach. J Mol Model 25:114. https://doi.org/10.1007/s00894-019-3995-6
32. Chaudhury S, Gray JJ (2009) Identification of structural mechanisms of HIV-1 protease

specificity using computational peptide docking: implications for drug resistance. Structure 17:1636–1648. https://doi.org/10.1016/j.str.2009.10.008

33. Jensen JH, Willemoës M, Winther JR, De Vico L (2014) In silico prediction of mutant HIV-1 proteases cleaving a target sequence. PLoS One 9:e95833. https://doi.org/10.1371/journal.pone.0095833

34. Watanabe C, Watanabe H, Fukuzawa K, Parker LJ, Okiyama Y, Yuki H, Yokoyama S, Nakano H, Tanaka S, Honma T (2017) Theoretical analysis of activity cliffs among benzofuranone-class Pim1 inhibitors using the fragment molecular orbital method with molecular mechanics Poisson–Boltzmann surface area (FMO+MM-PBSA) approach. J Chem Inf Model 57:2996–3010. https://doi.org/10.1021/acs.jcim.7b00110

35. Okimoto N, Otsuka T, Hirano Y, Taiji M (2018) Use of the multilayer fragment molecular orbital method to predict the rank order of protein–ligand binding affinities: a case study using tankyrase 2 inhibitors. ACS Omega 3:4475–4485. https://doi.org/10.1021/acsomega.8b00175

36. Morao I, Fedorov DG, Robinson R, Southey M, Townsend-Nicholson A, Bodkin MJ, Heifetz A (2017) Rapid and accurate assessment of GPCR–ligand interactions Using the fragment molecular orbital-based density-functional tight-binding method. J Comput Chem 38:1987–1990. https://doi.org/10.1002/jcc.24850

37. Söderhjelm P, Kongsted J, Ryde U (2010) Ligand affinities estimated by quantum chemical calculations. J Chem Theory Comput 6:1726–1737. https://doi.org/10.1021/ct9006986

Chapter 8

Computational Methods for Biochemical Simulations Implemented in GAMESS

Dmitri G. Fedorov, Hui Li, Vladimir Mironov, and Yuri Alexeev

Abstract

Computational methods for modeling biochemical processes implemented in GAMESS package are reviewed; in particular, quantum mechanics combined with molecular mechanics (QM/MM), semi-empirical, and fragmentation approaches. A detailed summary of capabilities is provided for the QM/MM implementation in QuanPol program and the fragment molecular orbital (FMO) method. Molecular modeling and visualization packages useful for biochemical simulations with GAMESS are described. GAMESS capabilities with corresponding references are tabulated for reader's convenience.

Key words GAMESS, Quantum mechanics (QM), Molecular mechanics (MM), Fragment molecular orbital (FMO), QuanPol, GUI

1 Introduction to Biochemical Simulations

Various methods for modeling biochemical processes have been developed over the years. Some of these methods are implemented in quantum chemistry package GAMESS [1, 2], described in this chapter together with visualization programs providing some level of GAMESS support. In this chapter, specifically GAMESS-US is described.

Quantum mechanical (QM) methods offer an accurate way to computing properties of chemical systems; in particular, they are appealing for modeling biochemical reactions. The major disadvantage of traditional QM methods such as Hartree-Fock (HF), density functional theory (DFT), Møller-Plesset perturbation theory (MP), and coupled cluster (CC) is that the computational time and memory requirements scale superlinearly with the system size, which makes them difficult to apply to biochemical simulations. There are various faster approaches such as force fields in molecular mechanics (MM) [3] or more sophisticated and modern methods parametrized to reproduce high-level QM results. In classical force fields, the potential energy of the system is approximated by the

Alexander Heifetz (ed.), *Quantum Mechanics in Drug Discovery*, Methods in Molecular Biology, vol. 2114, https://doi.org/10.1007/978-1-0716-0282-9_8,

sum of the individual terms describing bonded and nonbonded interactions between atoms. This is significantly less expensive computationally than solving the Schrödinger equation in QM. The accuracy of MM strongly depends on the force field used and the transferability of parameters.

As a compromise, QM and MM can be combined in QM/MM methods [4, 5]. The idea is to treat a small active site with an accurate but expensive QM method, while the rest of the system is treated with a cheaper MM force field. However, QM/MM may require large active sites for some reactions as in proton or electron hopping. Also, the QM/MM boundary may have a negative effect on the accuracy. QM/MM methods implemented in GAMESS are described below, including QuanPol built into GAMESS.

QM calculations can be accelerated in fragmentation methods [6–14] and other linear scaling approaches [15, 16]. In the fragment molecular orbital (FMO) method [17–21] implemented [22, 23] in GAMESS, a molecular system is divided into fragments. QM calculations are performed for fragments self-consistently in the Coulomb field from the whole system, and non-electrostatic interactions are added by performing explicit fragment pair calculations. FMO in GAMESS is efficiently parallelized [24–26], and it has been applied to many biochemical problems [27–33].

Finally, a list of graphical programs with a GAMESS support is provided. These programs can be used to do molecular modeling, as well as to create GAMESS input files and visualize outputs.

2 Parametrized QM Methods for Biochemical Simulations

A number of QM methods in GAMESS (Table 1) are parametrized to reproduce either experimental data or high-level QM calculations. The parametrization is usually less restrictive compared to force fields; there are no explicit atomic types, and QM treatment allows for breaking and making of covalent bonds. Polarization and charge transfer are accounted for.

Another route to accelerating QM calculations is offered by fragmentation. The elongation method (ELG) [48] implemented in GAMESS does not use fragments per se; instead, the system is gradually elongated, but variationally optimized groups of orbitals are analogous to fragments. The method is in particular suited to studying DNA [49]. Another approach [9], based on the divide-and-conquer idea, has been applied to a protein-ligand complex [50]. EFMO [51] has been used for mapping enzymatic reaction paths [52]. All of these methods are labeled as fragmentation approaches, summarized in Table 2. They can be combined with many QM methods (appropriate QM calculations are performed on subsystems). FMO is covered below, and more details on other methods can be found elsewhere [6].

Table 1
Parametrized methods in GAMESS suitable for biochemical simulations

Method	Description	Status	References
RM1	A semi-empirical method	Available	[34, 35]
PM6	A semi-empirical method	Not available[a]	[36]
DFTB	Density-functional tight-binding	Available	[37–39]
HF-3c	Hartree-Fock with three corrections	Available	[40–42]
EFP1 (rigid)	First-generation effective fragment potential	Available	[43, 44]
EFP1 (flexible)	First-generation effective fragment potential	Not available[a]	[45]
EFP2	Second-generation effective fragment potential	Available[b]	[46]

[a]Developed in private versions
[b]Recently extended to biosystems in another program [47]

Table 2
Fragmentation methods in GAMESS suitable for biochemical simulations

Method	Description	Status	References
FMO	Fragment molecular orbital	Available	[22, 23]
EFMO	Effective fragment molecular orbital	Available	[51, 53]
ELG	Elongation	Available	[48]
DC	Divide-and-conquer	Available	[9]
MTA	Molecular tailoring approach	Not available[a]	[54]
SMF	Systematic molecular fragmentation	Not available[a]	[55]

[a]The method needs additional software/scripts

3 QM/MM Methods Interfaced with GAMESS

In the hybrid quantum and molecular mechanics (QM/MM) approach, a small important part of a molecular system (an active site) up to a few hundreds of atoms is modeled by an accurate but expensive QM method. The rest of the system is described by a fast MM force field, reaching a compromise between accuracy and computational efficiency. If the MM treatment of the bulk of the system is appropriate, the accuracy of QM/MM is comparable to a QM calculation for the whole system. However, setting up a QM/MM calculation is not trivial, and expertise of the underlying chemical processes and good chemical intuition are needed to divide the system properly in QM and MM regions.

They are two groups of QM/MM methods, the additive and the subtractive schemes (Fig. 1) [56]. In the additive scheme, the

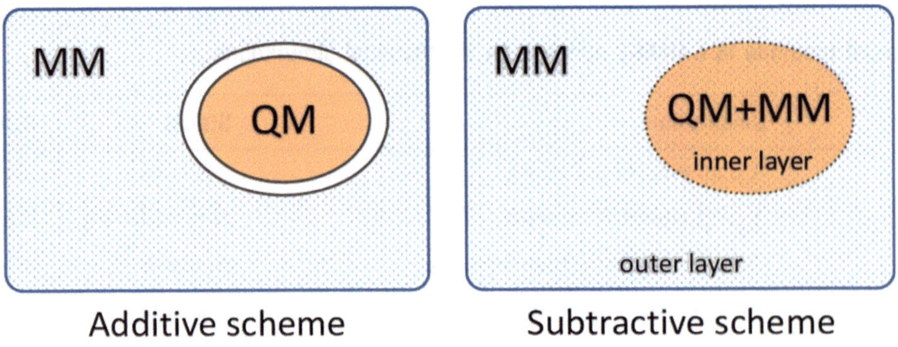

Fig. 1 Two QM/MM approaches: the additive and subtractive schemes

QM and MM parts are disjoint, and the total energy of the system (E_{tot}) is

$$E_{tot} = E_{QM} + E_{MM} + E_{QM/MM}, \tag{1}$$

where E_{QM} and E_{MM} are the energies of QM and MM regions, respectively, and $E_{QM/MM}$ is the interaction between them. The additive scheme describes the electronic embedding.

In the subtractive scheme, the system is partitioned into several nested layers. The outermost layer is the entire system modeled using MM. The innermost layer contains the active site described by both QM and MM methods separately. The total energy of the system is computed as

$$E_{tot}^{outer} = E_{MM}^{outer} + \left(E_{QM}^{inner} - E_{MM}^{inner}\right), \tag{2}$$

where *inner* and *outer* denote layers. This scheme can be readily extended to more than two layers, for example, by defining high- and low-level QM and MM regions (the total of three layers). Typically 2–3 layers are used. Examples of use of the subtractive QM/MM scheme applications are ONIOM [57] approach and its variations IMOMM [58] and SIMOMM [59]. The subtractive scheme describes the mechanical embedding.

GAMESS is a general-purpose QM package, and MM methods for QM/MM are mainly implemented using external MM software packages via interfaces. Some of the interfaces are integrated in the GAMESS and available out of the box. Other interfaces require substantial changes in the GAMESS source code often left to qualified users to do. QM/MM interfaces for GAMESS are listed in Table 3.

CHARMM package [60] was one of the first MM packages interfaced with GAMESS [61] for the additive QM/MM scheme. The drawback of this interface is that its installation is not automated and requires a manual modification of GAMESS source code.

Table 3
QM/MM methods using GAMESS for QM with external MM software packages

Program	Description	Status	References
CHARMM	Additive QM/MM scheme. This interface is an extension of the QUANTUM code	A patch is needed[a], see CHARMM website	[60, 61]
Gromacs	Script-based, data exchange through files. Mechanical embedding only	A patch is needed[a]	
Tinker	SIMOMM, MM = traditional force fields	Available[b]	[59]
Tinker/EFP	Electronic embedding with MM = EFP	A patch is needed[a]	[45]
QuanPol	MM = CHARMM, AMBER, and MMFF94. QM = HF, MCSCF, GVB, DFT, DFTB, MP2, TDDFT, and TDDFTB	Available[c]	[62]

[a]It has to be obtained separately from distributed GAMESS
[b]A specific version of Tinker adopted to GAMESS has to be separately obtained from GAMESS distribution (note that recent versions of Tinker cannot be used)
[c]Available in distributed GAMESS without further patches

An alternative to CHARMM is Tinker [63]. An old customized version of Tinker is distributed together with GAMESS. The distributed GAMESS/Tinker interface supports only the subtractive IMOMM scheme of QM/MM, whereas an additive QM/EFP scheme for GAMESS/Tinker [45] with an electrostatic embedding was implemented in a private version. Tinker can be used together with FMO (*see* Subheading 4) to reduce the computation cost of QM calculations.

GAMESS also has a native QM/MM code QuanPol (quantum chemistry polarizable force field program) [62]. It supports a large number of QM and MM methods, and its detailed description is presented in the next section. So far, QuanPol has not been interfaced with any fragment-based method in GAMESS.

3.1 Summary of QuanPol

QuanPol [62] can perform MM, QM/MM, or QM/MM/continuum solvent MD simulations, geometry optimization, saddle point search, and harmonic frequency calculation (Hessian), using HF, DFT, DFTB, GVB, MCSCF, MP2, TDDFT, or TDDFTB wave functions. CHARMM, AMBER, and MMFF94 force fields are implemented in QuanPol (Table 4).

In QuanPol, the total energy of a molecular system described with QM/MMpol/continuum solvent for HF, DFT, DFTB, GVB, and MCSCF methods can be schematically written as

$$E_{\mathrm{QM/MMpol/C}} = E_{\mathrm{QM}} + E_{\mathrm{MM}} + E_{\mathrm{chg}} + E_{\mathrm{rep}} + E_{\mathrm{disp}} + E_{\mathrm{pol}} + E_{\mathrm{sol}} \quad (3)$$

Here E_{QM} is the total QM energy (including nuclear repulsion) of the QM region calculated using the regular gas phase

Table 4
Summary of QuanPol capabilities

Method	Details
QM	HF, MCSCF, DFT, DFTB, TDDFT, TDDFTB, MP2
MM	CHARMM, AMBER, MMFF94, water models, user-defined force field; can use induced dipole polarizability for all listed QM method
QM-MM link	Automatic capping H atom method
Solvation model	Continuum model (COSMO/C-PCM) for QM and QM/MM
QM basis set	All basis sets with *s*, *p*, *d*, and *f* functions
Geometry optimization	Steepest descent and Hessian guided methods; transition state search; all MM and QM/MM methods have analytic gradients
IR spectrum	Hessian method for QM/MM; MM MD simulated IR spectrum
Free energy simulation	Free energy perturbation; umbrella sampling; thermodynamic integration
Boundary condition	Open boundary; periodic boundary; spheric boundary with solvation
MD	Radial distribution function; diffusion constants

Hamiltonian for these methods with the wave function polarized by MM atoms and continuum solvent. E_{MM} contains the covalent-potential terms and pairwise intermolecular interaction (nonpolarizable charge-charge and Lennard-Jones or their variants) potential terms. E_{chg} is the electrostatic interaction energy between QM electrons/nuclei and MM atomic point charges. E_{rep} and E_{disp} are the repulsion and dispersion interaction energies, respectively, between QM atoms and MM atoms. E_{pol} is the induced dipole polarization energy of the MM atoms in QM/MMpol or QM/MMpol/continuum. E_{sol} is the solvation free energy in QM/MMpol/continuum, obtained from induced surface charges.

In QuanPol the induced dipoles and induced surface charges are included into HF, DFT, DFTB, GVB, and MCSCF methods using a variational SCF scheme, in accordance with the SCF nature of these QM methods, and are solved iteratively and self-consistently together with the electron density in SCF calculations. After the SCF convergence, the wave function, induced dipoles, and induced surface charges are determined, and the induced dipole polarization energy and the induced surface charge solvation free energy are determined as well. The variational treatment is advantageous because analytic gradient can be readily evaluated.

Compared to various implementations of QM/MM and QM/MM/continuum methods in the literature, the QM/MMpol/continuum method implemented in the QuanPol program is probably the most general one. For example, when induced dipole polarizable force fields are used in QM/MMpol methods, the QM methods are usually HF and DFT or their semi-empirical variants. In QuanPol, the latest QM/MMpol style TDDFT (and TDDFTB) and MP2 methods are implemented with the analytic gradient, thus expanding the borderline of QM/MM methods. The QuanPol QM/MMpol/continuum method rigorously combines induced dipole polarizable force fields and induced surface charge solvation models into various QM calculations and has analytic gradients.

3.2 Covalent Boundary Treatment in QuanPol

In QuanPol a capping H atom method can be used to treat covalent bond cuts between QM and MM regions. For example, when a protein C_α-C_β bond is cut, an H atom is put in the place of the C_α to saturate the open bond of the side chain (QM region). The C_α atom retains its MM covalent parameters in the QM/MM calculation, but should not retain its force field charge, polarizability, Lennard-Jones potential, and/or effective Gaussian potentials. The force on the C_α atom is the sum of the QM force on the H atom and the MM force on the C_α atom. In all calculations, the H and C_α atoms are enforced to have the same coordinate, mass of carbon, force, and velocity. No additional degrees of freedom are introduced. QuanPol automatically uses an effective Gaussian repulsion potential for the capping H atom to produce a H-C bond length of ~1.5 Å. The capping H atom method can be used interchangeably for all QM methods.

3.3 Geometry Optimization and Harmonic Vibrational Frequencies

QuanPol can be used to perform pure MM and QM/MM geometry optimization (including transition state search) and Hessian vibration analysis to obtain frequencies and thermochemical properties. For pure MM system with thousands of atoms, optimization and frequency calculations can be done for the entire system. When the system is large, partial geometry optimization and Hessian calculation can be performed for selected atoms. For QM/MM systems, geometry optimization can be performed for 1000–2000 QM and MM atoms around the QM center, and subsequent Hessian calculations can be performed for only 20–30 QM atoms to obtain the local vibrational frequencies and thermochemical properties.

3.4 Molecular Dynamics Simulation

QuanPol contains subroutines to evaluate some key properties from MD simulation. Volume is meaningful only when periodic boundary condition (PBC) or spherical boundary condition (SBC) is used. Other implemented properties include radial distribution function, radial density profile (for spherical boundary condition only), root mean square displacement of atomic coordinates from the initial coordinates, radius of gyration (mass-weighted root mean square distances of the atoms from the center of mass), dielectric constant, and self-diffusion constant. The dipole moment of the QM atoms, the dipole moment of user-selected MM atoms, and the dipole moment of the entire QM/MM system are printed out at every MD step. The QuanPol Vibrational Spectrum Program (QPVIB) can be used to generate the time autocorrelation functions of these dipole moments and the corresponding infrared spectra.

3.5 Free Energy Perturbation Simulation

Both MM and QM/MM molecular dynamics-based free energy perturbation (FEP) simulation methods are implemented in QuanPol. The Gibbs free energy change (ΔG) from state A to state B (constant pressure and constant temperature) can be obtained from the internal potential energy change (ΔE) as calculated using pure MM methods or QM/MM methods. In general, the difference between state A and state B can be molecular geometry, electronic state, or even different types and numbers of atoms. FEP methods can be formulated to calculate a free energy profile along reaction coordinates (i.e., the potential of mean force, PMF), solvation free energy and binding free energy, relative reduction potential, and acid-base ionization constant.

Both single topology and dual topology MD simulation methods for FEP calculations are implemented. When single topology method is used, only one set of atoms are used to run the MD simulation. Some atoms are selected to be "perturbation atoms" and are given a second set of force field parameters (e.g., mass, charge, Lennard-Jones potential, bond stretching force constant, bond angle bending force constant, and others) or the same set of force field parameters but a second set of Cartesian coordinates. At every MD step, a single point energy calculation is performed using the second set of parameters or coordinates for the "perturbation atoms" to obtain the ΔE. Very often FEP simulations are performed for a hypothetic "alchemical" state lying between states A and B, with a mixing coefficient.

Several types of free energy changes can be calculated using QuanPol. For a pure MM system, one can set ΔE to be the difference in the charge-charge and Lennard-Jones interaction energy between the "perturbation atoms" and all other atoms in the system. The resulted free energy change thus is solvation free energy change. Both single and dual topology methods can be used to calculate solvation free energy change. In addition to the solvation potential energy of the "perturbation atoms," the covalent and noncovalent potential energy within the "perturbation atoms" can also be included into ΔE. In this case, one can calculate the free energy change associated with a geometric change of the "perturbation atoms." The resulted free energy change thus is the PMF.

3.5.1 Umbrella Sampling

Umbrella sampling is available in QuanPol, and there is a utility called "QuanPol Weighted Histogram Analysis Program" (QPWHA) to use the Weighted Histogram Analysis Method (WHAM) to analyze both 1D and 2D histograms and generate PMF curves.

3.5.2 Thermodynamic Integration (TI)

QuanPol can also use the thermodynamic integration (TI) scheme in MD simulation to evaluate the mean force (i.e., Boltzmann averaged) between two atoms that have a series of constrained distances over a user-defined reaction range (e.g., from 1.5 to 5.0 Å). The mean force, integrated over the displacement, gives the free energy change (i.e., potential of mean force) over the displacement. It works for both MM and QM/MM MD simulations. The two atoms can be MM or QM, in any combination.

3.6 Setting Up MM and QM/MM Calculations

In order to perform QM/MM calculations, prior pure MM calculations are always required. To set up pure MM calculations using QuanPol, a set of molecular Cartesian coordinates is required. Small molecules can be drawn manually by using visualization software or from a PDB file. QuanPol can use these Cartesian coordinates or PDB files to generate user-requested force field. For example, the MMFF94 force field can be generated for general organic molecules based on a set of Cartesian coordinates. CHARMM and AMBER force fields can only be generated for proteins and nucleic acids in PDB format. Once generated, the force field (i.e., the topology file) is in QuanPol format (the $FFDATA group) and can be saved and reused by QuanPol.

Geometry optimization and MD simulation can be run using the generated MM force field. For proteins and nucleic acids in aqueous solvent, the whole system can have more than 10,000 atoms. The system should be equilibrated with MD simulation. After equilibration, further MM MD simulations (including free energy simulation) or geometry optimization can be performed.

QM/MM calculation can be performed with QuanPol by defining a group of MM atoms be QM atoms (in the $DATA group). QuanPol will automatically run QM/MM calculations when both $FFDATA and $DATA groups are present. The QM calculation is controlled by GAMESS keywords. QuanPol can automatically set up the default capping H atom parameters for QM-MM covalent link atoms (one or multiple). In the simplest case, no user interference is necessary, but a user can always use the $QUANPO group keywords to tune the QM/MM settings. For a given QM/MM input file, different QM methods and basis sets can be used interchangeably (no adjustment of the MM part or QM/MM setting is required), and the results are comparable. For example, one can compare the QM/MM results obtained from DFTB to DFT or MP2 or TDDFTB to DFTB or MCSCF.

4 The Fragment Molecular Orbital Method

In FMO, a molecular system is divided into fragments. This fragmentation step can be done automatically in many graphical user interface (GUI) programs (see below) for proteins, polysaccharides (cellulose, heparin, etc.), and polynucleotides (RNA, DNA, etc.) or their combination. Also, water or other stand-alone molecules such as ligands can be assigned as fragments automatically. Manual fragmentation is necessary if there is a coordinated metal cation (the metal and the groups it is coordinated to are merged into one fragment), a large ligand (it is not mandatory to subdivide it, but sometimes useful), or large inorganic species. The user aided by GUI prepares an input file that contains Cartesian coordinates of all atoms, an assignment of each atom to a fragment, and a list of pairs of atoms between which a covalent bond has to be detached in order to define fragments (for proteins, this is a pair of carbon atoms in Cα–CO bonds). Typically, the user starts from a PDB file containing all atoms, and then the GUI makes an input file from it. However, the PDB file in most cases cannot be simply taken from a PDB database, because of missing atoms (hydrogens and sometimes others) and steric repulsions. Typically, one has to refine the structure using a solvated MM optimization or MD. The structure can also be further refined using FMO (see below). There are, however, some examples of high-quality entries in PDB database that can be used straight away (1UAO, 1L2Y, 1CRN, etc.). Perhaps, as the quality of submitted structures improves, QM calculations will be easier to begin.

FMO calculations are typically executed on a UNIX cluster consisting of a few or many [25] nodes. First, individual fragments (called monomers) are calculated in the electrostatic embedding due to the rest of fragments. Then, fragment pairs (called dimers) are calculated, also in the embedding. In the first monomer step, the many-body polarization is accounted for. In the second step, charge transfer, exchange repulsion, dispersion, and other effects are incorporated. Typically, the solvent is treated with a continuum model, polarizable continuum model (PCM) [64], or solvent model density (SMD) [65]. The solvent adds a solvent embedding to all fragment calculations. A cavity is constructed around the solute, and polarizable charges are placed on the surface representing the solvent. Non-electrostatic terms, such as the solute-solvent dispersion or the loss of entropy by the solvent due to the presence of the solute (as related to creating a cavity), are computed using parametrized models.

The total properties are computed using the many-body expansion [66]. If only monomers are computed, the method is called FMO1. In FMO2 dimers are also calculated, and in FMO3, in addition to that, trimers are computed. FMO1 calculations, neglecting interfragment QM effects such as charge transfer, are not commonly used; two examples of FMO1 applications are electronic excitations [67] and electronic density [68]. In biochemical systems, the FMO2 level is by far dominant, especially if one is interested in such properties as pair interactions. FMO3 is used sometimes to obtain a more accurate value of the total properties, for instance, to evaluate the relative stabilities of different isomers with a small energy difference [69]. For FMO2, the energy E for N fragments is computed as

$$E^{\mathrm{FMO2}} = \sum_{I}^{N} E_I + \sum_{I>J}^{N} \Delta E_{IJ} \tag{4}$$

where E_I is the internal energy of fragment I polarized by the embedding potential and ΔE_{IJ} is the energy of the interaction between fragments I and J. Many other properties can be obtained with a similar expansion, for example, dipole moments or atomic charges, reflecting the effect of charge transfer. The gradient is obtained by differentiating Eq. 4, and it is complicated by the dependence of the polarizable embedding on the electronic state of each fragment (the embedding is computed using the electron densities of fragments). The calculated gradient can be used for geometry optimization (energy minimization) [70] or MD simulations [71].

Many, but not all of the QM methods implemented in GAMESS can be used with FMO. There are also various add-on extensions such as solvent models, some of which can be used with FMO. A summary of methods is given in Table 5. This Table is meant as a glossary and is focused on extensions (such as solvent models), whereas a list of QM methods interfaced with FMO can be found elsewhere [21].

In biochemistry, FMO has been used to analyze contributions of residues to protein-ligand binding [27, 32], assess energetic stability of native vs. decoy protein structures [28], analyze topological factors in protein-protein interactions [33], map [52] and analyze [29] enzymatic reactions, optimize protein structures [97], and study protein structure fluctuations in 200–1000 ps FMO/MD [87, 98]. IR and Raman spectra of proteins can be simulated with FMO [92, 99].

5 Molecular Modeling and Visualization Software

They are many molecular visualization programs which can process GAMESS input and output file formats. A short outline of the most relevant visualization packages is presented in Table 6. The list is

Table 5
Summary of FMO development in GAMESS (an FMO glossary)

Method	Details	Status	References
FMO1	One-body expansion of FMO	Available	[67, 68]
FMO2	Two-body expansion of FMO	Available	[22]
FMO3	Three-body expansion of FMO	Available	[22]
MFMO	Multilayer FMO for reducing computational time (sometimes abbreviated MLFMO)	Available	[72]
HOP	Hybrid orbital projection, used for interfragment boundary	Available	[73]
BDA and BAA	A covalent bound on a fragment boundary is divided between a bond detached atom (BDA) and a bond attached atom (BAA)	Available	[73]
AFO	Adaptive frozen orbital, used for interfragment boundary	Available	[69]
PCM[*n*]	Polarizable continuum model (a continuum solvent model). *n* defines how solvent charges are induced	Available	[64]
SMD[*n*]	Solvent model density (a continuum solvent model)	Available	[65]
FMO/EFP	Effective fragment potential (an explicit solvent model). Only EFP1 can be used	Available	[74]
FMO/RISM	Reference interaction site model (a statistical solvent model)	Not available	[75]
FMO/MD	Molecular dynamics	Available	[71]
FMO/IRC	Intrinsic reaction coordinate (used to map a reaction path)	Available	[76]
FMO/FD(D)	Frozen domain (FD), optionally with the dimer approximation (FDD), to optimize active sites	Available	[76–78]
FMO/MM	A mechanical MM embedding (SIMOMM)	Available	[66, 70]
SCZV	Self-consistent Z-vector for accurate analytic gradients	Available	[79]
FMO/AP	Auxiliary polarization for using large basis sets	Available	[80]
FMO/ONIOM	Our own *N*-layered integrated molecular orbital and molecular mechanics for evaluating interactions	Not available[a]	[81]
PIE	Pair interaction energy ΔE_{IJ}, sometimes called interfragment interaction energy (IFIE)	Available	[22]
PIEDA	PIE decomposition analysis into components to identify fragments important for binding	Available	[82–84]
PL, PL0, PLH	Levels of polarization of fragments: full (PL), by isolated state density (PL0), and via Huckel (PLH)	Available	[68, 82]
ABA, SBA(C)	Asymmetric binding analysis (ABA), symmetric binding analysis (SBA) with separated cavitation (SBAC)	Available	[85]
FMO/SA	Subsystem analysis for analyzing binding energy	Available	[86]

(continued)

Table 5 (continued)

Method	Details	Status	References
FMO/FA	Fluctuation analysis to analyze interactions in MD	Available	[87]
IEA	Interaction energy analysis for analyzing solvent effects using EFP	Available	[88]
MEP	Molecular electrostatic potential	Available	[89, 90]
Hessian	Analytic second derivatives to compute harmonic vibrational frequencies	Available	[91]
IR spectra	Infrared spectra	Available	[91]
Raman spectra	Normal Raman spectra	Available	[92]
IMS	Ion mobility spectroscopy mobilities	Not available[a]	[68]
NMR	Nuclear magnetic resonance shifts	Not available[a]	[93]
FMO/F(X)	Total Fock matrix optionally with the exchange (X) added to the embedding (to obtain molecular orbitals and their energies)	Available[b]	[94]
FMO/LCMO (X)	Linear combination of molecular orbitals optionally with the exchange (X) added to the embedding (to obtain molecular orbitals and their energies)	Available	[95, 96]
FMO/DOS	Density of states for analyzing electronic structure of materials	Available[b]	[95]

[a]Developed in private versions or requires additional software
[b]It can be done using a stand-alone program distributed with GAMESS

not comprehensive and includes a selection of programs recently used in various applications.

Visualization packages for GAMESS can be divided into three groups. The packages in the first group, ChemCraft, Gabedit, MacMolPlt, and Molden, are primarily designed for quantum chemists, who study small- or medium-sized chemical systems using sophisticated quantum chemistry methods. Using these packages, one can visualize molecular orbitals, electron density, molecular electrostatic potential, normal vibrational modes, IR spectra, etc. Software packages in this group are usually not optimized to display large systems. All of them, except for Molden, can be used to prepare GAMESS input files. MacMolPlt has also a useful FMO fragmentation support, though at a basic level.

The second group of visualization software includes PyMol, VMD, Avogadro, and WinMostar, designed for macromolecule visualization from the beginning. PyMol and VMD can be viewed as domain-specific frameworks for high-level programming languages (Python and TCL, respectively), and the support of third-

Table 6
Graphical software for molecular modeling with GAMESS support

Program	Licensing	Language	Platform	FMO	QM/MM	References
Avogadro	GPL v2	C++	Cross-platform	No	Yes[a, b]	[100]
ChemCraft	Proprietary, commercial	Unknown	Windows, Linux	No	Yes[a, b]	[101]
Facio	Proprietary, freeware	Unknown	Windows[c], MacOS[d], UNIX[d]	Yes	No	[102, 103]
FU	BSD	Python, Fortran	Cross-platform	Yes	No	[85, 104]
Gabedit	MIT	C	Cross-platform	No	No	[105]
MacMolPlt	GPL v2	C++	Cross-platform	Yes	Yes[b, e]	[106]
Molden	Proprietary, open source, free for noncommercial use	Fortran	Cross-platform	No	No	[107]
PyMol	BSD-like, open source	Python, C++	Cross-platform	Via user scripts	Yes[b]	[108]
VMD	Open source with a free license	Tcl, C	Cross-platform	No	Yes[b]	[109]
WinMostar	Proprietary, commercial, free for students	Unknown	Windows	Via interface to FU	Yes[b, e]	[110]

[a]QM/EFP only
[bb]Visualization of XYZ files
[c]Native application
[d]Non-native application (requires emulators)
[e]Visualization of input files

party software file formats is provided in the form of scripts or plug-ins. This approach has an advantage that a new functionality can be added by the user. These programs have support for GAMESS output files and can be used to visualize some results of quantum chemistry calculations, like electronic orbitals and densities. However, using PyMol for GAMESS requires user-made plug-ins.

Avogadro and WinMostar softwares have features of both groups. They provide more functionality for quantum chemists than the second group, but their functionality cannot be easily extended. Both Avogadro and WinMostar have a visual generator of GAMESS input files. Compared to the first software group, they are better tuned for large chemical systems and include powerful tools for building macromolecules and structure analysis.

The third class of visualization software includes Facio and FU and has the feature of an advanced functionality for fragmentation, specifically for FMO. Facio has a well-developed FMO input file maker with a support for most FMO extensions; FU has a well-designed support for biochemical systems. Both programs can also visualize FMO results.

6 Summary

In this chapter, the computational methods for modeling biochemical processes implemented in GAMESS have been reviewed. A detailed summary of capabilities is provided for FMO and the QM/MM implementation in QuanPol. The relevant GAMESS features with corresponding references are tabulated for convenience.

Although GAMESS has been originally designed for simulations of atoms and small molecules, gradually, various methods have been implemented capable of handling large systems. GAMESS has a hardwired limit of 2000 QM atoms (that can be increased by a recompilation), but this limit does not apply to most models in fragmentation methods such as EFMO and FMO and QM/MM. Most types of FMO calculations can have an arbitrary number of atoms and fragments. Practically, of course, there is a limit of computational cost and memory. The largest FMO run reported so far had about 1.2 million atoms (FMO/MD using DFTB) [111]. However, obtaining Gibbs energy with FMO is a problem, whereas QuanPol can be used for estimating it.

A number of graphical interfaces for biochemical simulations in GAMESS are available, but there is still a lot of room for improvement. The important step for running biochemical simulations is to prepare a well-optimized structure, and QM methods are very sensitive to unnatural distortions of bonds and angles. In MM, one defines a list of bonds on springs that can be quickly relaxed, but a poor structure for QM methods implies broken or distorted bonds, for which simple single-reference methods such as RHF and DFT often fail to converge. The burden on preparing a good initial structure for QM methods lies with the user. Modeling software can prepare a reasonable initial structure and optimize it with semi-empirical methods or force fields. There is an ongoing work to enable force field optimization in GUI for GAMESS. Another common problem is a delay for adding GUI support for recently developed methods.

GAMESS is a suitable program for biochemical simulations. New methods are being implemented, the parallel efficiency is improving, and GAMESS support in visualization programs is expanding.

Acknowledgments

This work was supported by the Office of Science, US Department of Energy, under Contract DE-AC02-06CH11357. V.M. thanks Intel Parallel Computing Centers program for funding.

References

1. Schmidt MW, Baldridge KK, Boatz JA, Elbert ST, Gordon MS, Jensen JH, Koseki S, Matsunaga N, Nguyen KA, Su S, Windus TL, Dupuis M, Montgomery JA (1993) General atomic and molecular electronic structure system. J Comput Chem 14:1347–1363
2. Gordon MS, Schmidt MW (2005) Advances in electronic structure theory: GAMESS a decade later. In: Dykstra CE, Frenking G, Kim KS, Scuseria GE (eds) Theory and applications of computational chemistry, the first forty years. Elsevier, Amsterdam, pp 1167–1189
3. Lopes PE, Guvench O, MacKerell AD (2015) Current status of protein force fields for molecular dynamics simulations. In: Kukol A (ed) Molecular modeling of proteins. Humana, New York, NY, pp 47–71
4. Senn HM, Thiel W (2009) QM/MM methods for biomolecular systems. Angew Chem Int Ed 48:1198–1229
5. Lin H, Truhlar DG (2007) QM/MM: what have we learned, where are we, and where do we go from here? Theor Chem Accounts 117:185–199
6. Gordon MS, Pruitt SR, Fedorov DG, Slipchenko LV (2012) Fragmentation methods: a route to accurate calculations on large systems. Chem Rev 112:632–672
7. Otto P, Ladik J (1975) Investigation of the interaction between molecules at medium distances: I. SCF LCAO MO supermolecule, perturbational and mutually consistent calculations for two interacting HF and CH2O molecules. Chem Phys 8:192–200
8. Gao J (1997) Toward a molecular orbital derived empirical potential for liquid simulations. J Phys Chem B 101:657–663
9. Kobayashi M, Nakai H (2012) How does it become possible to treat delocalized and/or open-shell systems in fragmentation-based linear-scaling electronic structure calculations? The case of the divide-and-conquer method. Phys Chem Chem Phys 14:7629–7639
10. Collins MA, Bettens RPA (2015) Energy-based molecular fragmentation methods. Chem Rev 115:5607–5642
11. Raghavachari K, Saha A (2015) Accurate composite and fragment-based quantum chemical models for large molecules. Chem Rev 115:5643–5677
12. Liu J, Herbert JM (2016) Pair–pair approximation to the generalized many-body expansion: an alternative to the four-body expansion for ab initio prediction of protein energetics via molecular fragmentation. J Chem Theory Comput 12:572–584
13. Sahu N, Gadre SR (2016) Vibrational infrared and Raman spectra of polypeptides: fragments-in-fragments within molecular tailoring approach. J Chem Phys 144:114113
14. Fang T, Li Y, Li S (2017) Generalized energy-based fragmentation approach for modeling condensed phase systems. WIREs Comp Mol Sc 7:e1297
15. Kussmann J, Beer M, Ochsenfeld C (2013) Linear-scaling self-consistent field methods for large molecules. WIREs Comput Mol Sci 3:614–636
16. Akimov AV, Prezhdo OV (2015) Large-scale computations in chemistry: a bird's eye view of a vibrant field. Chem Rev 115:5797–5890
17. Kitaura K, Ikeo E, Asada T, Nakano T, Uebayasi M (1999) Fragment molecular orbital method: an approximate computational method for large molecules. Chem Phys Lett 313:701–706
18. Fedorov DG, Kitaura K (2007) Extending the power of quantum chemistry to large systems with the fragment molecular orbital method. J Phys Chem A 111:6904–6914
19. Fedorov DG, Nagata T, Kitaura K (2012) Exploring chemistry with the fragment molecular orbital method. Phys Chem Chem Phys 14:7562–7577
20. Tanaka S, Mochizuki Y, Komeiji Y, Okiyama Y, Fukuzawa K (2014) Electron-correlated fragment-molecular-orbital calculations for biomolecular and nano systems. Phys Chem Chem Phys 16:10310–10344
21. Fedorov DG (2017) The fragment molecular orbital method: theoretical development, implementation in GAMESS, and applications. WIREs Comput Mol Sci 7:e1322

22. Fedorov DG, Kitaura K (2004) The importance of three-body terms in the fragment molecular orbital method. J Chem Phys 120:6832–6840
23. Alexeev Y, Mazanetz MP, Ichihara O, Fedorov DG (2012) GAMESS as a free quantum-mechanical platform for drug research. Curr Top Med Chem 12:2013–2033
24. Fedorov DG, Olson RM, Kitaura K, Gordon MS, Koseki S (2004) A new hierarchical parallelization scheme: generalized distributed data interface (GDDI), and an application to the fragment molecular orbital method (FMO). J Comput Chem 25:872–880
25. Mironov V, Alexeev Y, Fedorov DG (2019) Multi-threaded parallelization of the energy and analytic gradient in the fragment molecular orbital method. Int J Quant Chem 119: e25937
26. Alexeev Y, Mahajan A, Leyffer S, Fletcher G, Fedorov DG (2012) Heuristic static load-balancing algorithm applied to the fragment molecular orbital method. In: Proceedings of the international conference on high performance computing. Networking, Storage and Analysis, IEEE, Salt Lake City, UT
27. Sawada T, Fedorov DG, Kitaura K (2010) Role of the key mutation in the selective binding of avian and human influenza hemagglutinin to sialosides revealed by quantum-mechanical calculations. J Am Chem Soc 132:16862–16872
28. Simoncini D, Nakata H, Ogata K, Nakamura S, Zhang KYJ (2015) Quality assessment of predicted protein models using energies calculated by the fragment molecular orbital method. Mol Inform 34:97–104
29. Ito M, Brinck T (2014) Novel approach for identifying key residues in enzymatic reactions: proton abstraction in ketosterbid isomerase. J Phys Chem B 118:13050–13058
30. Heifetz A, Trani G, Aldeghi M, MacKinnon CH, McEwan PA, Brookfield FA, Chudyk EI, Bodkin M, Pei Z, Burch JD, Ortwine DF (2016) Fragment molecular orbital method applied to lead optimization of novel interleukin-2 inducible T-cell kinase (ITK) inhibitors. J Med Chem 59:4352–4363
31. Pruitt SR, Steinmann C (2017) Mapping interaction energies in chorismate mutase with the fragment molecular orbital method. J Phys Chem A 121:1798–1808
32. Morao I, Fedorov DG, Robinson R, Southey M, Townsend-Nicholson A, Bodkin MJ, Heifetz A (2017) Rapid and accurate assessment of GPCR-ligand interactions using the fragment molecular orbital-based density-functional tight-binding method. J Comput Chem 38:1987–1990
33. Sladek V, Tokiwa H, Shimano H, Shigeta Y (2018) Protein residue networks from energetic and geometric data: are they identical? J Chem Theory Comput 14:6623–6631
34. Rocha GB, Freire RO, Simas AM, Stewart JJP (2006) RM1: a reparameterization of AM1 for H, C, N, O, P, S, F, Cl, Br, and I. J Comput Chem 27:1101–1111
35. Lima NBD, Rocha GB, Freir RO, Simas AM (2019) RM1 semiempirical model: chemistry, pharmaceutical research, molecular biology and materials science. J Braz Chem Soc 30:683–716
36. Kromann JC, Christensen AS, Steinmann C, Korth M, Jensen JH (2014) A third-generation dispersion and third-generation hydrogen bonding corrected PM6 method: PM6-D3H+. PeerJ 2:e449
37. Cui Q, Elstner M (2014) Density functional tight binding: values of semi-empirical methods in an ab initio era. Phys Chem Chem Phys 16:14368–14377
38. Christensen AS, Kubar T, Cui Q, Elstner M (2016) Semiempirical quantum mechanical methods for noncovalent interactions for chemical and biochemical applications. Chem Rev 116:5301–5337
39. Allec SI, Sun Y, Sun J, Chang CA, Wong BM (2019) Heterogeneous CPU+GPU-enabled simulations for DFTB molecular dynamics of large chemical and biological systems. J Chem Theory Comput 15:2807–2815
40. Sure R, Grimme S (2013) Corrected small basis set Hartree-Fock method for large systems. J Comput Chem 34:1672–1985
41. Kruse H, Grimme S (2012) Geometrical correction for the inter- and intra-molecular basis set superposition error in hartree-fock and density functional theory calculations for large systems. J Chem Phys 136:154101
42. Goerigk L, Collyer CA, Reimers JR (2014) Recommending Hartree-Fock theory with London-dispersion and basis-set-superposition corrections for the optimization or quantum refinement of protein structures. J Phys Chem B 118:14612–14626
43. Gordon MS, Freitag MA, Bandyopadhyay P, Jensen JH, Kairys V, Stevens WJ (2001) The effective fragment potential method: a QM-based MM approach to modeling environmental effects in chemistry. J Phys Chem A 105:293–307
44. Kairys V, Jensen JH (2000) QM/MM boundaries across covalent bonds: a frozen localized molecular orbital-based approach

for the effective fragment potential method. J Phys Chem A 104:6656–6665

45. Grigorenko BL, Nemukhin AV, Topol IA, Burt SK (2002) Modeling of biomolecular systems with the quantum mechanical and molecular mechanical method based on the effective fragment potential technique: proposal of flexible fragments. J Phys Chem A 106:10663–10672
46. Gordon MS, Smith QA, Xu P, Slipchenko LV (2013) Accurate first principles model potentials for intermolecular interactions. Ann Rev Phys Chem 64:553–578
47. Gurunathan PK, Acharya A, Ghosh D, Kosenkov D, Kaliman I, Shao Y, Krylov AI, Slipchenko LV (2016) Extension of the fragment potential method to macromolecules. J Phys Chem B 120:6562–6574
48. Aoki Y, Gu FL (2012) An elongation method for large systems toward bio-systems. Phys Chem Chem Phys 14:7640–7668
49. Orimoto U, Liu K, Aoki Y (2015) Elongation method for electronic structure calculations of random DNA sequences. J Comput Chem 36:2103–2113
50. Saparpakorn P, Kobayashi M, Hannongbua S, Nakai H (2013) Divide-and-conquer-based quantum chemical study for interaction between HIV-1 reverse transcriptase and MK-4965 inhibitor. Int J Quant Chem 113:510–517
51. Steinmann C, Fedorov DG, Jensen JH (2010) Effective fragment molecular orbital method: a merger of the effective fragment potential and fragment molecular orbital methods. J Phys Chem A 114:8705–8712
52. Steinmann C, Fedorov DG, Jensen JH (2013) Mapping enzymatic catalysis using the effective fragment molecular orbital method: towards all ab initio bio-chemistry. PLoS One 8:e60602
53. Bertoni C, Slipchenko LV, Misquitta AJ, Gordon MS (2017) Multipole moments in the effective fragment potential method. J Phys Chem A 121:2056–2067
54. Sahu N, Gadre SR (2014) Molecular tailoring approach: a route for ab initio treatment of large clusters. Acc Chem Res 47:2739–2747
55. Kobayashi R, Addicoat MA, Gilbert ATB, Amos RD, Collins MA (2019) The SMFA program for quantum chemistry calculations on large molecules. WIREs Comp Mol Sci 9: e1413
56. Cao L, Ryde U (2018) On the difference between additive and subtractive QM/MM calculations. Front Chem 6:89
57. Vreven T, Byun KS, Komáromi I, Dapprich S, Montgomery JA Jr, Morokuma K, Frisch MJ (2006) Combining quantum mechanics methods with molecular mechanics methods in ONIOM. J Chem Theory Comput 2:815–826
58. Maseras F, Morokuma K (1995) IMOMM: a new integrated ab initio + molecular mechanics geometry optimization scheme of equilibrium structures and transition states. J Comput Chem 16:1170–1179
59. Shoemaker JR, Burggraf LW, Gordon MS (1999) SIMOMM: an integrated molecular orbital/molecular mechanics optimization scheme for surfaces. J Phys Chem A 103:3245–3251
60. Brooks BR, Bruccoleri RE, Olafson BD, States DJ, Swaminathan S, Karplus M (1983) CHARMM: a program for macromolecular energy, minimization, and dynamics calculations. J Comput Chem 4:187–217
61. Field MJ, Bash PA, Karplus M (1990) A combined quantum mechanical and molecular mechanical potential for molecular dynamics simulations. J Comput Chem 11:700–733
62. Thellamurege NM, Si D, Cui F, Zhu H, Lai R, Li H (2013) QuanPol: a full spectrum and seamless QM/MM program. J Comput Chem 34:2816–2833
63. Rackers JA, Wang Z, Lu C, Laury ML, Lagardère L, Schnieders MJ, Piquemal JP, Ren P, Ponder JW (2018) Tinker 8: software tools for molecular design. J Chem Theory Comput 14:5273–5289
64. Fedorov DG, Kitaura K, Li H, Jensen J, Gordon M (2006) The polarizable continuum model (PCM) interfaced with the fragment molecular orbital method (FMO). J Comput Chem 27:976–985
65. Fedorov DG (2018) Analysis of solute-solvent interactions using the solvation model density combined with the fragment molecular orbital method. Chem Phys Lett 702:111–116
66. Fedorov DG, Asada N, Nakanishi I, Kitaura K (2014) The use of many-body expansions and geometry optimizations in fragment-based methods. Acc Chem Res 47:2846–2856
67. Chiba M, Fedorov DG, Kitaura K (2007) Time-dependent density functional theory with the multilayer fragment molecular orbital method. Chem Phys Lett 444:346–350
68. Alexeev Y, Fedorov DG, Shvartsburg AA (2014) Effective ion mobility calculations for macromolecules by scattering on electron clouds. J Phys Chem A 118:6763–6772
69. Fedorov DG, Jensen JH, Deka RC, Kitaura K (2008) Covalent bond fragmentation suitable to describe solids in the fragment molecular

orbital method. J Phys Chem A 112:11808–11816
70. Fedorov DG, Ishida T, Uebayasi M, Kitaura K (2007) The fragment molecular orbital method for geometry optimizations of polypeptides and proteins. J Phys Chem A 111:2722–2732
71. Komeiji Y, Mochizuki Y, Nakano T, Fedorov DG (2009) Fragment molecular orbital-based molecular dynamics (FMO-MD), a quantum simulation tool for large molecular systems. J Mol Struc THEOCHEM 898:2–7
72. Fedorov DG, Ishida T, Kitaura K (2005) Multilayer formulation of the fragment molecular orbital method (FMO). J Phys Chem A 109:2638–2646
73. Nagata T, Fedorov DG, Kitaura K (2010) Importance of the hybrid orbital operator derivative term for the energy gradient in the fragment molecular orbital method. Chem Phys Lett 492:302–308
74. Nagata T, Fedorov DG, Kitaura K, Gordon MS (2009) A combined effective fragment potential-fragment molecular orbital method. I. The energy expression and initial applications. J Chem Phys 131:024101
75. Yoshida N (2014) Efficient implementation of the three-dimensional reference interaction site model method in the fragment molecular orbital method. J Chem Phys 140:214118
76. Nakata H, Fedorov DG, Nagata T, Kitaura K, Nakamura S (2015) Simulations of chemical reactions with the frozen domain formulation of the fragment molecular orbital method. J Chem Theory Comput 11:3053–3064
77. Fedorov DG, Alexeev Y, Kitaura K (2011) Geometry optimization of the active site of a large system with the fragment molecular orbital method. J Phys Chem Lett 2:282–288
78. Nakata H, Fedorov DG (2016) Efficient geometry optimization of large molecular systems in solution using the fragment molecular orbital method. J Phys Chem A 120:9794–9804
79. Nagata T, Brorsen K, Fedorov DG, Kitaura K, Gordon MS (2011) Fully analytic energy gradient in the fragment molecular orbital method. J Chem Phys 134:124115
80. Fedorov DG, Kitaura K (2014) Use of an auxiliary basis set to describe the polarization in the fragment molecular orbital method. Chem Phys Lett 597:99–105
81. Asada N, Fedorov DG, Kitaura K, Nakanishi I, Merz KM Jr (2012) An efficient method to evaluate intermolecular interaction energies in large systems using overlapping multicenter ONIOM and the fragment molecular orbital method. J Phys Chem Lett 3:2604–2610
82. Fedorov DG, Kitaura K (2007) Pair interaction energy decomposition analysis. J Comput Chem 28:222–237
83. Fedorov DG, Kitaura K (2012) Energy decomposition analysis in solution based on the fragment molecular orbital method. J Phys Chem A 116:704–719
84. Green MC, Fedorov DG, Kitaura K, Francisco JS, Slipchenko LV (2013) Open-shell pair interaction energy decomposition analysis (PIEDA): formulation and application to the hydrogen abstraction in tripeptides. J Chem Phys 138:074111
85. Fedorov DG, Kitaura K (2017) Modeling and visualization for the fragment molecular orbital method with the graphical user interface FU, and analyses of protein-ligand binding. In: Gordon MS (ed) Fragmentation: toward accurate calculations on complex molecular systems. Wiley, Hoboken, NJ, pp 119–139
86. Fedorov DG, Kitaura K (2016) Subsystem analysis for the fragment molecular orbital method and its application to protein-ligand binding in solution. J Phys Chem A 120:2218–2231
87. Fedorov DG, Kitaura K (2018) Pair interaction energy decomposition analysis for density functional theory and density-functional tight-binding with an evaluation of energy fluctuations in molecular dynamics. J Phys Chem A 122:1781–1795
88. Nagata T, Fedorov DG, Sawada T, Kitaura K (2012) Analysis of solute-solvent interactions in the fragment molecular orbital method interfaced with effective fragment potentials: theory and application to a solvated griffithsin-carbohydrate complex. J Phys Chem A 116:9088–9099
89. Mazanetz MP, Chudyk E, Fedorov DG, Alexeev Y (2016) Applications of the fragment molecular orbital method to drug research. In: Zhang W (ed) Computer aided drug discovery. Springer, New York, NY, pp 217–255
90. Fedorov DG, Brekhov A, Mironov V, Alexeev Y (2019) Molecular electrostatic potential and electron density of large systems in solution computed with the fragment molecular orbital method. J Phys Chem A 123 (29):6281–6290
91. Nakata H, Nagata T, Fedorov DG, Yokojima S, Kitaura K, Nakamura S (2013) Analytic second derivatives of the energy in the fragment molecular orbital method. J Chem Phys 138:164103

92. Nakata H, Fedorov DG, Yokojima S, Kitaura K, Nakamura S (2014) Simulations of Raman spectra using the fragment molecular orbital method. J Chem Theory Comput 10:3689–3698
93. Gao Q, Yokojima S, Kohno T, Ishida T, Fedorov DG, Kitaura K, Fujihira M, Nakamura S (2007) Ab initio NMR chemical shift calculations on proteins using fragment molecular orbitals with electrostatic environment. Chem Phys Lett 445:331–339
94. Fedorov DG, Kitaura K (2009) The role of the exchange in the embedding electrostatic potential for the fragment molecular orbital method. J Chem Phys 131:171106
95. Tsuneyuki S, Kobori T, Akagi K, Sodeyama K, Terakura K, Fukuyama H (2009) Molecular orbital calculation of biomolecules with fragment molecular orbitals. Chem Phys Lett 476:104–108
96. Fedorov DG, Kitaura K (2017) Many-body expansion of the Fock matrix in the fragment molecular orbital method. J Chem Phys 147:104106
97. Nishimoto Y, Fedorov DG (2016) The fragment molecular orbital method combined with density-functional tight-binding and the polarizable continuum model. Phys Chem Chem Phys 18:22047–22061
98. Vuong VQ, Nishimoto Y, Fedorov DG, Sumpter BG, Niehaus TA, Irle S (2019) The fragment molecular orbital method based on long-range corrected density-functional tight-binding. J Chem Theory Comput 15:3008–3020
99. Nakata H, Fedorov DG (2019) Simulations of infrared and Raman spectra in solution using the fragment molecular orbital method. Phys Chem Chem Phys 21:13641–13652
100. Hanwell MD, Curtis DE, Lonie DC, Vandermeersch T, Zurek E, Hutchison GR (2012) Avogadro: an advanced semantic chemical editor, visualization, and analysis platform. J Cheminf 4:17
101. Chemcraft – graphical software for visualization of quantum chemistry computations. https://www.chemcraftprog.com. Accessed Jun 2019
102. Suenaga M (2005) Facio: new computational chemistry environment for PC GAMESS. J Comput Chem, Jpn 4:25–32
103. Suenaga M (2008) Development of GUI for GAMESS / FMO calculation. J Comput Chem, Jpn 7:33–53
104. FU. URL https://sourceforge.net/projects/fusuite/. Accessed Jun 2019
105. Allouche AR (2011) Gabedit – a graphical user interface for computational chemistry softwares. J Comput Chem 32:174–182
106. Bode BM, Gordon MS (1998) MacMolPlt: a graphical user interface for GAMESS. J Mol Graph Model 16:133–138
107. Schaftenaar G, Vlieg E, Vriend G (2017) Molden 2.0: quantum chemistry meets proteins. J Comput Aided Mol Des 31:789–800
108. The PyMOL Molecular Graphics System, Version 2.0 Schrödinger, LLC
109. Humphrey W, Dalke A, Schulten K (1996) VMD – visual molecular dynamics. J Mol Graph 14:33–38
110. Wimostar: structure modeller and visualizer for free chemistry simulations. https://winmostar.com/en/index.php. Accessed Jun 2019
111. Nishimoto Y, Fedorov DG (2018) Adaptive frozen orbital treatment for the fragment molecular orbital method combined with density-functional tight-b inding. J Chem Phys 148:064115

Chapter 9

Accurate Scoring in Seconds with the Fragment Molecular Orbital and Density-Functional Tight-Binding Methods

Inaki Morao, Alexander Heifetz, and Dmitri G. Fedorov

Abstract

The accurate evaluation of receptor-ligand interactions is an essential part of rational drug design. While quantum mechanical (QM) methods have been a promising means by which to achieve this, traditional QM is not applicable for large biological systems due to its high computational cost. Here, the fragment molecular orbital (FMO) method has been combined with the density-functional tight-binding (DFTB) method to compute energy calculations of biological systems in seconds. FMO-DFTB outperformed GBVI/WSA in identifying a set of 10 binders versus a background of 500 decoys applied to human k-opioid receptor. The significant increase in the speed and the high accuracy achieved with FMO-DFTB calculations allows FMO to be applied in areas of drug discovery that were not previously accessible to traditional QM methodologies. For the first time, it is now possible to perform FMO calculations in a high-throughput manner.

Key words Fragment molecular orbital method (FMO), Density-functional tight-binding method (DFTB), Structure-based drug discovery (SBDD), Structure-activity relationship (SAR)

1 Introduction

The rationalization of potency and selectivity in the structure-based drug design (SBDD) process requires a deep understanding of the binding interactions between ligand and its target protein [1]. However, visual inspection and force field (FF)-based molecular mechanics calculations (MM) cannot always explain the full complexity of the molecular interactions. Recently, several notable reports have been published [2–5] that emphasize the crucial role of a large number of "underappreciated" or non-obvious intramolecular interactions involved in biomolecular recognition. These interactions include CH/π [6, 7], halogen/π [8], cation/π [9], and non-classical hydrogen bonds [10] which, typically, are not properly parameterized in the currently available force fields (FF) [3]. Furthermore, the role of hydrophobic interactions, vital

Alexander Heifetz (ed.), *Quantum Mechanics in Drug Discovery*, Methods in Molecular Biology, vol. 2114,
https://doi.org/10.1007/978-1-0716-0282-9_9,

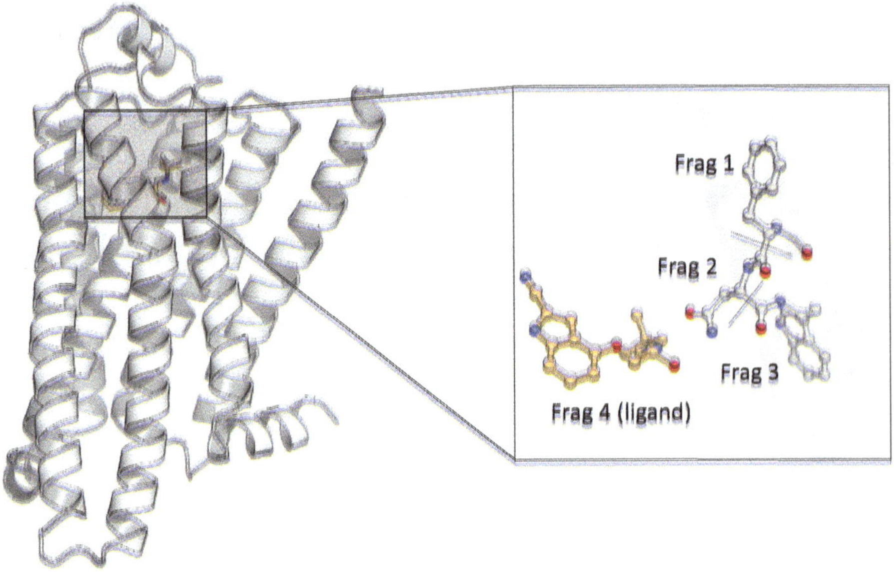

Fig. 1 FMO fragmentations are achieved by dividing the receptor protein into fragments, each of which corresponds to an amino acid (Frag 1–3). Similarly, the ligand can be represented by single or multiple fragments as necessary (Frag 4)

for receptor stability [11], still has no reliable non-QM predictive method for its quantification [5, 12].

The use of quantum mechanical (QM) methods can represent charge fluctuations and dynamic polarization, which are essential in assessing molecular interactions. However, despite the many advantages that QM can bring, traditional QM methods are not feasible for large biological systems, such as proteins, due to their high computational cost.

As described in detail at Chapter 3 of this book, the fragment molecular orbital (FMO) quantum mechanical (QM) method [1, 7, 13, 14] accelerates traditional QM methods, by dividing the system into smaller pieces called fragments and performing QM calculations on these fragments (Fig. 1). For example, in proteins, each residue can be represented by a fragment. Similarly, the ligand can be represented by single or multiple fragments as necessary. A key advantage of FMO is that it can provide the individual contribution of each receptor-ligand pair interaction energy (PIE) to the total interaction energy (TIE). TIE is a sum of PIEs for all residues; it is a guestimate of the total protein-ligand binding energy in enthalpy-driven cases. FMO is able to detect and quantify protein-ligand interactions that are difficult to identify using non-QM approaches, providing precise information on the individual contribution of each residue in the protein target to ligand binding [15]. This information is essential for medicinal chemists in order to increase favorable interactions or to generate new interactions as part of the rational design during lead optimization. In spite of the increased speed of FMO compared to traditional QM methods due to the fragmentation scheme, it is still relatively slow (takes a few hours to complete for a single ligand-protein complex) which makes it impractical to use in a high-throughput manner.

The cost of doing traditional MP2 calculations of fragments scales as N^5, where N is the number of basis functions in the fragment, due to the transformation of two-electron integrals. Semiempirical quantum mechanical (SQM) methods have been used for decades to bridge the gap between QM and FF methods providing a great compromise between computational time and accuracy albeit the drastic integral approximations made by parameterization. They typically consist of valence-only minimal basis SCF methods derived from either Hartree-Fock theory (PM6, OM2, etc.) or Kohn-Sham density functional theory (DFT). DFTB is based on a series expansion of electron density from DFT; for instance, DFTB3 is based on the third-order expansion. As a consequence, FMO can be accelerated by 1000-fold when it is combined with the density-functional tight-binding (DFTB) approach [15–17]. FMO-DFTB energy calculations complete in seconds and possess comparable accuracy of much more expensive methods like FMO-MP2 in the systems analyzed so far.

As we previously demonstrated [16] FMO-DFTB is a rapid, accurate, and reliable method for the assessment of receptor-ligand interactions and TIE calculations [16, 18]. The interactions detected by FMO-DFTB are consistent with the experimental data and with those detected by traditional FMO-MP2 method. The application of FMO-DFTB will be of great utility for the design and evaluation of new compounds, helping considerably to reduce the effort and cost of chemical synthesis needed for drug discovery programs. The high correlation found between experimental affinity and receptor-ligand $TIE^{FMO\text{-}DFTB}$ indicates that FMO-DFTB can be used to determine the binding affinities of new targets and, therefore, provides a means of predicting experimental outcomes.

FMO-DFTB is similar to previously reported FMO-MP2 [16] making it a very attractive method to support SBDD in hit-to-lead and in lead optimization stages. However, the significant increase in the speed with which FMO-DFTB calculations can be performed allows FMO-DFTB to be applied in areas of drug discovery that were not previously accessible to FMO-MP2 or other QM methodologies. It opens an avenue to use QM calculations in a high-throughput manner. For example, hundred/thousands of docked ligands can be refined and re-ranked with FMO-DFTB in the presence of the surrounded protein and water molecules in order to prioritize hits of virtual screening (VS) campaigns. FMO-DFTB can be applied in the modelling and structural optimization of receptor-ligand complexes or in molecular dynamic (MD) simulations [19]. In the future, FMO-DFTB can be also used in ADMET prediction.

2 Method

FMO code version 5.1 is distributed inside ab initio quantum chemistry package GAMESS [20]. We used the third-order DFTB3 method with 3ob parameters [21] and the Møller-Plesset second-order perturbation theory (MP2), and, for treating solvent effects, we combined both calculations with the polarizable continuum model (PCM). MP2 was used with the 6-31G* basis set, whereas the UFF dispersion model was used for DFTB3. The structures were taken from the previous study.

We consider any interaction with an absolute PIE $\geq$3.0 kcal/mol to be significant. The water molecules in this work were extracted from the crystal structures (if resolved in these) and treated explicitly to explore their contribution to the receptor-ligand binding.

In this section of the chapter, we would like to illustrate the potential use of FMO-DFTB approach as a QM-based scoring function to distinguish between known binders and chemically similar analogues. For example, such approach can be taken in structure-based virtual screenings. A set of 10 potent binders of human κ-opioid receptor (KOR, PDB entry 4DJH [22]) were extracted from the literature [23]. A decoy of 500 chemically similar analogues were extracted from commercial vendor library by applying Blaze (Cresset) search technology [24] in the presence of the KOR. A Blaze similarity score for these 500 compounds was >0.65 to the known KOR binders (1.00 mean 100% identity to the query molecule). The use of decoys similar to the known compounds provided a challenging validation exercise, success in which would indicate which scoring function (FMO-DFTB or force field based) is preferable in this case. The library of 500 compounds was run through Blaze (Cresset) in the presence of the protein. The overlaid poses were re-optimized within the protein active site and scored and ranked by FMO-DFTB and GBVI/WSA force field-based scoring function (as implemented in docking module of molecular operating environment (MOE) of computational chemistry group (CCG)).

100% of the known KOR binders were found among the top 20 and top 90 when ranked by FMO-DFTB and GBVI/WSA, respectively. The corresponding receiver operating characteristic (ROC) curves for both methodologies are shown in Fig. 2. This "enrichment" test demonstrated that the FMO-DFTB used as scoring was of high quality, as proven by its ability to discriminate between known binders and very similar decoys. In this case, and in concordance with our in-house results, the performance of the QM-based scoring was found to be better than GBVI/WSA force field-based scoring.

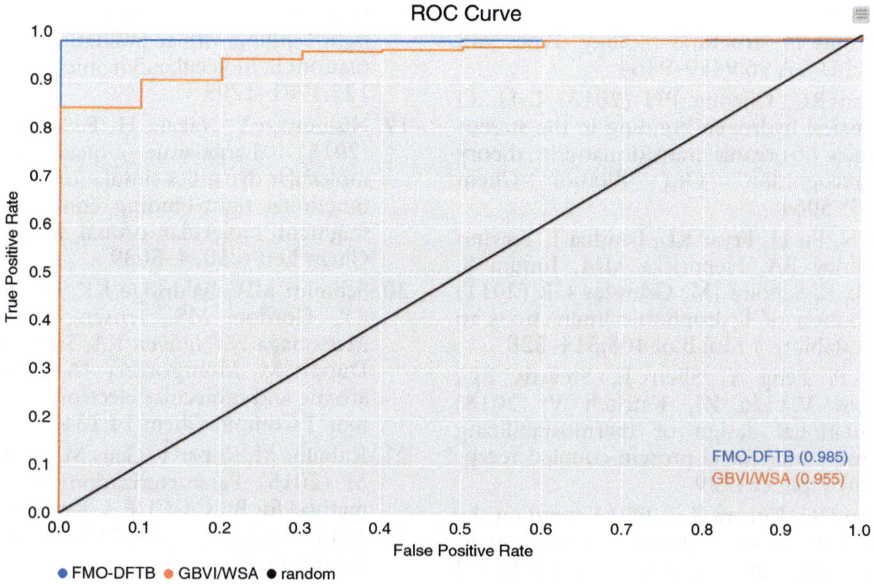

Fig. 2 ROC curve plotting the true-positive rate against the false-positive rate as its discrimination threshold is varied for the in silico screening of 10 known KOR binders embedded in a library of 500 similar compounds. The blue curve corresponds to FMO-DFTB method and the orange line to GBVI/WSA scoring

Acknowledgments

I.M. would like to thank Dr. Balínt Aradi and the Bremen Center for Computational Materials Science for the kind permission to use the Slater-Koster files to run DFTB calculations. D.G.F. was supported by JSPS KAKENHI Grant 19H02682.

References

1. Heifetz A, Chudyk EI, Gleave L, Aldeghi M, Cherezov V, Fedorov DG, Biggin PC, Bodkin MJ (2016) The fragment molecular orbital method reveals new insight into the chemical nature of GPCR-ligand interactions. J Chem Inf Model 56:159–172
2. Tong Y, Mei Y, Li YL, Ji CG, Zhang JZ (2010) Electrostatic polarization makes a substantial contribution to the free energy of avidin-biotin binding. J Am Chem Soc 132:5137–5142
3. Raha K, Peters MB, Wang B, Yu N, Wollacott AM, Westerhoff LM, Merz KM Jr (2007) The role of quantum mechanics in structure-based drug design. Drug Discov Today 12:725–731
4. Beratan DN, Liu C, Migliore A, Polizzi NF, Skourtis SS, Zhang P, Zhang Y (2015) Charge transfer in dynamical biosystems, or the treachery of (static) images. Acc Chem Res 48:474–481
5. Bissantz C, Kuhn B, Stahl M (2010) A medicinal chemist's guide to molecular interactions. J Med Chem 53:5061–5084
6. Ozawa T, Okazaki K, Kitaura K (2011) CH/pi hydrogen bonds play a role in ligand recognition and equilibrium between active and inactive states of the beta2 adrenergic receptor: an ab initio fragment molecular orbital (FMO) study. Bioorg Med Chem 19:5231–5237
7. Fedorov DG, Nagata T, Kitaura K (2012) Exploring chemistry with the fragment molecular orbital method. Phys Chem Chem Phys 14:7562–7577
8. Lu Y-X, Zou J-W, Wang Y-H, Yu Q-S (2007) Substituent effects on noncovalent halogen/π interactions: theoretical study. Int J Quantum Chem 107:1479–1486

9. Gallivan JP, Dougherty DA (1999) Cation-pi interactions in structural biology. Proc Natl Acad Sci U S A 96:9459–9464
10. Johnston RC, Cheong PH (2013) C-H...O non-classical hydrogen bonding in the stereomechanics of organic transformations: theory and recognition. Org Biomol Chem 11:5057–5064
11. Pace CN, Fu H, Fryar KL, Landua J, Trevino SR, Shirley BA, Hendricks MM, Iimura S, Gajiwala K, Scholtz JM, Grimsley GR (2011) Contribution of hydrophobic interactions to protein stability. J Mol Biol 408:514–528
12. Popov P, Peng Y, Shen L, Stevens RC, Cherezov V, Liu ZJ, Katritch V (2018) Computational design of thermostabilizing point mutations for G protein-coupled receptors. elife 7:pii: e34729
13. Fedorov DG, Kitaura K (2007) Extending the power of quantum chemistry to large systems with the fragment molecular orbital method. J Phys Chem A 111:6904–6914
14. Kitaura K, Ikeo E, Asada T, Nakano T, Uebayasi M (1999) Fragment molecular orbital method: an approximate computational method for large molecules. Chem Phys Lett 313:701–706
15. Nishimoto Y, Fedorov DG, Irle S (2014) Density-functional tight-binding combined with the fragment molecular orbital method. J Chem Theory Comput 10:4801–4812
16. Morao I, Fedorov DG, Robinson R, Southey M, Townsend-Nicholson A, Bodkin MJ, Heifetz A (2017) Rapid and accurate assessment of GPCR-ligand interactions Using the fragment molecular orbital-based density-functional tight-binding method. J Comput Chem 38:1987–1990
17. Vuong VQ, Nishimoto Y, Fedorov DG, Sumpter BG, Niehaus TA, Irle S (2019) The fragment molecular orbital method based on long-range corrected density-functional tight-binding. J Chem Theory Comput 15:3008–3020
18. Fedorov DG, Kitaura K (2018) Pair interaction energy decomposition analysis for density functional theory and density-functional tight-binding with an evaluation of energy fluctuations in molecular dynamics. J Phys Chem A 122:1781–1795
19. Nishimoto Y, Nakata H, Fedorov DG, Irle S (2015) Large-scale quantum-mechanical molecular dynamics simulations using density-functional tight-binding combined with the fragment molecular orbital method. J Phys Chem Lett 6:5034–5039
20. Schmidt MW, Baldridge KK, Boatz JA, Elbert ST, Gordon MS, Jensen JH, Koseki S, Matsunaga N, Nguyen KA, Su S, Windus TL, Dupuis M, Montgomery JA (1993) General atomic and molecular electronic structure system. J Comput Chem 14:1347–1363
21. Kubillus M, Kubar T, Gaus M, Rezac J, Elstner M (2015) Parameterization of the DFTB3 method for Br, Ca, Cl, F, I, K and Na in organic and biological systems. J Chem Theory Comput 11(1):332–342
22. Wu H, Wacker D, Mileni M, Katritch V, Han GW, Vardy E, Liu W, Thompson AA, Huang XP, Carroll FI, Mascarella SW, Westkaemper RB, Mosier PD, Roth BL, Cherezov V, Stevens RC (2012) Structure of the human kappa-opioid receptor in complex with JDTic. Nature 485:327–332
23. Kormos CM, Gichinga MG, Runyon SP, Thomas JB, Mascarella SW, Decker AM, Navarro HA, Carroll FI (2016) Design, synthesis, and pharmacological evaluation of JDTic analogs to examine the significance of replacement of the 3-hydroxyphenyl group with pyridine or thiophene bioisosteres. Bioorg Med Chem 24:3842–3848
24. Clarke E, Jarvis CI, Goncalves MB, Kalindjian SB, Adams DR, Brown JT, Shiers JJ, Taddei DMA, Ravier E, Barlow S, Miller I, Smith V, Borthwick AD, Corcoran JPT (2018) Design and synthesis of a potent, highly selective, orally bioavailable, retinoic acid receptor alpha agonist. Bioorg Med Chem 26:798–814

Chapter 10

Protein Molecular Dynamics Simulations with Approximate QM: What Can We Learn?

Stephan Irle, Van Q. Vuong, Mouhmad H. Elayyan, Marat R. Talipov, and Steven M. Abel

Abstract

Classical force fields are essential for computer simulations of proteins and are typically parameterized to reproduce secondary and tertiary structure of isolated proteins. However, while protein-protein interactions are ubiquitous in nature, they are not considered in parameterization efforts and are far less understood than isolated proteins. A better characterization of intermolecular interactions is widely recognized as a key to revolutionizing drug and therapeutic developments with high-throughput computational screening. Urgently needed is a critical assessment of the performance of modern protein force fields against first-principles electronic structure methods and experiments. In a daring step toward this goal, we here describe a comparison of peptide folding dynamics as predicted by a molecular mechanics force field on the one hand and by an approximate electronic structure quantum mechanical (QM) method based on density-functional tight-binding (DFTB) on the other. We further compare the dynamics from straightforward DFTB simulations with a near-linear scaling version of DFTB for massively parallel computation based on the fragment molecular orbital (FMO-DFTB) method. We illustrate differences between the phenomenology of the folding dynamics from these three methods for a small model peptide, as well as charge polarization and dynamic fluctuations, point out possible correlations and implications for force field developers, and discuss the lessons learned that might become applicable to future predictive high-throughput computer screening for personalized neoantigen cancer therapy.

Key words Molecular dynamics simulations, Density-functional tight-binding, Near-linear scaling methods, Neoantigen therapy

1 Introduction

Protein-protein interactions are a cornerstone of cellular processes including receptor-mediated sensing, signal transduction, growth, and motility [1, 2]. Fast and accurate computational modeling of protein-protein interactions is a critical gap in many therapeutic applications. As an illustrative example, T cells are known to orchestrate adaptive immunity and have emerged as a therapeutically important agent in immunotherapies [3]. T cells use a surface receptor (TCR) to scan short peptides (typically 8–11 amino

Alexander Heifetz (ed.), *Quantum Mechanics in Drug Discovery*, Methods in Molecular Biology, vol. 2114,
https://doi.org/10.1007/978-1-0716-0282-9_10, © Springer Science+Business Media, LLC, part of Springer Nature 2020

acids) that are presented by surface proteins (MHC) on other cells. When they identify a "non-self" peptide (activation is correlated with binding affinity), they can either kill the identified cell directly or orchestrate other immune responses. Interestingly, cancers typically have tens to thousands of mutations in their genome, many of which change the sequences of proteins. When processed and presented by MHC, the aberrant proteins can appear to T cells as signatures of "non-self." Such peptide sequences are known as neoantigens and can lead to T cells identifying and mounting an attack on cancer cells. Since tumors in individual patients likely carry different mutations, each patient is likely to have different neoantigens. Thus, therapies will likely need to be personalized. Pipelines for targeting tumor neoantigens have been proposed [4, 5], but they rely on computational screening to characterize the binding of peptides to MHC and the ability of TCRs to distinguish the neoantigen from the original protein sequence. This is acknowledged as a key area of development for practical development of personalized cancer immunotherapies. Accurate descriptions of protein-peptide binding that can be carried out in a high-throughput manner will provide a means to achieve this goal [6]. Of crucial importance in this context is the treatment of entropic contributions to the binding free energies, which are nontrivial and require treatment beyond simple approximations.

Protein-peptide interactions are typically modeled using classical molecular mechanics (MM) force fields such as CHARMM36m [7] or AMBER ff14SB [8]. These are parameterized to reproduce available experimental structural data of single proteins. While this reference data may contain polypeptides (dimers, trimers, etc.) characterized by quaternary structure [9], the intermolecular interactions between the monomer units are usually not explicitly considered in the optimization of the force fields. The parameterization typically aims to accurately predict secondary (α-helix vs. β-sheets and irregular loops) and tertiary (three-dimensional backbone) protein structure from a given amino acid sequence in molecular dynamics (MD) simulations. However, the entropic term needed in the calculation of binding free energies is not immediately parameterizable, as it would require complete knowledge of the full potential energy surface of the protein.

MM force fields contain two types of potential energy terms: bonded terms describe interactions between covalently connected atoms, and nonbonded terms describe interactions such as Pauli repulsion, van der Waals attraction, and electrostatic interactions. In a successful parameterization, both bonded and nonbonded terms work in concert to achieve accurate structure prediction. However, minute changes are known to greatly influence the resulting preference of a force field for either α-helix or β-sheet, as shown in Fig. 1a (adapted from Ref. 10). This figure shows the seven lowest-energy conformations of the G-peptide, a well-studied

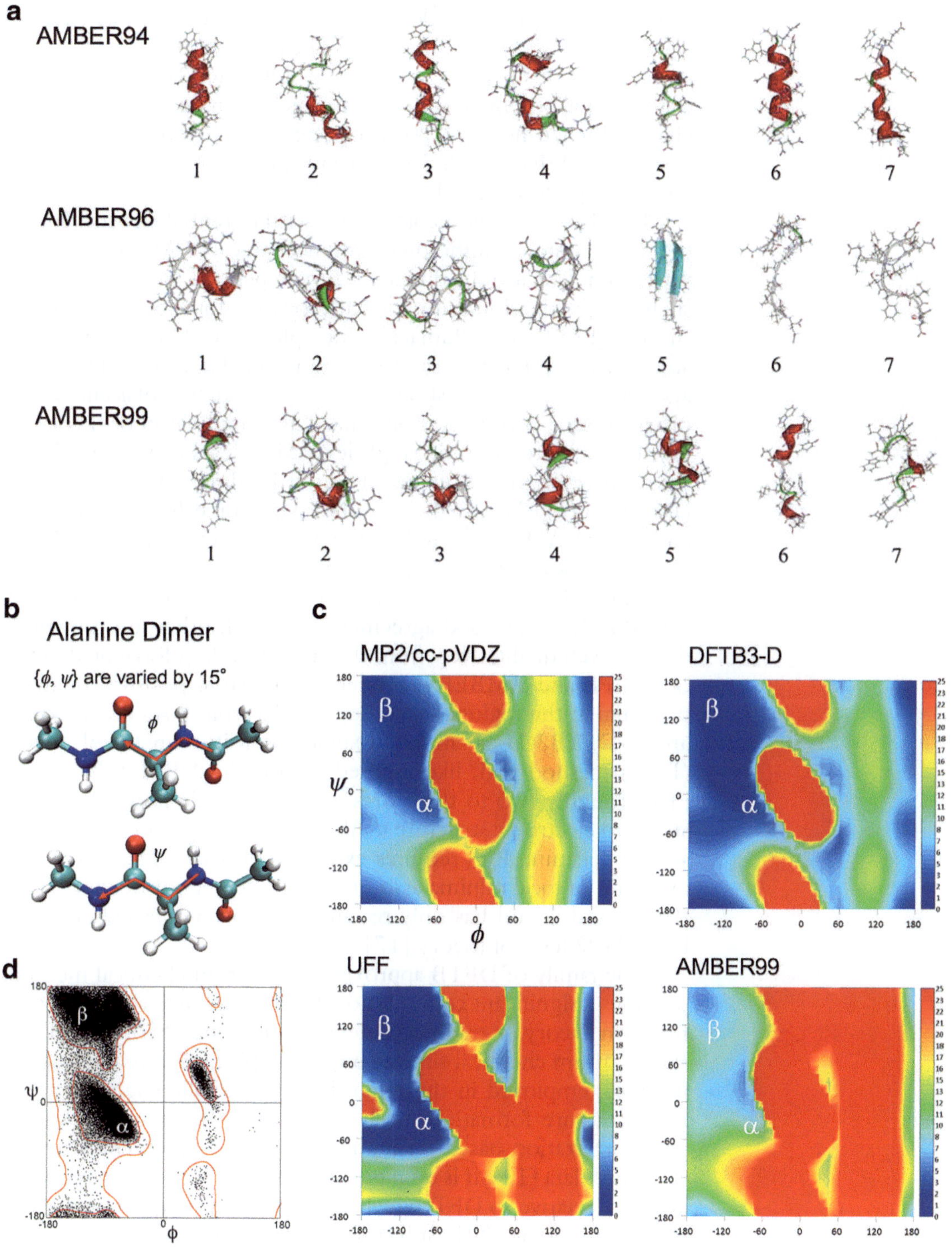

Fig. 1 (**a**) Lowest energy structures of the G-peptide from simulated annealing runs using different force fields. α-Helices are highlighted in red, β-sheets in cyan color. Adapted from Ref. 10. (**b**) Alanine dimer for potential energy calculations with definition of ϕ and ψ torsional angles. (**c**) ϕ—potential energy plots for the alanine dimer molecule in gas phase, energy units are in kcal/mol. (**d**) Ramachandran ϕ—plot; each point denotes a structure in the PDB database

"toy" peptide for force field developers, predicted by three different versions of the AMBER force field. The differences in the secondary structure of this peptide are striking. Reference 10 illustrates how the torsional potentials of these force fields can be tuned to entirely change their α-helix vs. β-sheet preferences. Moreover, it is well documented that the choice of atomic partial charges greatly influences this preference [11].

For these reasons, a comparison with potential energies from a highly accurate, ab initio quantum mechanical (QM) perspective is warranted. The alanine dimer is a simple peptide model that allows such an investigation using the backbone torsional potential energy surface (PES). In a Ramachandran plot, the ϕ and ψ torsional angles (*see* Fig. 1b) are used to distinguish helix from sheet secondary structure. Figure 1c shows the PES as a function of ϕ and ψ with other internal coordinates optimized at a single level of theory for (1) the standard ab initio, high-level MP2 level of theory, (2) the third-order density-functional tight-binding plus dispersion (DFTB3-D) [12], (3) the universal MM force field (UFF) [13], and (4) the AMBER99 MM force field. Figure 1d shows a Ramachandran plot with each point on the map taken from a PDB structure [14]. The PESs from the QM methods (MP2 and DFTB3-D) are in good agreement with each other, but obviously deviate even qualitatively from the corresponding force field methods (UFF and AMBER99), which show large positive, repulsive potential in the regions separating α-helix from β-sheets and other areas. DFTB3-D is an approximate density functional theory (DFT) method and, like its predecessor, the DFTB2-D method [15], roughly two to three orders of magnitude faster than most standard electronic structure methods [16]. A rigorous study of the simple dialanine PES requires extensive MD sampling due to the existence of local minima associated with degrees of freedom other than ϕ and ψ and has only recently been accomplished at the ab initio MP2 level of theory [17].

The family of DFTB approximate quantum chemical methods provides a significant computational advantage over ab initio wavefunction theory or traditional density functional theory (DFT) for the quantum chemical simulation of complex systems and was, for instance, employed in the unraveling of the fullerene and carbon nanostructure formation mechanism [18, 19]. By way of careful parameterization based on highly accurate, coupled-cluster-based reference data [12], it is even possible for DFTB to achieve superior accuracy relative to DFT, which is traditionally plagued by the self-interaction error [20]. Still, the computational bottleneck of the DFTB methods, which is the diagonalization of the electronic Hamiltonian matrix, scales cubically with system size. This unfavorable scaling behavior imposes a limit of roughly 1000 atoms on the range of accessible molecular system sizes for which DFTB can be reasonably applied to. However, a near-linear scaling, massively parallel version called FMO-DFTB [21, 22] is available in the GAMESS-US code [23] and offers a practical route to performing protein MD simulations on parallel computer platforms [21]. The error associated with the many-body expansion inherent to FMO was found minimal and convergent with increasing fragment size [21, 22].

We here report the first Born-Oppenheimer MD (BOMD) simulations for the folding of a capped alanine decamer (ala10) with the straightforward DFTB2 and FMO-DFTB2 QM methods in the gas phase and compare dynamics and charge distribution against corresponding results from a corresponding MM AMBER ff14SB force field-based MD simulation. The study provides insight into the differences of the dynamics and therefore different entropy contributions, between MM and QM beyond a standard potential energy comparison, in particular for the role of charge polarization and dynamic charge fluctuations during the MD simulation of the folding event. This demonstration represents a starting point for future, more systematic, and detailed investigations into the question of what QM methods might be able to teach MM force field developers concerning the entropic contributions needed for accurate binding free energies.

2 Computational Methodology

AMBER ff14SB force field-based MD simulations were performed using the AMBER18 code [23]. DFTB2 and FMO-DFTB2 calculations were performed using the GAMESS-US code [24] with the mio [25] parameterization. We first constructed an extended ala10 model system with acetyl cap at the N-terminus and methyl amine cap at the C-terminus and performed a geometry optimization at all three levels of theory. As previously reported [26], the geometry performance of DFTB2 for peptide systems is very good, with deviations of less than 0.02 Å from experimental bond length values. AMBER predicts for the polyalanine model system an α-helix as most stable structure over the β-turn, and full DFTB2 and FMO-DFTB2 concur with this prediction [21]. Subsequently, we proceeded to perform a number of MD simulations in the NVT canonical ensemble at a target temperature of 300.0 K up to 100 ps, using random initial velocities. In the AMBER MD simulations, we employed a Langevin thermostat, whereas in GAMESS-US, a Nose-Hoover chain thermostat was employed. The initial velocities were generated to match the target temperature. We performed time integration using the velocity Verlet algorithm with a time step of 1.0 fs in both AMBER and GAMESS-US codes for all three levels of theory. Analyses on structures and atomic charges were performed in intervals of 1 ps.

3 Results and Discussion

Since an exhaustive study of the free energy landscape between extended form, α-helix, and β-turn is beyond the scope of this demonstration study and a brute-force scan impossible even on modern supercomputers, we here report only results for a selected folding trajectory in all three cases. The end points of the three trajectories for AMBER, DFTB2, and FMO-DFTB2 methods are shown in Fig. 2 after a simulation of 100 ps at 300 K. We note that, unlike the AMBER-MM approach, for which we find that the folding "success" rate to an α-helix is practically 100%, DFTB2 and FMO-DFTB2 are less prone to produce a folded α-helix after 100 ps. In addition to the trajectories reported here, which arrive at an α-helix after 100 ps, we have also observed the folding of the ala10 model system into a β-turn or even unsuccessfully folded structures during the simulation time. This fact by itself indicates already a significant difference between MM and QM potentials and the associated dynamics. For the sake of method comparison, we discuss here only trajectories that all arrive at an α-helix structure.

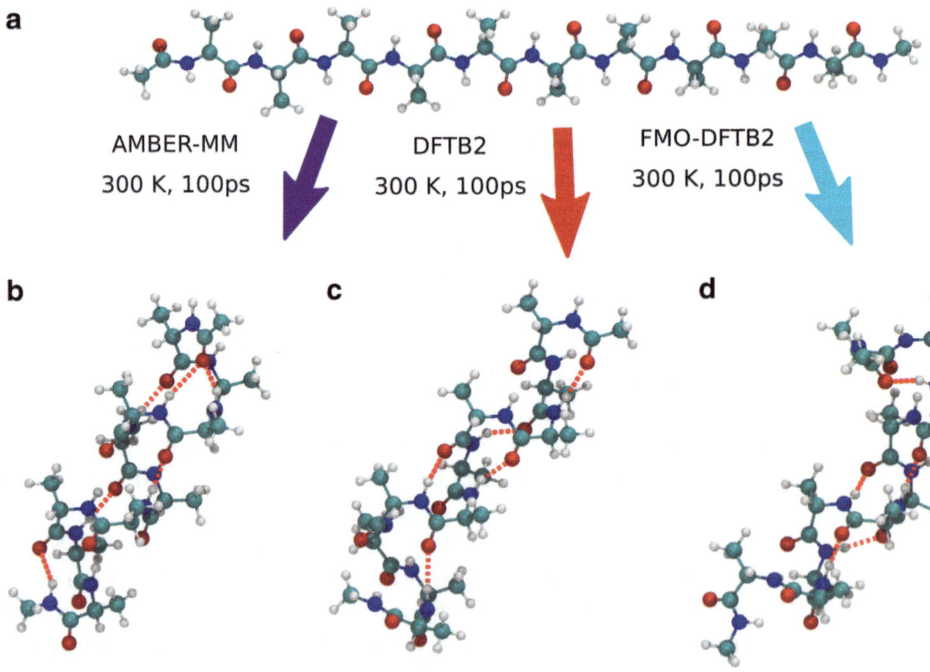

Fig. 2 (**a**) Initial geometry of ala10. (**b–d**) Final geometries of ala10 after 100 ps simulation under the NVT ensemble with the AMBER-MM, DFTB2, and FMO-DFTB2 potentials, respectively. Red dotted bands indicate hydrogen bonds

3.1 Phenomenological Description of Trajectories

Figure 3a displays the root mean square deviation (RMSD) of geometries from the initial extended, individually optimized geometries for the three trajectories over the full simulation period. The RMSD value for the AMBER-MM trajectory increases generally monotonously to a maximum value of ~7.5 Å during the first 50 ps and then remains steady with minor fluctuations of around ±0.5 Å. Both QM trajectories start out with a large geometry displacement and reach an RMSD value of >9 Å at about 7 ps before a "relaxation" occurs during the following 7 ps (14 ps, FMO-DFTB2) and 10 ps (17 ps, DFTB2). At about 30 ps, both QM trajectories reach an average RMSD value of ~7 Å but with rather large intermittent fluctuations of around ±0.5 Å. Visual inspection of the AMBER trajectories reveals that the helix formation starts from the N-terminus of the chain until the first 20 ps, followed by the completion of the full helix at about 40 ps, with the final hydrogen bond (H-bond) formed at the C-terminus. The process resembles that of a "zipper" mechanism where basically all H-bonds are "snapping" in place in an orderly manner. The peptide maintains its overall linear structure throughout the process, and hence no "wild" RMSD fluctuations are observed. Due to the limited backbone dynamics visible in AMBER MM simulations, we can expect minor entropic free energy contributions in these simulations. Figure 3b displays the number of H-bonds as a function of time for the three trajectories in integer intervals. The criterion for determining hydrogen bonding was that the X···H bond length $r_H < 3.5$ Å and the X-H···Y angle was $\alpha < 30°$. The acetyl and methyl amine caps are involved in the hydrogen bond formation network, and therefore a maximum of nine H-bonds can be sometimes observed. In both QM trajectories, on the other hand, the linear chain first bends into a "U" shape, only to extend again during at about 15 ps, and then individual H-bonds in a more

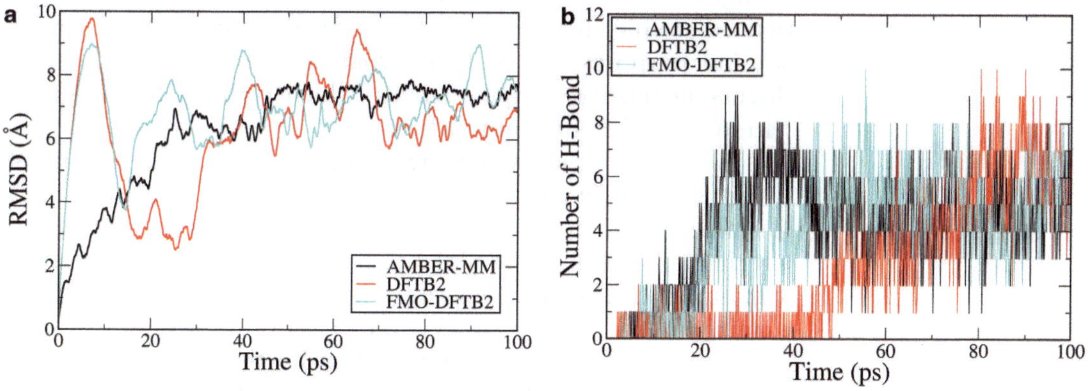

Fig. 3 (**a**) Root mean square deviation (RMSD) of geometries from initial linear optimized geometries in units of Å as a function of time. (**b**) Number of hydrogen bonds as a function of time for the three trajectories AMBER, DFTB2, and FMO-DFTB2

random succession are formed. The formation process is less orderly as in the AMBER trajectory, and intermittent major interruptions of the H-bond network can be observed. In Fig. 3b this can be seen as the peaks, both minimum and maximum number of H-bonds, reached by the DFTB2 and FMO-DFTB2 trajectories, from only one H-bond present to up to ten H-bonds. The structure remains volatile, congruent with the large fluctuations in the RMSD plots of Fig. 3a. Due to the larger backbone dynamics in comparison to AMBER MM simulations, we can expect larger entropic free energy contributions in these QM simulations. From all of the above observations, we surmise that the steep torsional potentials in AMBER (*see* Fig. 1c) "drove" the system toward reaching the α-helix in a much shorter time period as in the corresponding DFTB2 and FMO-DFTB2 trajectories. We note that DFTB is formulated within the framework of a two-center approximation for the electronic Hamiltonian and does not contain explicit four-center potential energy terms, as present in most MM force fields.

3.2 Analysis of Charge Evolution

The AMBER ff14SB force field contains point charge values for individual atom types, as in most common MM force fields. For instance, the O atom in alanine carries a charge of $-0.57\ e^-$, whereas the H atom attached to the N atom is positive with $+0.27\ e^-$. These charge values are never updated as the geometry changes during an MD simulation. Polarizable force fields do possess a mechanism for charge update, for instance, on the basis of the Drude oscillator model [27], but are less commonly used and computationally more expensive than traditional MM force fields. DFTB2 and FMO-DFTB2 obtain atomic charges from a Mulliken population analysis of the molecular orbitals, obtained at each geometry during a Born-Oppenheimer MD simulation, and therefore naturally exhibit charge fluctuation as a function of geometry changes. There is a large variety of wavefunction- and density-derived atomic charge schemes, and for the present brief demonstration purposes, we wish to avoid an overcomplication when discussing relative charge changes, which typically follow similar patterns in different charge schemes.

In order to understand the effect of charge polarization and dynamic charge fluctuations in the peptide folding QM trajectories, it is necessary to first study a model system for H-bond formation, which in the context of ala10 is best realized by an H-bonded alanine dimer, shown in Fig. 4a. The geometry of this dimer was obtained at the B3LYP/6-31G(d,p) level of theory, and we have performed Mulliken population analysis for the atoms in this complex, as well as for the atoms in corresponding isolated alanine molecules at infinite separation. The different Mulliken charges shown in Fig. 4b represent the changes in Mulliken atomic charges for the atoms in the hydrogen-bonded alanine dimer with respect

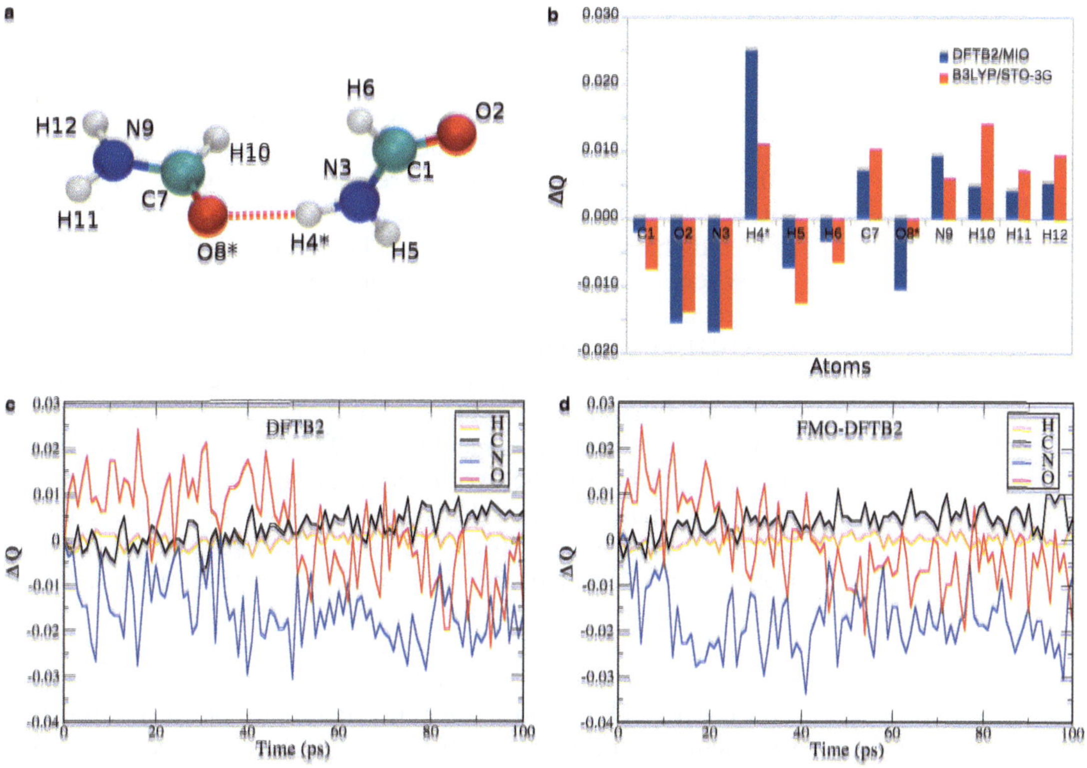

Fig. 4 (**a**) B3LYP/6-31G(d,p) optimized geometry of the hydrogen-bonded alanine dimer with atomic labels. (**b**) Different Mulliken atomic charges for the atoms in the hydrogen-bonded alanine dimer with respect to those in the isolated individual alanine molecules, illustrating the effect of charge polarization due to H-bond formation. (**c**) Different Mulliken atomic charges averaged over H, C, N, and O atoms of ala10 as a function of time for the DFTB2 trajectory. (**d**) Different Mulliken atomic charges averaged over H, C, N, and O atoms of ala10 as a function of time for the FMO-DFTB2 trajectory

to those in the isolated individual alanine molecules, illustrating the effect of charge polarization due to the formation of the H-bond. Atoms labeled with a star, namely, O8∗ and H4∗, are mediating the H-bond, which is marked by a red dotted line in Fig. 4a. In this system, both DFTB2 and B3LYP/STO-3G levels of theory[1] exhibit similar effects of charge polarization in direction and on the same order of magnitude. The donor H4∗ atom becomes more positive when forming the H-bond, whereas the acceptor O8∗ atom becomes more negative. We further observe a negative charge polarization on the donor N3 atom which carries the H atom, since charge polarization in molecular systems typically propagates in an oscillatory fashion and in concord with the electronegativities of the chemical elements. The magnitude of charge polarization is on the

[1] Since DFTB is a minimum basis set method, it is appropriate to compare with the minimum STO-3G basis set for the DFT calculation. Larger basis sets such as the 6-31G(d,p) exhibit an order of magnitude larger polarization effects, but generally follow the same direction.

order of ± 0.02 e^- only, which is only a few percent relative to the total Mulliken charge at both levels of theory. For instance, the total DFTB2 Mulliken charge on the oxygen atom in alanine is -0.49 e^-, closer to the AMBER oxygen charge value than predicted by B3LYP/STO-3G where the corresponding charge is only -0.24 e^-. Nevertheless, the aforementioned charge polarization pattern ($O\delta^- \cdots H\delta^+ – N\delta^-$) accurately reflects the traditionally expected charge fluctuation due to H-bond formation.

When analyzing the charge polarization and dynamic charge changes occurring in the ala10 folding dynamics of the DFTB2 and FMO-DFTB2 trajectories, it is inconvenient to analyze the charge evolution on individual atoms. Instead, we average different Mulliken atomic charges for the H, C, N, and O chemical elements of the peptide and plot this average different Mulliken atomic charges as a function of time for the two trajectories (*see* Fig. 4c, d). We recognize the negative average charge polarization on O atoms from the second half of the trajectories, for which we concluded the formation of an H-bonded, dynamic α-helix structure, as well as the negative charge polarization on the N atoms which carry the donor H atoms. Clearly, the average charge polarization observed in the two QM trajectories conforms with the traditional polarization picture ($O\delta^- \cdots H\delta^+ – N\delta^-$) in this monomolecular simulation of a neutral peptide and with the charge patterns observed in the H-bonded alanine dimer model system. We note that the relatively small scale of charge polarization in DFTB2 (as well as B3LYP/STO-3G minimal basis set calculations) has been discussed and addressed in the literature by way of chemical-potential equalization (CPE) in the third-order DFTB3 method [28].

4 Conclusions and Outlook

The comparison between MM and QM folding simulations of a small model peptide in gas phase revealed the following noteworthy observations:

1. AMBER MM/MD simulations quickly and perfectly fold ala10 into α-helix structure, following a "zipper"-like mechanism driven by the torsional potentials, large repulsive barriers in PES.
2. DFTB2 and FMO-DFTB2 MD simulations for 100 ps do not readily result in the formation of an α-helix structure even though thermodynamically the DFTB2/mio and FMO-DFTB2/mio methods predict a lower energy over the β-turn or extended structure.

3. DFTB2 and FMO-DFTB2 MD simulations reveal that atomic charges are affected by H-bond formation upon folding, in a manner anticipated from traditional charge polarization concepts.

Concerning the rapid and efficient folding observed in the AMBER MM/MD simulations, we note that the torsional potentials of force fields are often characterized by strongly repulsive torsional potentials, as seen, for instance, in Fig. 1c. It is not so much the appropriate potential energy well that drives the dynamics, but the barriers that separate the different minima (-α-helix vs. β-turn vs. extended form). Potential energy surfaces from QM methods, including ab initio MP2 and DFTB, show much lower barriers and more moderate landscapes. It is therefore understandable that in folding dynamics, MM simulations "find" the correct minimum faster than the QM-based simulations. In the gas phase, one can assume that the QM-based, moderate landscape represents the correct physics. The free energy contributions from protein dynamics simulations will be strongly affected by the differences in barriers between α-helix and β-sheet areas, and MM force field developers working on the predictive methodologies for protein binding free energies need to be aware of these traditionally overlooked areas of the MM potential energy landscapes.

Charge polarization and dynamic charge fluctuations in the DFTB2 and FMO-DFTB2 folding simulations were found to be relatively small, on the order of a few percent relative to the total atomic charges, following previous observations [28]. We therefore expect that the DFTB2 and FMO-DFTB2 folding dynamics presented here are largely determined by the aforementioned torsional potentials as well as by Coulombic stabilization of the H-bonds. At this stage it is difficult to assess whether a fixed atomic charge scheme as employed in most MM force fields is appropriate, given larger charge polarization effects in QM methods that employ larger than minimum basis sets [28]. We further recall the well-known principle in force field parameterization that a modification of charge should go hand in hand with changes in torsional potentials [29]. If charge fluctuations become larger, as can be expected when charged residues are present, fixed MM charge schemes might steer the folding simulation to unphysical dynamics. This might also happen in the simulation of protein-protein interactions, as in these systems small charge fluctuations can quickly add up due to the large size of the interaction interface. For such systems, relevant for the personalized neoantigen cancer treatments, it will be imperative to employ the long-range corrected (LC) LC-DFTB2 and FMO-LC-DFTB2 methods [22] which are capable of dealing with zwitterionic charged species and to explore differential neural network corrections to address deficiencies of the DFTB methods related to their torsional potential energies and use of a minimal basis set [30]. Further, more detailed simulations along these lines are necessary in order for approximate QM MD simulations to inform MM developers on the correct adjustment of their potentials for high-throughput screening in the search of personalized neoantigens.

Acknowledgments

S.I. acknowledges support by the Laboratory Directed Research and Development (LDRD) Program of Oak Ridge National Laboratory (ORNL). ORNL is managed by UT-Battelle LLC, for DOE under Contract No. DE-AC05-00OR22725. M.H.E. acknowledges support from the DOE Science Undergraduate Laboratory Internship (SULI) program. V.Q.V. acknowledges support from an Energy Science and Engineering Fellowship of the Bredesen Center for Interdisciplinary Research and Graduate Education at the University of Tennessee, Knoxville. M.T. acknowledges support from ORNL's summer research appointment program for faculty of Historically Black Colleges and Universities (HBCUs) and other Minority Educational Institutions (MEIs). Calculations were performed at the High Performance Computer Center of Tennessee Technological University and at the Center for Nanophase Materials Sciences, which is a US Department of Energy Office of Science User Facility.

References

1. Keskin O, Tuncbag N, Gursoy A (2016) Predicting protein–protein interactions from the molecular to the proteome level. Chem Rev 116:4884–4909
2. Blundell TL, Burke DF, Chirgadze D, Dhanaraj V, Hyvönen M, Innis CA, Parisini E, Pellegrini L, Sayed M, Sibanda BL (2000) Protein-protein interactions in receptor activation and intracellular signalling. Biol Chem 381:955–959
3. Jackson HJ, Rafiq S, Brentjens RJ (2016) Driving CAR T-cells forward. Nat Rev Clin Oncol 13:370–383
4. Schumacher TN, Schreiber RD (2015) Neoantigens in cancer immunotherapy. Science 348:69–74
5. Yarchoan M, Johnson BA, Lutz ER, Laheru DA, Jaffee EM (2017) Targeting neoantigens to augment antitumour immunity. Nat Rev Cancer 17:209–222
6. Pierce BG, Hellmanm LM, Hossain M, Singh NK, Vander Kooi CW, Weng Z, Baker BM (2014) Computational design of the affinity and specificity of a therapeutic T cell receptor. PLoS Comput Biol 10:e1003478
7. Huang J, Rauscher S, Nawrocki G, Ran T, Feig M, de Groot BL, Grubmüller H, MacKerell AD Jr (2017) CHARMM36m: an improved force field for folded and intrinsically disordered proteins. Nat Methods 14:1–73
8. Maier JA, Martinez C, Kasavajhala K, Wickstrom L, Hauser KE, Simmerling C (2015) ff14SB: improving the accuracy of protein side chain and backbone parameters from ff99SB. J Chem Theor Comput 11:3696–3713
9. Berg JM, Tymoczko JL, Stryer L (2002) Quaternary structure: polypeptide chains can assemble into multisubunit structures. In: Freeman WH (ed) Biochemistry, 5th edn. Palgrave MacMillan, New York, NY
10. Sakae Y, Okamoto Y (2010) Controlling the secondary-structure-forming tendencies of proteins by a backbone torsion-energy term. Mol Simulat 36:138–158
11. Cerutti DS, Swope WC, Rice JE, Case DA (2014) ff14ipq: a self-consistent force field for condensed-phase simulations of proteins. J Chem Theory Comput 10:4515–4534

12. Gaus M, Cui Q, Elstner M (2011) DFTB3: extension of the self-consistent-charge density-functional tight-binding method (SCC-DFTB). J Chem Theory Comput 7:931–948
13. Rappe AK, Casewit CJ, Colwell KS, Goddard WA III, Skiff WM (1992) UFF, a full periodic table force field for molecular mechanics and molecular dynamics simulations. J Am Chem Soc 114:10024–10035
14. Lovell SC, Davis IW, Arendall WB, De Bakker PIW, Word JM, Prisant MG, Richardson JS, Richardson DC (2003) Structure validation by Cα geometry: ϕ,ψ and Cβ deviation. Proteins Struct Funct Genet 50:437–450
15. Elstner M, Porezag D, Jungnickel G, Elsner J, Haugk M, Frauenheim T, Suhai S, Seifert G (1998) Self-consistent-charge density-functional tight-binding method for simulations of complex materials properties. Phys Rev B 58:7260–7268
16. Cui Q, Elstner M (2014) Density functional tight binding: values of semi-empirical methods in an ab initio era. Phys Chem Chem Phys 16:14368–14377
17. Mironov V, Alexeev Y, Mulligan VK, Fedorov DG (2019) A systematic study of minima in alanine dipeptide. J Comput Chem 40:297–309
18. Irle S, Zheng G, Wang Z, Morokuma K (2006) The C_{60} formation puzzle "solved": QM/MD simulations reveal the shrinking hot giant road of the dynamic fullerene self-assembly mechanism. J Phys Chem B 110:14531–14545
19. Irle S, Page AJ, Saha B, Wang Y, Chandrakumar KRS, Nishimoto Y, Qian HJ, Morokuma K (2012) Atomistic mechanism of carbon nanostructure self-assembly as predicted by nonequilibrium QM/MD simulations. In: Leszczynski J, Shukla MK (eds) Practical aspects of computational chemistry II: an overview of the last two decades and current trends. Springer-European Academy of Sciences, New York, NY
20. Lutsker V, Aradi B, Niehaus TA (2015) Implementation and benchmark of a long-range corrected functional in the density functional based tight-binding method. J Chem Phys 143:184107–184120
21. Nishimoto Y, Fedorov DG, Irle S (2014) Density-functional tight-binding combined with the fragment molecular orbital method. J Chem Theory Comput 10:4801–4812
22. Vuong VQ, Nishimoto Y, Fedorov DG, Sumpter BG, Niehaus TA, Irle S (2019) The fragment molecular orbital method based on long-range corrected density-functional tight-binding. J Chem Theory Comput 15:3008–3020
23. Case DA et al (2018) AMBER. University of California, San Francisco, CA, p 2018
24. Gordon MS, Schmidt MS (2005) Advances in electronic structure theory: GAMESS a decade later. In: Dykstra CE, Frenking G, Kim KS, Scuseria GE (eds) Theory and applications of computational chemistry: the first forty years. Elsevier, Amsterdam
25. The mio-1-1 parameter set for DFTB2, http://www.dftb.org. Accessed 23 Sep 2019
26. Elstner M (2006) The SCC-DFTB method and its application to biological systems. Theor Chem Accounts 116:316–325
27. PEM Lopes, Huang J, Shim J, Luo Y, Li H, Roux B, Mackerell Jr AD (2016) Polarizable force field for peptides and proteins based on the classical drude oscillator J Chem Theory Comput 9:12
28. Christensen AS, Elstner M, Qui C (2013) Improving intermolecular interactions in DFTB3 using extended polarization from chemical-potential equalization. J Chem Phys 143:084123
29. Maier JA, Martinez C, Kasavajhala K, Wickstrom L, Hauser KE, Simmerling C (2013) ff14SB: improving the accuracy of protein side chain and backbone parameters from ff99SB. J Chem Theory Comput 9:5430–5449
30. Zhu J, Vuong VQ, Sumpter BG, Irle S (2019) Artificial neural network correction for density-functional tight-binding molecular dynamics simulations. MRS Commun. https://doi.org/10.1557/mrc.2019.80

Chapter 11

Analyzing GPCR-Ligand Interactions with the Fragment Molecular Orbital (FMO) Method

Alexander Heifetz, Tim James, Michelle Southey, Inaki Morao, Dmitri G. Fedorov, Mike J. Bodkin, and Andrea Townsend-Nicholson

Abstract

G-protein-coupled receptors (GPCRs) have enormous physiological and biomedical importance, and therefore it is not surprising that they are the targets of many prescribed drugs. Further progress in GPCR drug discovery is highly dependent on the availability of protein structural information. However, the ability of X-ray crystallography to guide the drug discovery process for GPCR targets is limited by the availability of accurate tools to explore receptor-ligand interactions. Visual inspection and molecular mechanics approaches cannot explain the full complexity of molecular interactions. Quantum mechanics (QM) approaches are often too computationally expensive to be of practical use in time-sensitive situations, but the fragment molecular orbital (FMO) method offers an excellent solution that combines accuracy, speed, and the ability to reveal key interactions that would otherwise be hard to detect. Integration of GPCR crystallography or homology modelling with FMO reveals atomistic details of the individual contributions of each residue and water molecule toward ligand binding, including an analysis of their chemical nature. Such information is essential for an efficient structure-based drug design (SBDD) process. In this chapter, we describe how to use FMO in the characterization of GPCR-ligand interactions.

Key words G-protein-coupled receptors (GPCR), Chemical interactions, Pair interaction energy (PIE), Drugs, Receptor, Computational, Modelling, Quantum mechanics (QM), Fragment molecular orbital method (FMO), Computer-aided drug design (CADD), Structure-based drug design (SBDD), General atomic and molecular electronic structure system (GAMESS), Pair interaction energy decomposition analysis (PIEDA)

1 Introduction

1. G-protein-coupled receptor-ligand interactions are fundamental to almost all processes occurring in living organisms. Not surprisingly, therefore, 34% of all drugs approved by the FDA (US Food and Drug Administration) act on this group of proteins [1]. What is surprising is that, collectively, these drugs only target around 13% of the 800 known GPCRs

Alexander Heifetz (ed.), *Quantum Mechanics in Drug Discovery*, Methods in Molecular Biology, vol. 2114, https://doi.org/10.1007/978-1-0716-0282-9_11,

[2]. Thus, there is huge potential in terms of the number of new therapeutic targets within this family of receptors.

2. Further progress in drug discovery for GPCRs is highly dependent upon gaining an in-depth understanding of the structure-activity relationships (SAR) and the individual interactions between the receptor and a small molecule (drug candidate) [2–5]. The efficiency and cost-effectiveness of the drug-design process can be increased by the availability of structural data regarding the target protein and by the reliability of the computational tools used for data exploration [3–5]. However, even with the crystal structure in hand, "visual inspection" and the force field-based molecular mechanics (MM) calculations often used for the rationalization of ligand-protein potency cannot always explain the full complexity of the molecular interactions [6].

3. There is increasing evidence [6–9] that there are a number of non-intuitive interactions, such as CH/π [10, 11], halogen/π [12], cation/π [13], non-classical H-bonds [14], and others, which play important roles in protein-ligand binding but that are not sufficiently parameterized in most popular force fields (FF) [8]. Quantum mechanical methods have been employed to improve the reliability of the exploration of protein-ligand interactions [15, 16]. Historically, the application of high-level ab initio QM calculations was limited to molecular systems consisting of a small number of atoms, usually low molecular weight organic molecules. Recent and continuing advances in computer science have enabled the method to be applied to much larger biological molecules such as GPCRs [17].

4. The fragment molecular orbital method [11, 16, 18] offers a considerable increase in computational speed over traditional QM methods [19]. One of the key advantages of the FMO approach is that the output from these calculations includes a list of the interactions formed between the ligand and the receptor along with a chemically intuitive breakdown of these interactions [17]. Such information is essential for medicinal chemists to be able to adopt a rational approach to the modification of lead compounds in order to enhance favorable interactions.

5. FMO works by partitioning a system into smaller fragments (Fig. 1). For example, in a receptor, each residue can be represented as a fragment. Similarly, the ligand can be represented by a single or multiple fragments, as necessary. By performing QM calculations on the fragments, one is able to attain a high level of accuracy with increased efficiency.

6. As illustrated in Fig. 1, the pair interaction energy (PIE) between any two fragments calculated by FMO is the sum of

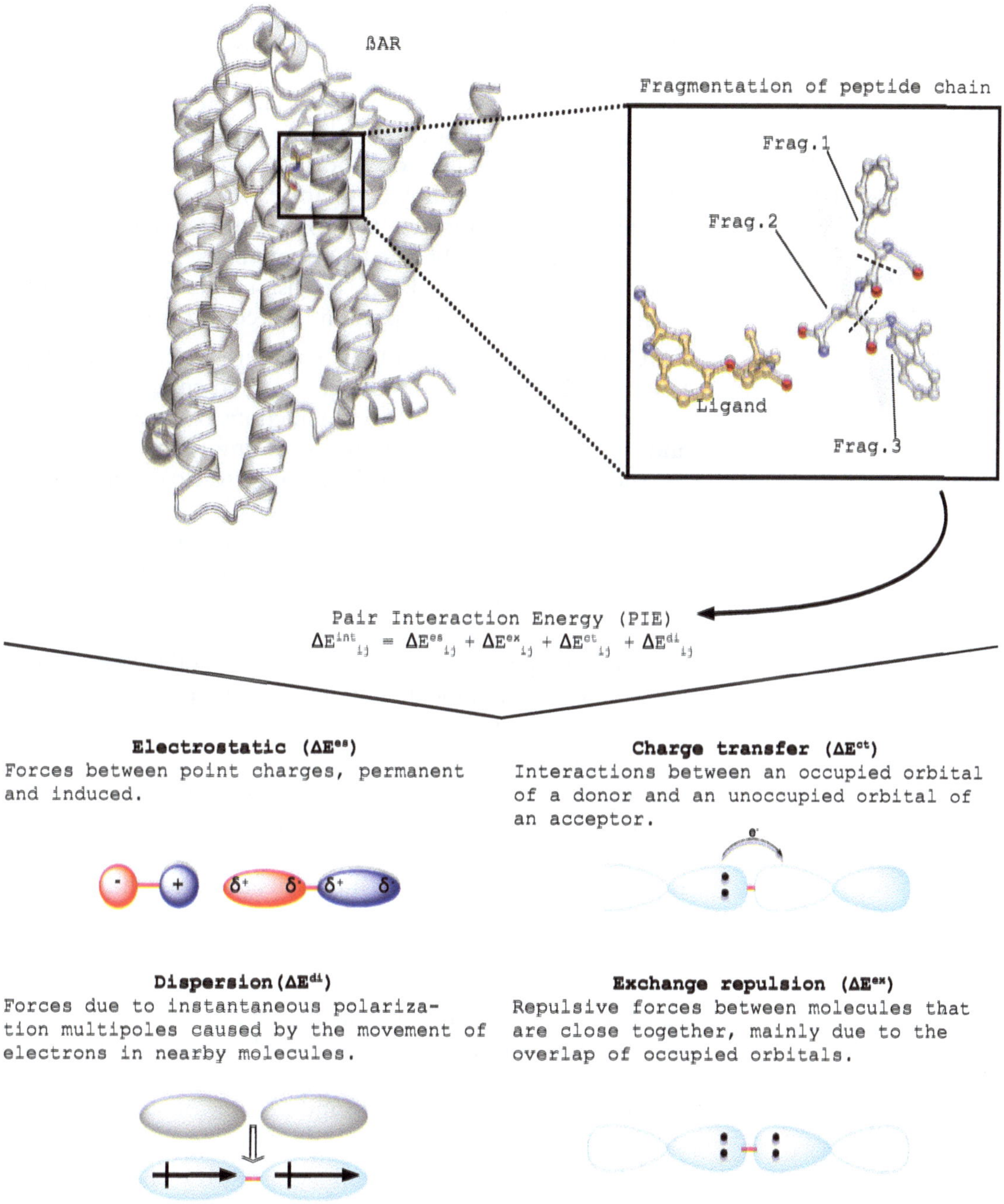

Fig. 1 Workflow for FMO calculations and details on each of the PIE terms that are computed [17]. The electrostatic component arises from the Coulomb interaction between polarized charge distributions of fragments. The exchange repulsion term is derived from the interaction between fragments situated in close proximity and is always repulsive; it is due to the Pauli repulsion and is related to the overlap of two occupied orbitals. The charge transfer term arises from the interaction between occupied orbitals of a donor and unoccupied orbitals of an acceptor. The dispersion arises as the interaction between instantaneous dipole moments of two fragments; it is hydrophobic (nonpolar) in nature and is obtained in PIEDA from the correlation energy of electrons

four energy terms—electrostatic, exchange repulsion, charge transfer, and dispersion—and is calculated using pair interaction energy decomposition analysis (PIEDA) [20]. The electrostatic and charge transfer terms are predominant in salt bridge, hydrogen bond, and polar interactions, while the dispersion term generally corresponds to interactions that are predominantly hydrophobic in nature. The role of hydrophobic interactions is integral to biomolecular recognition, but there is still no reliable non-QM predictive method for its quantification [6]. Exchange repulsion is a high-level QM term that quantifies the repulsion between electrons [17].

7. FMO is highly useful tool for structure-based drug design [11, 21, 22], as it provides an accurate and comprehensive list of strong, weak, and repulsive interactions between the ligand and its surrounding residues. Such information is highly instructive in terms of modifications, scaffold replacement (scaffold hopping), linking (specifically in the case of fragment-based drug discovery), or extension of chemical moieties to form stronger or new interactions with the protein or alternatively to remove repulsions. FMO can also be applied in the analysis of ligand-water-protein networks to distinguish between energetically favorable and unfavorable water molecules, thus enabling the design of ligands that can interact with or displace certain waters.
8. Application of the FMO method in the hit-to-lead and lead optimization stages of drug discovery is a highly valuable approach for the design, evaluation, and filtering of targets for synthesis as it can significantly decrease the effort and cost of chemical synthesis (for more details, *see* Chapter 3 of this book).
9. As previously demonstrated [23], significant correlation between receptor-ligand affinities and FMO energy terms [24] indicates that the latter can be efficiently used as descriptors in QSAR modelling to predict the binding affinities of new molecules. FMO has been successfully applied in the discovery of novel Hsp90 inhibitors [21] and in many of our confidential drug discovery programs.
10. Reported examples [24] comparing FMO and MM have shown that the FMO method clearly outperformed force field-based scoring functions and demonstrated a high correlation with experimentally measured values of protein-ligand affinity [23]. The key differentiator between the FMO and MM methods is that the former accounts for polarization and charge transfer effects [11, 25]. The description of electrostatics in most force fields is based on static charges that neglect polarization, and in systems such as proteins, this is an

approximation to the actual state. In addition, van der Waals forces, despite being generally well parameterized on average, are not capable of detecting the directional nature of the dispersion terms involving halogens [26].

11. In this chapter, we describe a number of examples of the application of FMO to the exploration of receptor-ligand interactions.

2 Methods

1. Several QM packages, including GAMESS [27], ABINIT-MP [28], and PAICS [29], contain modules for performing FMO calculations. A typical FMO-MP2 calculation on a ligand-receptor complex takes approximately 4 h on a 36-CPU cluster to complete, significantly faster than the weeks to a month (or more) for traditional QM approaches that have been used for estimating binding energies. Recent developments in methodology and integration with the density-functional tight-binding (DFTB) approach [30] have further increased the throughput.
2. In FMO calculations, a biological system is partitioned into fragments (Fig. 1) [11, 16]. Each residue is typically characterized as a fragment, and the interaction energies reported herein correspond to actual amino acid residues as opposed to residue fragments. The ligand can be represented as a whole or subdivided; some ligands are very large, and dividing such molecules into multiple fragments has the benefit of both reducing the computational cost and providing a more detailed analysis. An in-depth description of the FMO strategy and methodology can be found in previously published reports [11, 16, 20], including detailed mathematical formulae that are beyond the scope of this manuscript.
3. An FMO calculation consists of the following steps: (a) fragmentation (i.e., assigning atoms in a system to a fragment); (b) fragment self-consistent field (SCF) calculations in their embedding polarizable potential, so that fragments mutually polarize each other in a self-consistent fashion whereby intra-fragment charge transfer and other quantum effects are accounted for; (c) fragment pair SCF calculations, bringing in inter-fragment charge transfer; and (d) total property (energy, gradient, etc.) evaluation. By performing QM computations on fragments, one can achieve high efficiency, often resulting in linear scaling. The FMO method has been parallelized for PC clusters [19]. By default we use the MP2 method (second-order Møller-Plesset perturbation theory [31]) with the 6-31G* basis set. This basis set is often considered a good compromise between speed and accuracy [24].

4. As illustrated in Fig. 1, the pair interaction energy ($\Delta E_{ij}^{\text{int}}$, or PIE) between fragments i and j is a sum of four terms: electrostatics ($\Delta E_{ij}^{\text{es}}$), exchange repulsion ($\Delta E_{ij}^{\text{ex}}$), charge transfer ($\Delta E_{ij}^{\text{ct}}$), and dispersion ($\Delta E_{ij}^{\text{di}}$):

$$\Delta E_{ij}^{\text{int}} = \Delta E_{ij}^{\text{es}} + \Delta E_{ij}^{\text{ex}} + \Delta E_{ij}^{\text{ct}} + \Delta E_{ij}^{\text{di}} \tag{1}$$

5. The PIE is not a difference between the energies of the receptor-ligand complex and the sum of the "free" protein and ligand, but rather represents the "strength" of the interaction between the ligand and protein residues in the complex. The Δ denotes the differences in total QM energy of a fragment pair ij and two individual fragments i and j, both computed in the receptor-ligand complex. In the present work, fragment i is the ligand and the other n fragments are receptor residues and water molecules. In the equations below, there is no self-interaction (the sums exclude $j = i$). The total PIE calculated by the FMO method describes the stability of the receptor-ligand complex. This stability correlates with, but is not the same as, the binding energy [24]. The difference lies in the polarization factors—the ligand is polarized by the protein and vice versa [19]. Based on previous reports [11], we recommend that only interactions with an absolute PIE greater than or equal to 3.0 kcal/mol should be considered as significant.

3 Notes

1. We originally applied the FMO method to investigate the interactions in 18 class A GPCR-ligand crystal structures, which represent different branches of the GPCR family and include both active and inactive states [23].
2. The purpose of this investigation was to explore the chemical nature of GPCR-ligand interactions and to test whether FMO can provide insights beyond the conventional "visual inspection" or MM-based assessments [16]. We were also interested in exploring which of the three potentially attractive PIE terms (electrostatics, charge transfer, and dispersion) is dominant in the receptor-ligand binding and if there is a consensus among the whole class A GPCR family.
3. To illustrate the typical outcome of an FMO calculation, we give five examples (Figs. 2, 3, 4, 5, and 6) taken from an earlier report [23]. The results are shown as panels where the left-hand figures (a) show 3D images of the receptor-ligand complexes. The carbon atoms of the ligands are shown in light orange, and those of the receptors are colored according to

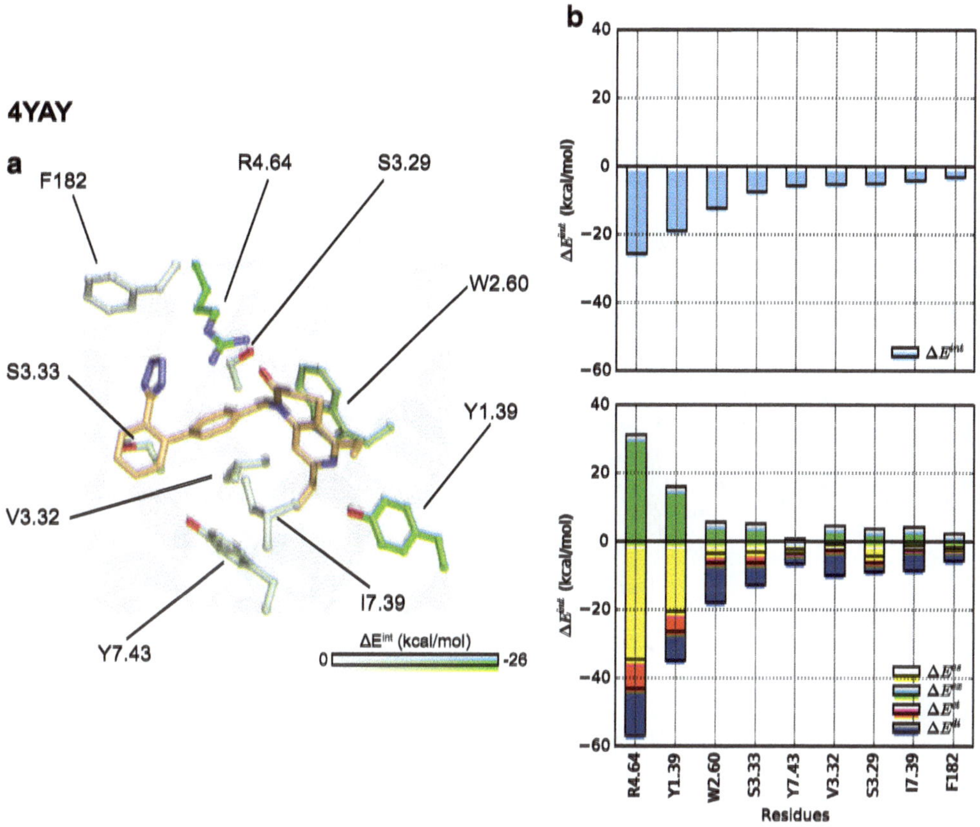

Fig. 2 Human angiotensin II type 1 receptor (AT$_1$R) in complex with ZD7155 (PDB entry 4YAY [32])—AT$_1$R is primarily involved in blood pressure regulation. Several antihypertensive drugs have been developed as AT$_1$R blockers. However, the structural basis for AT$_1$R ligand binding and function had previously been poorly understood. The recently developed method of serial femtosecond crystallography at an X-ray free-electron laser was successfully applied to determine the crystal structure of the human AT$_1$R-ZD7155 complex. FMO calculations detected three major interactions involved in ZD7155 binding, formed by the residues Tyr35$^{1.39}$, Trp84$^{2.60}$, and Arg167$^{4.64}$ [32]. The PIEDA analysis showed that the interactions formed by Tyr35$^{1.39}$ and Arg167$^{4.64}$ are mainly electrostatic albeit with a significant portion of dispersion. This balance of terms is characteristic of salt bridges or hydrogen bonding [11]. On the other hand, the interaction with Trp84$^{2.60}$ is mainly driven by dispersion, which fits more with π-stacking or hydrophobic forces [11]. These results are consistent with published SDM data [32], which showed a salt bridge with Arg167$^{4.64}$, hydrogen bond with Tyr35$^{1.39}$, and face-to-face π-stack with Trp84$^{2.60}$ which are critical for ZD7155 binding. The other six interactions formed by residues Ser105$^{3.29}$, Val108$^{3.32}$, Ser109$^{3.33}$, Phe182^{ECL2}, Ile288$^{7.39}$, and Tyr292$^{7.43}$ are also consistent with the published analysis of this crystal structure

the PIE values calculated by FMO. Nitrogen atoms are shown in blue, oxygen in red, sulfur in yellow, and chlorine in light green. The right-hand plots (b) show the sorted PIE values of the most strongly interacting residues on the top and the PIEDA of these key interactions on the bottom. PIE terms, electrostatics, dispersion, charge transfer, and exchange repulsion, are color-coded yellow, blue, red, and green, respectively.

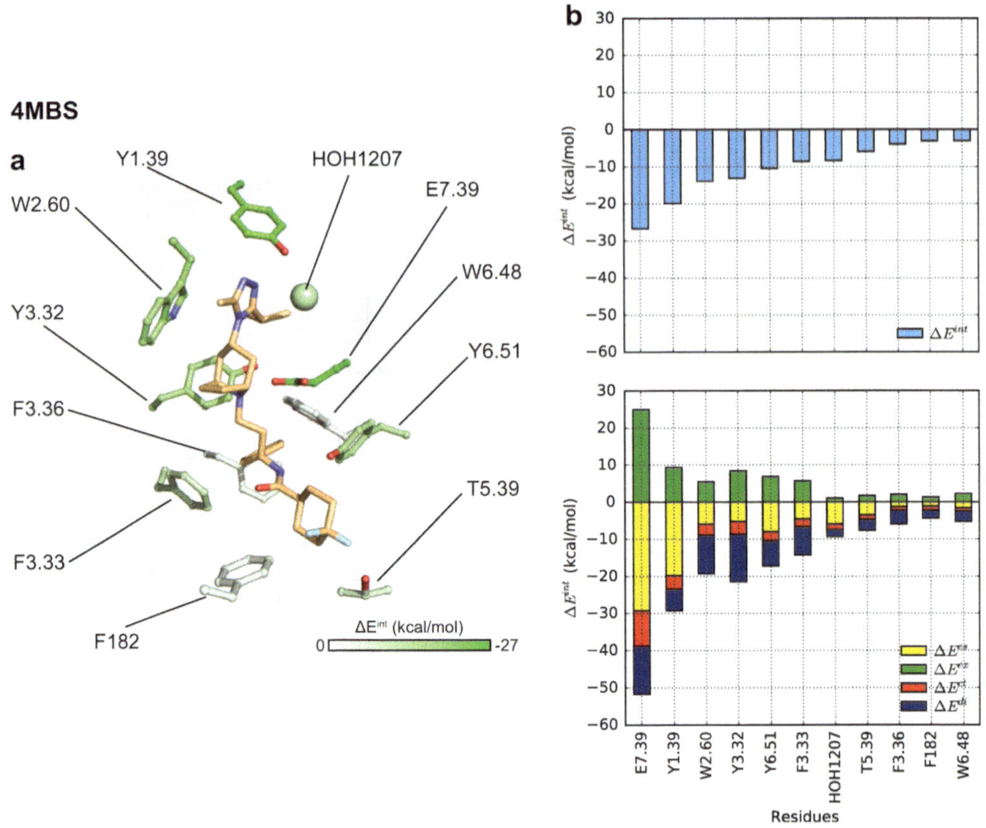

Fig. 3 Human chemokine CCR_5 receptor in complex with maraviroc (PDB entry 4MBS [33])—FMO calculations detected 11 significant interactions with ten residues—Tyr37$^{1.39}$, Trp86$^{2.60}$, Tyr108$^{3.32}$, Phe109$^{3.33}$, Phe112$^{3.36}$, Phe182^{ECL2}, Thr195$^{5.39}$, Trp248$^{6.48}$, Tyr251$^{6.51}$, and Glu283$^{7.39}$—and one water molecule, HOH1207. These interactions were consistent with published analysis of the crystal structure [4, 33]. It is interesting to note that, in contrast to other class A GPCRs, in this receptor the salt bridge with the ligand is formed by the negatively charged residue located at position 7.39 and not at 3.32. While all of these interactions were suggested in the analysis of the crystal structure, additional interactions with residues Gln194$^{5.38}$, Ile198$^{5.42}$, Thr2596$^{.59}$, and Met287$^{7.43}$, postulated from published reports, were also characterized as significant by FMO

4. PIEDA calculations across all 18 GPCR-ligand systems showed [23] the utility of FMO and highlighted the importance of hydrophobic interactions, non-classical hydrogen bonds [14], interactions with the backbone atoms, and water molecules for GPCR-ligand binding. According to this analysis, electrostatic and hydrophobic contributions to the ligand binding are approximately equal across all systems, on average [23].

5. We also observed [23] general trends in ligand binding for different proteins belonging to the same family. For example, the residues in positions 3.32, 3.33, 6.48, 6.51, 6.52, 7.39, and 7.43 (located on TM3, TM6, and TM7) make considerable contributions to receptor-ligand binding and are quite

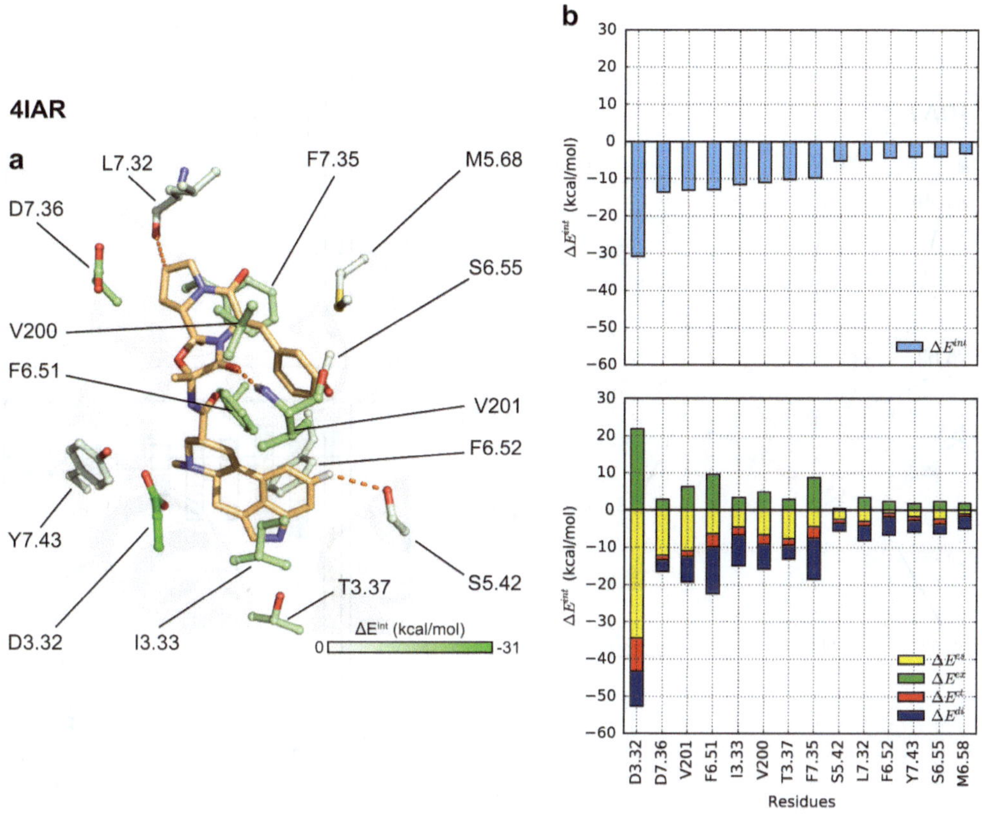

Fig. 4 Human serotonin 5-HT$_{1B}$ receptor in complex with ergotamine (PDB entry 4IAR [34])—FMO calculations highlighted 13 significant interactions with residues Asp129$^{3.32}$, Ile130$^{3.33}$, Thr134$^{3.37}$, Val200^{ECL2}, Val201^{ECL2}, Ser212$^{5.42}$, Phe330$^{6.51}$, Phe331$^{6.52}$, Ser334$^{6.55}$, Met337$^{6.58}$, Leu348$^{7.32}$, Phe351$^{7.35}$, Asp352$^{7.36}$, and Tyr359$^{7.43}$. The vast majority of these interactions were consistent with published analysis of the crystal structure [35]. However, three non-classical hydrogen bonds [14] formed by the ligand with Ser212$^{5.42}$, Phe330$^{6.51}$, and Leu348$^{7.32}$ (backbone carbonyl) and two hydroxy-interactions [36] with residues Ser334$^{6.55}$ and Phe351$^{7.35}$ were newly detected by FMO. FMO also emphasized the two interactions of Asp352$^{7.36}$, which forms both strong electrostatic attraction with ergotamine and, via its backbone atoms, a non-classical hydrogen bond with the ligand (marked as a dashed yellow line). An additional interesting interaction detected by FMO is the classical hydrogen bond formed between the backbone of Val201^{ECL2} and the ligand (also marked as a dashed yellow line)

conserved (>70%, in the 18 tested cases). The residues in positions 3.33, 6.48, 6.51, 6.52, and 7.43 form interactions of a mainly hydrophobic nature, while the residues in positions 3.32 and 7.39 form mainly electrostatic interactions. Published mutations in these positions frequently affect selectivity [40] and ligand binding affinity [41]. This observation is also consistent with a previous report [40] that residues in these positions highly frequently make contact with diverse ligands across nearly all class A GPCRs. Residues in the other positions are less frequently involved in interactions only with specific ligands [40]. These key positions in TM3, TM6, and TM7 form a

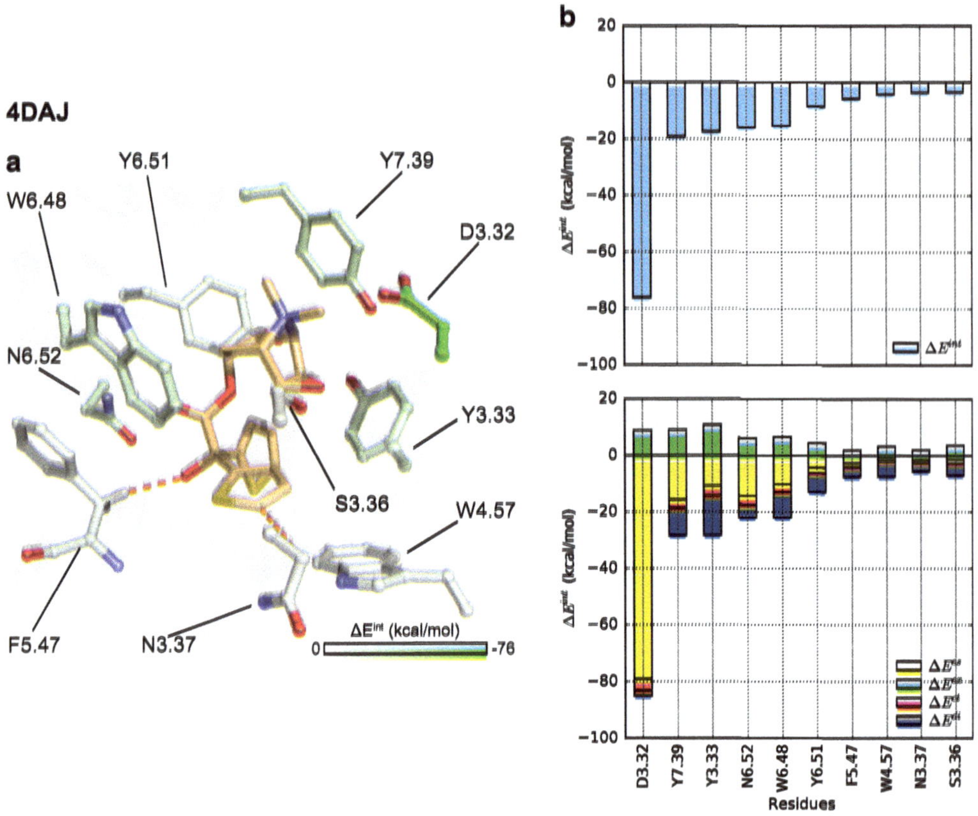

Fig. 5 Rat muscarinic M_3 muscarinic acetylcholine receptor in complex with tiotropium (PDB entry 4DAJ [37])—FMO calculations detected ten significant interactions with residues Asp147$^{3.32}$, Tyr148$^{3.33}$, Ser151$^{3.36}$, Asn152$^{3.37}$, Trp199$^{4.57}$, Phe239$^{5.47}$, Trp503$^{6.48}$, Tyr506$^{6.51}$, Asn507$^{6.52}$, and Tyr529$^{7.39}$. The majority of these interactions were suggested previously [4, 35, 38]. However, as in the M_2:QNB complex, the interactions with Asn152$^{3.37}$ and Phe239$^{5.47}$ were newly detected by FMO. The backbone nitrogen of Asn152$^{3.37}$ and a hydrogen from the methylene linker of Phe239$^{5.47}$ form hydrophobic and non-classical hydrogen bonds [14], respectively, with the ligand (marked as dashed yellow lines)

consensus core of the ligand-binding pocket with any variation in the amino acids occupying the topologically equivalent positions contributing to ligand specificity in different GPCRs. This is in spite of the observation that the amino acids present in these positions are not conserved as been previously reported for 18 tested systems [23]. According to PIEDA, the predominant contribution in the majority of these conserved positions is, in fact, hydrophobic [23].

6. Overall, we have demonstrated how the FMO method can provide new insight into the nature of key chemical interactions required for receptor-ligand recognition. We anticipate that this approach can be used to provide insight into many protein-ligand interactions and specifically to the design of new compounds.

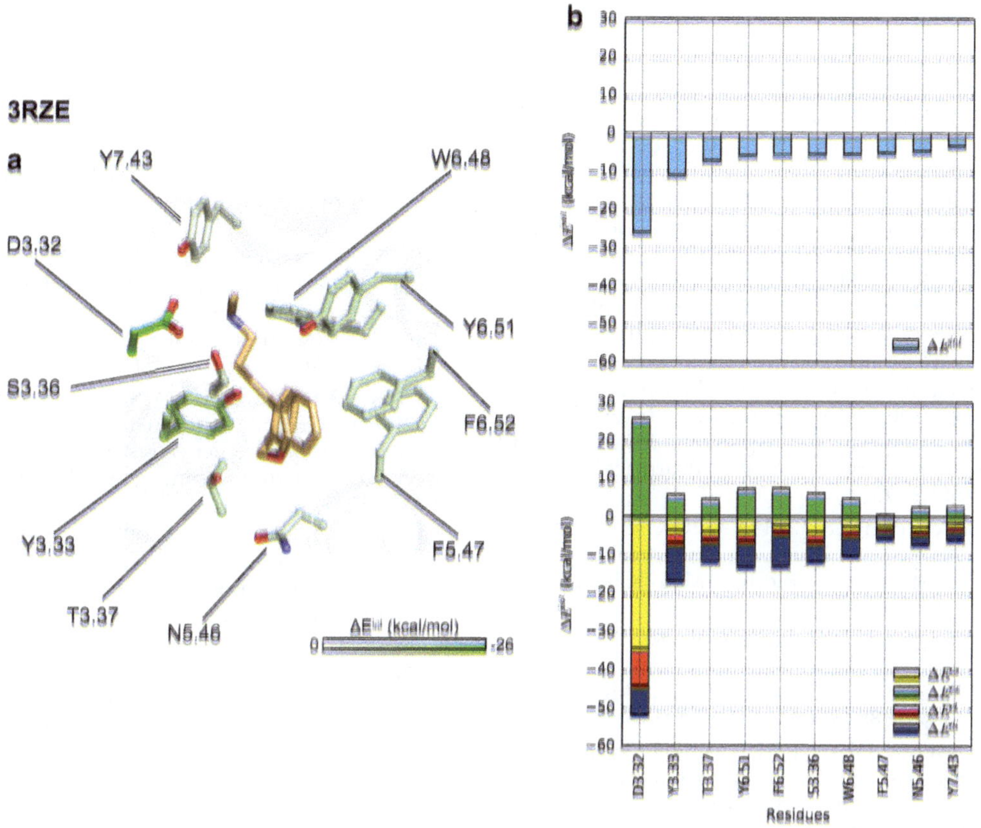

Fig. 6 Human histamine H_1 receptor in complex with doxepin (PDB entry 3RZE [39])—FMO calculations highlighted ten significant interactions with residues Asp107$^{3.32}$, Tyr108$^{3.33}$, Ser111$^{3.36}$, Thr112$^{3.37}$, Asn198$^{5.46}$, Phe199$^{5.47}$, Trp428$^{6.48}$, Tyr431$^{6.51}$, Phe432$^{6.52}$, and Tyr458$^{7.43}$. The majority of these interactions are consistent with previously reported analysis [39]. Several hydrophobic interactions postulated from a previous report [39] (with Ile115$^{3.40}$, Trp158$^{4.56}$, Thr194$^{5.43}$, and Phe435$^{6.55}$) were not detected as significant by FMO. The importance of hydrophobic interactions between doxepin and H_1, emphasized here by FMO, is consistent with published observations that the tricyclic moiety of doxepin adopts a deep binding position occupying a highly conserved hydrophobic pocket [4]

Acknowledgments

Some sections of this chapter are reprinted (adapted) with permission from the *Journal of Chemical Information and Modeling* 2016 Jan 25;56(1):159–72 [23]. Copyright (2019) American Chemical Society.

A.H. and A.T.N. would like to acknowledge the support of EU H2020 CompBioMed project (http://www.compbiomed.eu/, 675451).

References

1. Hauser AS, Attwood MM, Rask-Andersen M, Schioth HB, Gloriam DE (2017) Trends in GPCR drug discovery: new agents, targets and indications. Nat Rev Drug Discov 16:829–842
2. Heifetz A, Schertler GF, Seifert R, Tate CG, Sexton PM, Gurevich VV, Fourmy D, Cherezov V, Marshall FH, Storer RI, Moraes I, Tikhonova IG, Tautermann CS, Hunt P, Ceska T, Hodgson S, Bodkin MJ, Singh S, Law RJ, Biggin PC (2015) GPCR structure, function, drug discovery and crystallography: report from Academia-Industry International Conference (UK Royal Society) Chicheley Hall, 1–2 September 2014. Naunyn Schmiedebergs Arch Pharmacol 388 (8):883–903
3. Tautermann CS (2014) GPCR structures in drug design, emerging opportunities with new structures. Bioorg Med Chem Lett 24:4073–4079
4. Shonberg J, Kling RC, Gmeiner P, Lober S (2015) GPCR crystal structures: medicinal chemistry in the pocket. Bioorg Med Chem 23(14):3880–3906
5. Jazayeri A, Dias JM, Marshall FH (2015) From G protein-coupled receptor structure resolution to rational drug design. J Biol Chem 290:19489–19495
6. Bissantz C, Kuhn B, Stahl M (2010) A medicinal chemist's guide to molecular interactions. J Med Chem 53:5061–5084
7. Tong Y, Mei Y, Li YL, Ji CG, Zhang JZ (2010) Electrostatic polarization makes a substantial contribution to the free energy of avidin-biotin binding. J Am Chem Soc 132:5137–5142
8. Raha K, Peters MB, Wang B, Yu N, Wollacott AM, Westerhoff LM, Merz KM Jr (2007) The role of quantum mechanics in structure-based drug design. Drug Discov Today 12:725–731
9. Beratan DN, Liu C, Migliore A, Polizzi NF, Skourtis SS, Zhang P, Zhang Y (2015) Charge transfer in dynamical biosystems, or the treachery of (static) images. Acc Chem Res 48:474–481
10. Ozawa T, Okazaki K, Kitaura K (2011) CH/pi hydrogen bonds play a role in ligand recognition and equilibrium between active and inactive states of the beta2 adrenergic receptor: an ab initio fragment molecular orbital (FMO) study. Bioorg Med Chem 19:5231–5237
11. Fedorov DG, Nagata T, Kitaura K (2012) Exploring chemistry with the fragment molecular orbital method. Phys Chem Chem Phys 14:7562–7577
12. Lu Y-X, Zou J-W, Wang Y-H, Yu Q-S (2007) Substituent effects on noncovalent halogen/π interactions: Theoretical study. Int J Quantum Chem 107:1479–1486
13. Gallivan JP, Dougherty DA (1999) Cation-pi interactions in structural biology. Proc Natl Acad Sci U S A 96:9459–9464
14. Johnston RC, Cheong PH (2013) C-H...O non-classical hydrogen bonding in the stereomechanics of organic transformations: theory and recognition. Org Biomol Chem 11:5057–5064
15. Yu N, Li X, Cui G, Hayik SA, Merz KM 2nd (2006) Critical assessment of quantum mechanics based energy restraints in protein crystal structure refinement. Protein Sci 15:2773–2784
16. Fedorov DG, Kitaura K (2007) Extending the power of quantum chemistry to large systems with the fragment molecular orbital method. J Phys Chem A 111:6904–6914
17. Phipps MJ, Fox T, Tautermann CS, Skylaris CK (2015) Energy decomposition analysis approaches and their evaluation on prototypical protein-drug interaction patterns. Chem Soc Rev 44:3177–3211
18. Kitaura K, Ikeo E, Asada T, Nakano T, Uebayasi M (1999) Fragment molecular orbital method: an approximate computational method for large molecules. Chem Phys Lett 313:701–706
19. Alexeev Y, Mazanetz MP, Ichihara O, Fedorov DG (2012) GAMESS as a free quantum-mechanical platform for drug research. Curr Top Med Chem 12:2013–2033
20. Fedorov DG, Kitaura K (2007) Pair interaction energy decomposition analysis. J Comput Chem 28:222–237
21. Barker JJ, Barker O, Courtney SM, Gardiner M, Hesterkamp T, Ichihara O, Mather O, Montalbetti CA, Muller A, Varasi M, Whittaker M, Yarnold CJ (2010) Discovery of a novel Hsp90 inhibitor by fragment linking. ChemMedChem 5:1697–1700
22. Sawada T, Fedorov DG, Kitaura K (2010) Binding of influenza A virus hemagglutinin to the sialoside receptor is not controlled by the homotropic allosteric effect. J Phys Chem B 114:15700–15705
23. Heifetz A, Chudyk EI, Gleave L, Aldeghi M, Cherezov V, Fedorov DG, Biggin PC, Bodkin MJ (2016) The fragment molecular orbital method reveals new insight into the chemical nature of GPCR-ligand interactions. J Chem Inf Model 56:159–172

24. Mazanetz MP, Ichihara O, Law RJ, Whittaker M (2011) Prediction of cyclin-dependent kinase 2 inhibitor potency using the fragment molecular orbital method. J Cheminform 3:2
25. Fedorov DG, Kitaura K (2012) Energy decomposition analysis in solution based on the fragment molecular orbital method. J Phys Chem A 116:704–719
26. El Kerdawy A, Murray JS, Politzer P, Bleiziffer P, Hesselmann A, Gorling A, Clark T (2013) Directional noncovalent interactions: repulsion and dispersion. J Chem Theory Comput 9:2264–2275
27. Schmidt MW, Baldridge KK, Boatz JA, Elbert ST, Gordon MS, Jensen JH, Koseki S, Matsunaga N, Nguyen KA, Su S, Windus TL, Dupuis M, Montgomery JA Jr (1993) General atomic and molecular electronic structure system. J Comput Chem 14:1347–1363
28. Tanaka S, Mochizuki Y, Komeiji Y, Okiyama Y, Fukuzawa K (2014) Electron-correlated fragment-molecular-orbital calculations for biomolecular and nano systems. Phys Chem Chem Phys 16:10310–10344
29. Ishikawa T, Kuwata K (2012) RI-MP2 gradient calculation of large molecules using the fragment molecular orbital method. J Phys Chem Lett 3:375–379
30. Nishimoto Y, Fedorov DG (2017) Three-body expansion of the fragment molecular orbital method combined with density-functional tight-binding. J Comput Chem 38:406–418
31. Fedorov DG, Kitaura K (2004) Second order Moller-Plesset perturbation theory based upon the fragment molecular orbital method. J Chem Phys 121:2483–2490
32. Zhang H, Unal H, Gati C, Han GW, Liu W, Zatsepin NA, James D, Wang D, Nelson G, Weierstall U, Sawaya MR, Xu Q, Messerschmidt M, Williams GJ, Boutet S, Yefanov OM, White TA, Wang C, Ishchenko A, Tirupula KC, Desnoyer R, Coe J, Conrad CE, Fromme P, Stevens RC, Katritch V, Karnik SS, Cherezov V (2015) Structure of the angiotensin receptor revealed by serial femtosecond crystallography. Cell 161(4):833–844
33. Tan Q, Zhu Y, Li J, Chen Z, Han GW, Kufareva I, Li T, Ma L, Fenalti G, Li J, Zhang W, Xie X, Yang H, Jiang H, Cherezov V, Liu H, Stevens RC, Zhao Q, Wu B (2013) Structure of the CCR5 chemokine receptor-HIV entry inhibitor maraviroc complex. Science 341:1387–1390
34. Wang C, Jiang Y, Ma J, Wu H, Wacker D, Katritch V, Han GW, Liu W, Huang XP, Vardy E, McCorvy JD, Gao X, Zhou XE, Melcher K, Zhang C, Bai F, Yang H, Yang L, Jiang H, Roth BL, Cherezov V, Stevens RC, Xu HE (2013) Structural basis for molecular recognition at serotonin receptors. Science 340:610–614
35. Kruse AC, Hu J, Kobilka BK, Wess J (2014) Muscarinic acetylcholine receptor X-ray structures: potential implications for drug development. Curr Opin Pharmacol 16:24–30
36. Dougherty DA (2013) The cation-pi interaction. Acc Chem Res 46:885–893
37. Kruse AC, Hu J, Pan AC, Arlow DH, Rosenbaum DM, Rosemond E, Green HF, Liu T, Chae PS, Dror RO, Shaw DE, Weis WI, Wess J, Kobilka BK (2012) Structure and dynamics of the M3 muscarinic acetylcholine receptor. Nature 482:552–556
38. Tautermann CS, Kiechle T, Seeliger D, Diehl S, Wex E, Banholzer R, Gantner F, Pieper MP, Casarosa P (2013) Molecular basis for the long duration of action and kinetic selectivity of tiotropium for the muscarinic M3 receptor. J Med Chem 56:8746–8756
39. Shimamura T, Shiroishi M, Weyand S, Tsujimoto H, Winter G, Katritch V, Abagyan R, Cherezov V, Liu W, Han GW, Kobayashi T, Stevens RC, Iwata S (2011) Structure of the human histamine H1 receptor complex with doxepin. Nature 475:65–70
40. Venkatakrishnan AJ, Deupi X, Lebon G, Tate CG, Schertler GF, Babu MM (2013) Molecular signatures of G-protein-coupled receptors. Nature 494:185–194
41. Dore AS, Robertson N, Errey JC, Ng I, Hollenstein K, Tehan B, Hurrell E, Bennett K, Congreve M, Magnani F, Tate CG, Weir M, Marshall FH (2011) Structure of the adenosine A(2A) receptor in complex with ZM241385 and the xanthines XAC and caffeine. Structure 19:1283–1293

Chapter 12

Characterizing Rhodopsin-Arrestin Interactions with the Fragment Molecular Orbital (FMO) Method

Alexander Heifetz and Andrea Townsend-Nicholson

Abstract

Arrestin binding to G protein-coupled receptors (GPCRs) plays a vital role in receptor signaling. Recently, the crystal structure of rhodopsin bound to activated visual arrestin was resolved using XFEL (X-ray free electron laser). However, even with the crystal structure in hand, our ability to understand GPCR-arrestin binding is limited by the availability of accurate tools to explore receptor-arrestin interactions. We applied fragment molecular orbital (FMO) method to explore the interactions formed between the residues of rhodopsin and arrestin. FMO enables ab initio approaches to be applied to systems that conventional quantum mechanical (QM) methods would be too compute-expensive. The FMO calculations detected 35 significant interactions involved in rhodopsin-arrestin binding formed by 25 residues of rhodopsin and 28 residues of arrestin. Two major regions of interaction were identified: at the C-terminal tail of rhodopsin (D330-S343) and where the "finger loop" (G69-T79) of arrestin directly inserts into rhodopsin active core. Out of these 35 interactions, 23 were mainly electrostatic and 12 hydrophobic in nature.

Key words Arrestin, Rhodopsin, G protein-coupled receptors (GPCR), Chemical interactions, Pair interaction energy (PIE), Quantum mechanics (QM), Fragment molecular orbital method (FMO), Biased ligands, Signaling

1 Introduction

1. G protein-coupled receptors (GPCRs) are a large family of membrane receptors that mediates transmembrane signaling through G proteins and arrestins. Arrestins are a small family of proteins with four human isoforms [1] that play a key role in homologous desensitization of GPCRs, as well as regulating several other vital cellular signaling pathways [1–3]. Remarkably, only two arrestins regulate the signaling of >800 GPCRs [2]. Several G protein-independent pathways have been identified, with the β-arrestin-dependent signaling pathway the best characterized, to date [4].
2. GPCRs exist in multiple, distinct conformational states and form population ensembles comprised of several of these

Alexander Heifetz (ed.), *Quantum Mechanics in Drug Discovery*, Methods in Molecular Biology, vol. 2114, https://doi.org/10.1007/978-1-0716-0282-9_12,

different active and inactive states. Ligands tend to bind to one of these states, increasing its population, which in turn activates (or deactivates) a particular signaling pathway via an effector (e.g., G proteins or β-arrestins). Ligands that preferentially activate a specific GPCR signaling pathways are known as "biased ligands." Further study is required to determine how, at the atomic level, GPCRs adopt specific conformations that lead to a preference for a particular effector, whether a G protein or an arrestin.

3. Recently, the crystal structure of rhodopsin bound to activated visual arrestin has been solved using XFEL (X-ray free electron laser) (PDB code 5W0P) [5] (Fig. 1a). However, even with the crystal structure in hand, the "visual inspection" and force field-based molecular mechanics (MM) calculations often used for structural exploration cannot always explain the full complexity of the molecular interactions between GPCR and arrestin [6].

4. Fragment molecular orbital (FMO) method [7–9] readily brings the power of QM to molecular biochemical research and drug discovery. FMO provides insights into the chemical nature of interactions that are normally difficult to detect with non-QM methods. The FMO approach has a proven track record in the deep analysis of crystal or modelled GPCR structures and in characterizing the interactions between receptors and ligands [10]. FMO calculations are reasonably straightforward to set up and can be performed on relatively small PC clusters within an acceptable timeframe. FMO analysis can result in two considerable benefits: (a) complex QM theories are condensed into four simple and intuitively clear quantities, and (b) calculations become much faster than those performed using traditional QM approaches [11].

5. FMO works by partitioning a large system such as a protein into small fragments, for example, each residue in a protein can be represented as a fragment (Fig. 1a). By performing QM calculations on fragments, one can make the computational cost scaling with respect to system size nearly linear. One of the key features of the FMO approach is that it can provide a list of the interactions formed between the fragments of the system under study, yielding a chemically intuitive breakdown of these interactions (Fig. 1b) [12].

6. The pair interaction energy (PIE) between any two fragments calculated by FMO is the sum of four energy terms: electrostatic, exchange repulsion, charge transfer, and dispersion, provided by pair interaction energy decomposition analysis (PIEDA) [13] (see Fig. 1b). The electrostatic and charge transfer terms are predominant in salt bridge, hydrogen bond, and

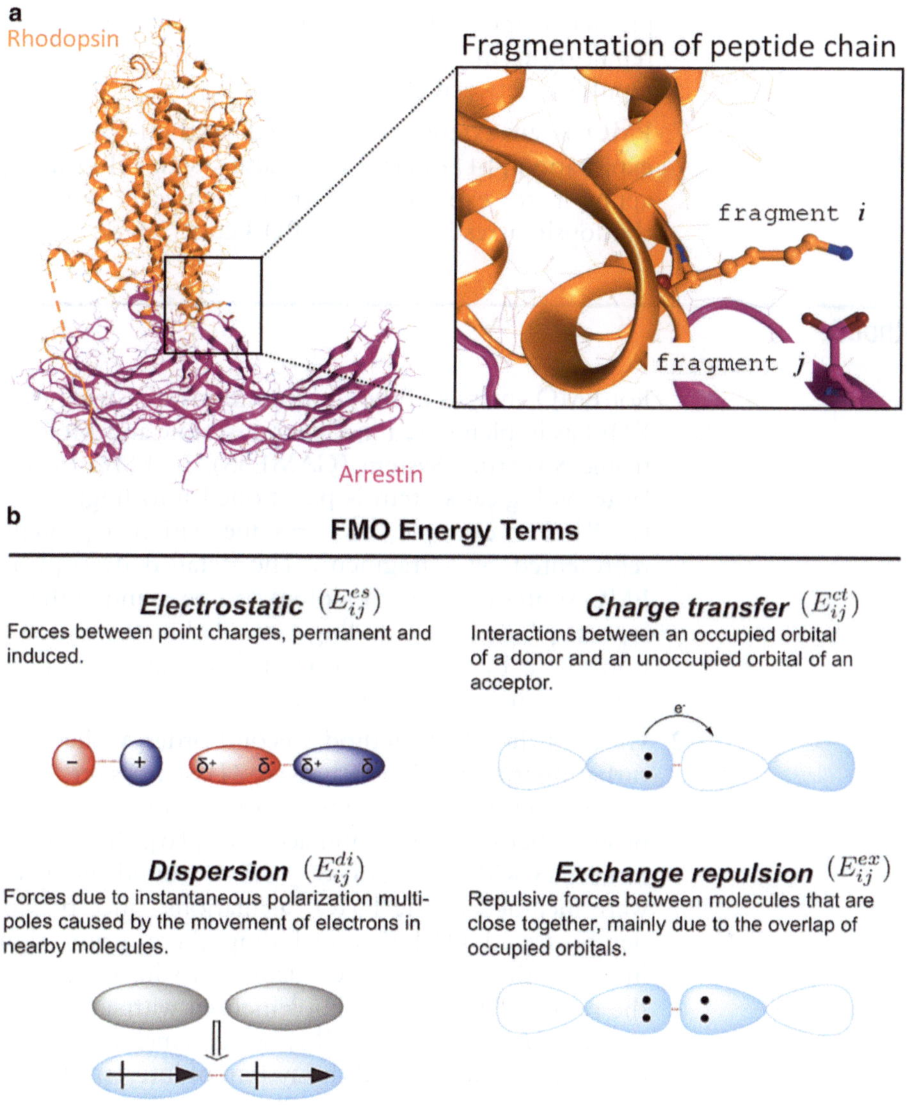

Fig. 1 (**a**) Crystal structure of the rhodopsin-arrestin complex (PDB code 5W0P) [5]. (**b**) Details of each of the PIE terms that are computed using the FMO workflow [12]. The electrostatic component arises from the Coulomb interaction between polarized charge distributions of fragments. The exchange repulsion term is derived from the interaction between fragments situated in close proximity and is always repulsive; it is due to the Pauli repulsion and is related to the overlap of two occupied orbitals. The charge transfer term arises from the interaction between occupied orbitals of a donor and unoccupied orbitals of an acceptor. The dispersion arises as the interaction between instantaneous dipole moments of two fragments; it is hydrophobic (nonpolar) in nature and is obtained in PIEDA from the correlation energy of electrons

polar interactions, while the dispersion term generally corresponds to interactions that are predominantly hydrophobic in nature. The role of hydrophobic interactions is integral for biomolecular recognition, but there is still no reliable predictive method for its quantification [6]. The exchange repulsion term is a high-level QM term which quantifies the repulsion between electrons [12].

7. FMO is an extensively validated method for the structural exploration [14] of large biological systems. In this report, we apply it to explore interactions between residues of the rhodopsin-arrestin complex (PDB code 5W0P, [5]).

2 Methods

1. For FMO analysis of the rhodopsin-arrestin structure, we used FMO as implemented in General Atomic and Molecular Electronic Structure System (GAMESS). In FMO calculations, a large biological system is partitioned into fragments (Fig. 1) [7, 8]. For example, each residue within a protein can be represented by a fragment. The detailed description of the FMO strategy and methodology can be found in the published reports [7, 8, 13], including a detailed mathematical formulation that is beyond the scope of this chapter and described in detail in Chapter 11 of this book.
2. We used the MP2 method (second-order Møller-Plesset perturbation theory [15]) with the 6-31G* basis set. This basis set is most commonly used and is often considered the best compromise between speed and accuracy [16]. Residues and water molecules within a radius of $\leq$4.5 Å around the ligand atoms were included in the FMO calculations, since previous work demonstrated [10] that including these atoms significantly increases the speed of the calculation without compromising the results. FMO can be implemented with a polarizable continuum solvent model (PCM), which ameliorates the effect of charged residues around the system of interest [17].
3. The residues of arrestin are numbered as for the rhodopsin-arrestin crystal structure (PDB code 5W0P) [5]. Arrestin elements are mapped based on the recent publication of Gurevich et al. [18] (Fig. 2).

3 Notes

1. Despite the extensive efforts described in the literature [1–3, 5, 18], the specific molecular interactions, in terms of interaction energy and chemical nature, between rhodopsin and arrestin have not yet been characterized. We applied the FMO method to investigate the interactions between rhodopsin and arrestin as evidenced in the crystal structure 5W0P [5].

Fig. 2 Sequence of arrestin as extracted from rhodopsin-arrestin crystal structure (PDB code 5W0P) [5]. Arrestin elements are colored as follows: green, "finger loop," G69-T79; yellow, "157-loop," D156-I165; cyan, "C-loop," V249-S253; and pink, "back loop," R319-M322

2. The FMO calculations detected 35 significant interactions involved in rhodopsin-arrestin binding (Fig. 3a); these were formed by 25 residues of rhodopsin and 28 residues of arrestin. Of these 35 interactions, 23 had mainly electrostatic nature and 12 hydrophobic (Fig. 3a). Two major regions of interactions were identified: the C-terminal tail of rhodopsin (D330-S343) and the finger loop of arrestin (Fig. 3b) that directly inserts into rhodopsin active core.
3. FMO detected that the C-tail of rhodopsin forms 16 interactions with arrestin. This FMO result is in agreement with a previous report [5] suggesting that the C-terminal tail of rhodopsin plays a role as one of the anchor points for rhodopsin-arrestin binding. FMO detected that the phosphorylated residues pThr 336 (Tpo336) and pSer 338 (Sep338) of rhodopsin form an extensive network of salt bridges (Fig. 4) with the

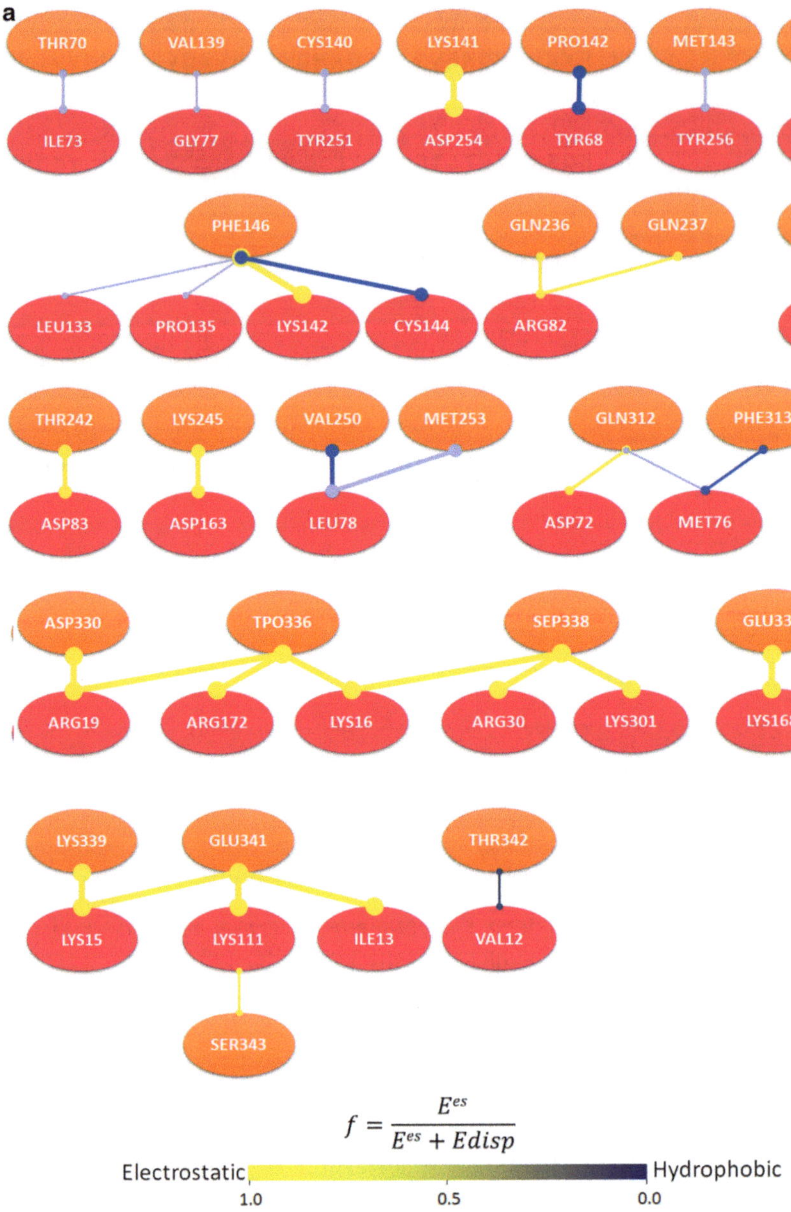

Fig. 3 (**a**) Network of 35 conserved inter-TM interactions involved in rhodopsin-arrestin binding for crystal structure 5W0P [5], formed by 25 residues of rhodopsin and 28 residues of arrestin. The ellipsoids represent residues. The lines between a pair of circles indicate the presence of interaction energy, and the thickness of the line is proportional to the size of the PIE. Only interactions with absolute value of $>$3.0 kcal/mol are shown. Lines are colored according to their chemical factor ($f = E^{\text{electrostatic}}/E^{\text{dispersion}} + E^{\text{electrostatic}}$): from 100% electrostatic (yellow) to 100% hydrophobic (dark blue). (**b**) The GPCR is shown in ribbon form, and the location of the 35 interactions on the surface of arrestin is highlighted in magenta

b

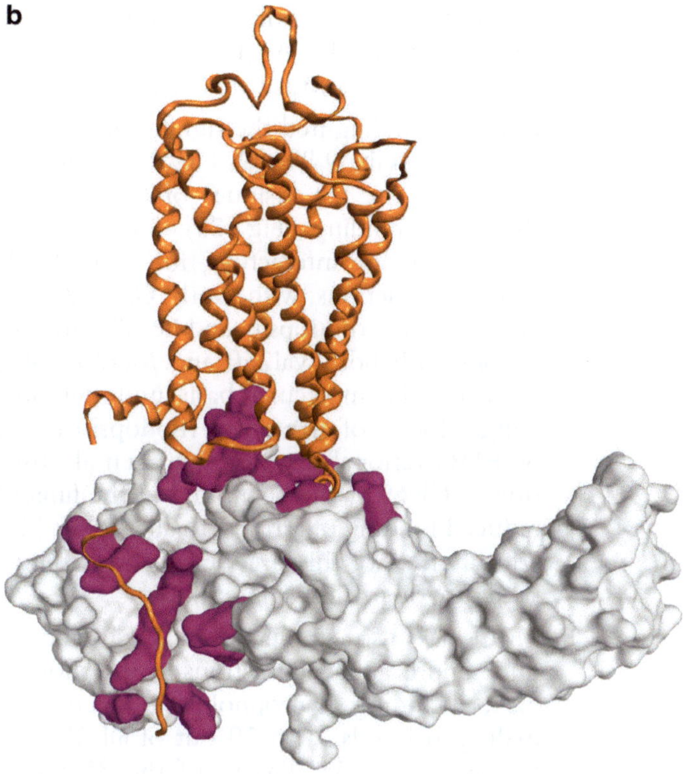

Fig. 3 (continued)

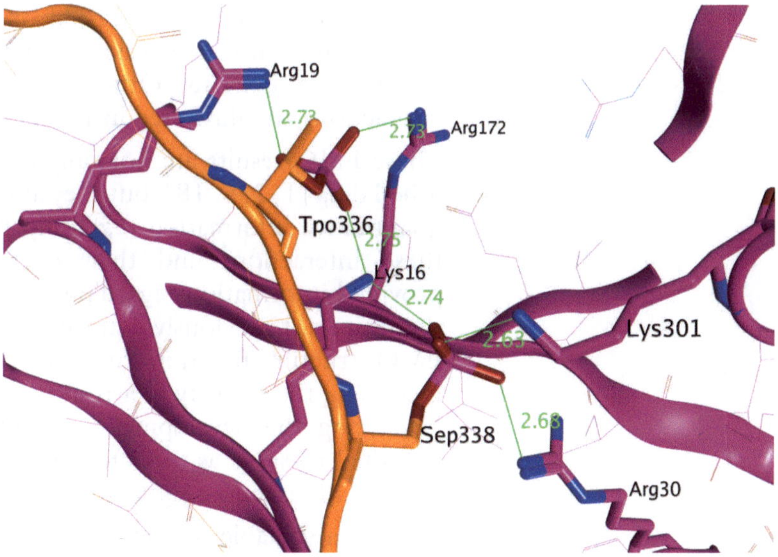

Fig. 4 Six salt bridges formed between phosphorylated residues Tpo336 and Sep338 of rhodopsin and arrestin

positively charged residues of arrestin: Tpo336 forms three salt bridges with Lys16, Arg19, and Arg172, while Sep338 forms three salt bridges with Lys16, Arg30, and Arg301.

4. FMO has highlighted the significant role of the "finger loop" of arrestin (Gly69-Thr79, Fig. 2) in rhodopsin binding, with six out of the ten residues of this helical loop involved in rhodopsin binding (Fig. 3a). Some of these residues form more than one interaction, for example, Met76 of arrestin forms interactions with both Gln312 and Phe313 of the C-terminus of rhodopsin, and Leu78 forms hydrophobic interactions with both Val250 and Met253 of rhodopsin's TM6 domain. The molecular basis underpinning the role of the "finger loop" of arrestin in rhodopsin binding as established by FMO rationalizes the experimental observation that deletion of Gly69 at the beginning of the finger loop dramatically reduced binding of arrestin to all GPCRs [18], reinforcing the critical effect of this residue on the stability of the secondary structure of arrestin's finger loop.

5. FMO has also emphasized the significant contribution of underappreciated interactions in the rhodopsin-arrestin complex, including hydrophobic interactions and non-classical hydrogen bonds, with 20 out of all 35 interactions being of this type (Fig. 5). Twelve of the 35 interactions were of a hydrophobic nature (Fig. 3a). Additional examples of underappreciated interactions include those formed by rhodopsin's Phe146 (located on intracellular loop 2). Phe146 forms interactions with four different arrestin residues (Fig. 5a) as follows: a hydrogen bond with the NH backbone of Lys142, a hydrophobic interaction with Pro35, a CH-π interaction with Cys144, and a non-classical hydrogen bond with the backbone carbonyl of Leu133. Other examples of underappreciated interactions are illustrated in Fig. 5a–5d.

6. These FMO results are not only in agreement with the published data [1, 2, 5, 18], but they also have provided additional quantitative information regarding the energetic strengths of these interactions and their chemical nature. FMO also provided information regarding underappreciated interactions that were not previously detected by any other method. These FMO structure-function studies reveal several different ways of increasing the capacity of arrestins to quench GPCR signaling, facilitating the development of enhanced arrestins with the potential to serve as tools for gene therapy of disorders associated with excessive signaling of mutant GPCRs [2].

7. FMO is a valuable resource that can be applied to aid further understanding of GPCR function. The regular application of FMO to GPCR studies can be used to expedite the generation

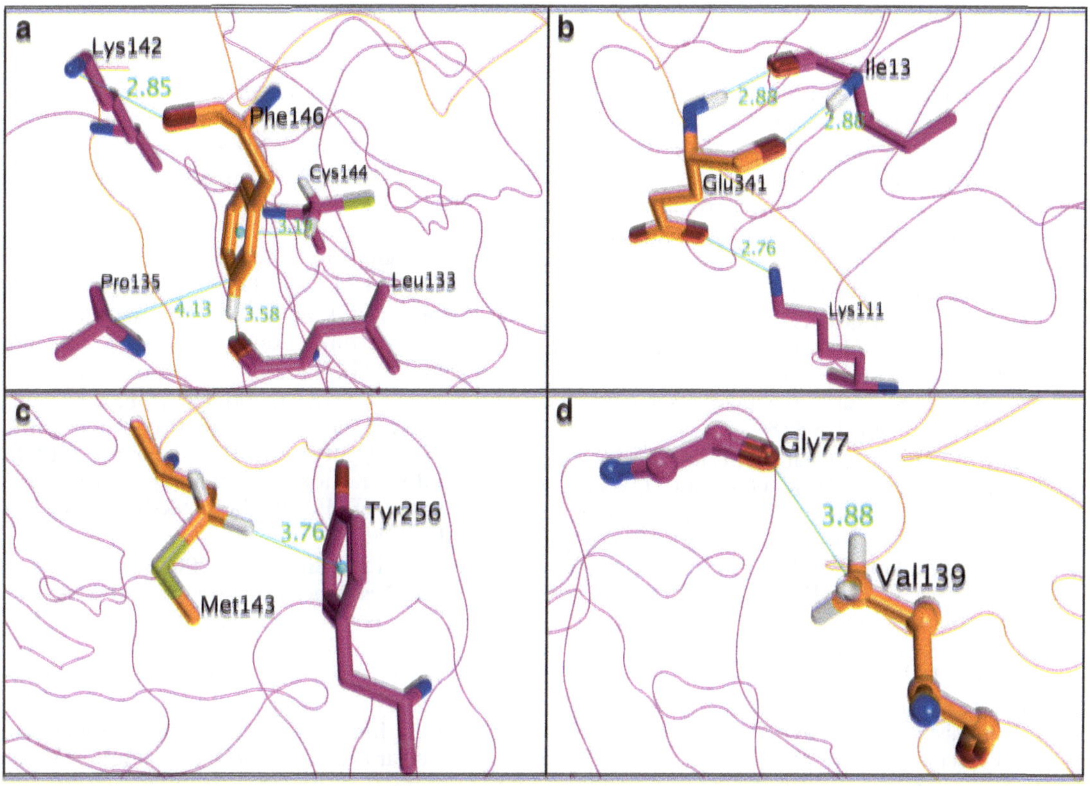

Fig. 5 (**a**) Phe146 (rhodopsin) forms interactions with four different arrestin residues: a hydrogen bond with the NH backbone of Lys142, a hydrophobic interaction with Pro35, a CH-π interaction with Cys144, and a non-classical hydrogen bond with the backbone carbonyl of Leu133. (**b**) Backbone atoms of Glu341 (rhodopsin) form two hydrogen bonds with the backbone atoms of Ile13 (arrestin), while the side chain of Glu341 forms salt bridge Lys111, (**c**) CH-π interaction of Met143 (rhodopsin) with Tyr256 (arrestin), and (**d**) non-classical hydrogen bond formed between Val139 (rhodopsin) and Gly77 (arrestin)

of more effective GPCR-biased drugs [19, 20]. This utility of FMO is not restricted to GPCRs, however, as it can be applied to the structural exploration of other protein families and biosystems.

Acknowledgments

A.H. and A.T.N. would like to acknowledge the support of EU H2020 CompBioMed project (http://www.compbiomed.eu/, 675451).

References

1. Chen Q, Iverson TM, Gurevich VV (2018) Structural basis of arrestin-dependent signal transduction. Trends Biochem Sci 43:412–423
2. Gurevich VV, Gurevich EV (2019) Arrestin mutations: some cause diseases, others promise cure. Prog Mol Biol Transl Sci 161:29–45
3. Gurevich VV, Gurevich EV (2013) Structural determinants of arrestin functions. Prog Mol Biol Transl Sci 118:57–92
4. Tan L, Yan W, McCorvy JD, Cheng J (2018) Biased ligands of G protein-coupled receptors (GPCRs): structure-functional selectivity relationships (SFSRs) and therapeutic potential. J Med Chem 61:9841–9878
5. Zhou XE, He Y, de Waal PW, Gao X, Kang Y, Van Eps N, Yin Y, Pal K, Goswami D, White TA, Barty A, Latorraca NR, Chapman HN, Hubbell WL, Dror RO, Stevens RC, Cherezov V, Gurevich VV, Griffin PR, Ernst OP, Melcher K, Xu HE (2017) Identification of phosphorylation codes for arrestin recruitment by G protein-coupled receptors. Cell 170:457–469.e13
6. Bissantz C, Kuhn B, Stahl M (2010) A medicinal chemist's guide to molecular interactions. J Med Chem 53:5061–5084
7. Fedorov DG, Nagata T, Kitaura K (2012) Exploring chemistry with the fragment molecular orbital method. Phys Chem Chem Phys 14:7562–7577
8. Fedorov DG, Kitaura K (2007) Extending the power of quantum chemistry to large systems with the fragment molecular orbital method. J Phys Chem A 111:6904–6914
9. Kitaura K, Ikeo E, Asada T, Nakano T, Uebayasi M (1999) Fragment molecular orbital method: an approximate computational method for large molecules. Chem Phys Lett 313:701–706
10. Heifetz A, Chudyk EI, Gleave L, Aldeghi M, Cherezov V, Fedorov DG, Biggin PC, Bodkin MJ (2016) The fragment molecular orbital method reveals new insight into the chemical nature of GPCR-ligand interactions. J Chem Inf Model 56:159–172
11. Alexeev Y, Mazanetz MP, Ichihara O, Fedorov DG (2012) GAMESS as a free quantum-mechanical platform for drug research. Curr Top Med Chem 12:2013–2033
12. Phipps MJ, Fox T, Tautermann CS, Skylaris CK (2015) Energy decomposition analysis approaches and their evaluation on prototypical protein-drug interaction patterns. Chem Soc Rev 44:3177–3211
13. Fedorov DG, Kitaura K (2007) Pair interaction energy decomposition analysis. J Comput Chem 28:222–237
14. Chudyk EI, Sarrat L, Aldeghi M, Fedorov DG, Bodkin MJ, James T, Southey M, Robinson R, Morao I, Heifetz A (2018) Exploring GPCR-ligand interactions with the fragment molecular orbital (FMO) method. Methods Mol Biol 1705:179–195
15. Fedorov DG, Kitaura K (2004) Second order Moller-Plesset perturbation theory based upon the fragment molecular orbital method. J Chem Phys 121:2483–2490
16. Mazanetz MP, Ichihara O, Law RJ, Whittaker M (2011) Prediction of cyclin-dependent kinase 2 inhibitor potency using the fragment molecular orbital method. J Cheminform 3:2
17. Li H, Fedorov DG, Nagata T, Kitaura K, Jensen JH, Gordon MS (2010) Energy gradients in combined fragment molecular orbital and polarizable continuum model (FMO/PCM) calculation. J Comput Chem 31:778–790
18. Zheng C, Tholen J, Gurevich VV (2019) Critical role of the finger loop in arrestin binding to the receptors. PLoS One 14:e0213792
19. Sloop KW, Emmerson PJ, Statnick MA, Willard FS (2018) The current state of GPCR-based drug discovery to treat metabolic disease. Br J Pharmacol 175(21):4060–4071
20. Hauser AS, Attwood MM, Rask-Andersen M, Schioth HB, Gloriam DE (2017) Trends in GPCR drug discovery: new agents, targets and indications. Nat Rev Drug Discov 16:829–842

Chapter 13

Characterizing Protein-Protein Interactions with the Fragment Molecular Orbital Method

Alexander Heifetz, Vladimir Sladek, Andrea Townsend-Nicholson, and Dmitri G. Fedorov

Abstract

Proteins are vital components of living systems, serving as building blocks, molecular machines, enzymes, receptors, ion channels, sensors, and transporters. Protein-protein interactions (PPIs) are a key part of their function. There are more than 645,000 reported disease-relevant PPIs in the human interactome, but drugs have been developed for only 2% of these targets. The advances in PPI-focused drug discovery are highly dependent on the availability of structural data and accurate computational tools for analysis of this data. Quantum mechanical approaches are often too expensive computationally, but the fragment molecular orbital (FMO) method offers an excellent solution that combines accuracy, speed and the ability to reveal key interactions that would otherwise be hard to detect. FMO provides essential information for PPI drug discovery, namely, identification of key interactions formed between residues of two proteins, including their strength (in kcal/mol) and their chemical nature (electrostatic or hydrophobic). In this chapter, we have demonstrated how three different FMO-based approaches (pair interaction energy analysis (PIE analysis), subsystem analysis (SA) and analysis of protein residue networks (PRNs)) have been applied to study PPI in three protein-protein complexes.

Key words Protein-protein interactions, PPI, QM, Molecular recognition, Quantum mechanics, FMO, Fragment molecular orbital, Protein residue networks, Subsystem analysis, Binding energy, PIE-PRN, PRN, Desolvation penalty, Pair interaction energy, PIE, Network analysis of protein residue networks, Efficiency centrality

1 Introduction

1. Living organisms rely on the specific recognition of pairs of proteins in practically every biological process [1–3]; hence, the importance of understanding molecular recognition and analysis of protein-protein interactions (PPIs) cannot be overstated. Analyses of experimentally determined structures indicate that intermolecular recognition is facilitated by a myriad of noncovalent interactions that, together, promote specificity at different levels.

Alexander Heifetz (ed.), *Quantum Mechanics in Drug Discovery*, Methods in Molecular Biology, vol. 2114,
https://doi.org/10.1007/978-1-0716-0282-9_13,

2. There are more than 645,000 reported disease-relevant PPIs in the human interactome [4]. However, only 2% of these had been targeted with drugs by 2011. Most of the remaining disease-relevant PPIs in protein complexes, such as transcription factors and many other signalling proteins, have been widely considered 'undruggable' and remain elusive, underexplored and yet to be fully understood [4–6]. Modulating disease-relevant protein-protein interactions using small-molecule inhibitors remains a challenging task owing to their highly dynamic and expansive interfacial areas (flat, featureless and relatively large). However, advances in PPI-focused drug discovery technology have been reported, and a few drugs are already on the market, with some potential drug-like candidates already in clinical trials [7].
3. Technological progress has played a central role in the identification of small-molecule modulators of PPIs that have to date reached clinical production [7]. The use of structural biology to determine 'hotspots' in PPIs' binding interfaces has been an important strategy in discovering small-molecule modulators [4, 5]. Despite the large sizes of PPI interfaces, only a small subset of amino acid residues that comprise the hotspot contributes most of the binding free energy. These 'hotspot' regions are potential targets for drug discovery [4]. An experimental way of identifying these hotspots in PPIs has been the combination of alanine-scanning mutagenesis and X-ray crystallography [8]. However even with the availability of high-resolution crystal structures, 'visual inspection' and the force field-based molecular mechanics (MM) calculations typically used for structural exploration cannot explain the full complexity of the intramolecular interactions [9]. Recently, several notable reports have been published [9–12] that emphasize the crucial role of 'underappreciated' or non-obvious intramolecular interactions involved in biomolecular recognition. These interactions include CH/π [13, 14], halogen/π [15], cation/π [16] and non-classical hydrogen bonds [17], features that are often not properly parameterized in the currently available force fields (FF) [11]. Furthermore, hydrophobic interactions, vital for receptor stability [18], still have no reliable non-QM predictive method for their quantification [9, 19].
4. Quantum mechanical (QM) methods have always been considered to be a reliable approach for the exploration of molecular interactions [20, 21]. However, despite their many advantages, traditional QM approaches are generally not feasible for large biological systems like GPCRs, due to their high computational cost [22]. We therefore have employed the fragment molecular orbital (FMO) QM approach [14, 21, 23, 24] in the current study.

5. The fragment molecular orbital method [14, 21, 23] offers a considerable computational speed-up over traditional QM methods [25]. By performing QM calculations on fragments, one can make the computational cost scaling with respect to system size nearly linear.
6. A specific advantage of FMO is that it provides a quantitative breakdown of the interactions formed between pairs of fragments (residues), including their strength (in kcal/mol) and chemical nature (electrostatic or hydrophobic) [14]. FMO offers an excellent solution that combines accuracy, speed and the ability to reveal key interactions that would be hard to detect otherwise [24].
7. The pair interaction energy (PIE; *see* Subheading 2.1) between any two fragments calculated by FMO can be expressed as the sum of several energy terms such as electrostatic, exchange repulsion, charge transfer, dispersion and solvent screening accomplished in the pair interaction energy decomposition analysis (PIEDA) [26, 27]. The electrostatic and charge transfer terms are predominant in salt bridge, hydrogen bond and polar interactions, whilst the dispersion term generally corresponds to interactions which are predominantly hydrophobic in nature. A strong electrostatic interaction is compensated for by the solvent screening, so that the sum of these two terms is often comparable to the dispersion interaction. The role of hydrophobic interactions is integral for biomolecular recognition, but there is still no reliable predictive method for its quantification [9]. The exchange repulsion term is a non-electrostatic QM term which quantifies the Pauli repulsion between electrons [22]. FMO is an extensively validated method for the structural exploration [28] of large biological systems.
8. As described in the literature [29] and as can be observed from Fig. 1, protein interfaces are characterized by complementarity of the shape and the chemical character of the interacting surfaces. FMO provides detail analysis of the interactions formed by the residues located on these interfaces.
9. In this chapter, we demonstrate how FMO can be applied to characterize the strength and chemical nature of protein-protein interactions. We applied three different FMO-based approaches (pair interaction energy analysis (PIE analysis), subsystem analysis (SA) and analysis of protein residue networks (PRNs)) to explore PPIs between three protein complexes (Fig. 1): trypisin/inhibitor (PDB code 1AVW [30]), barnase/barstar (PDB code 1BRS [31]) and subtilisin/eglin-C (PDB code 2SEC [32]).

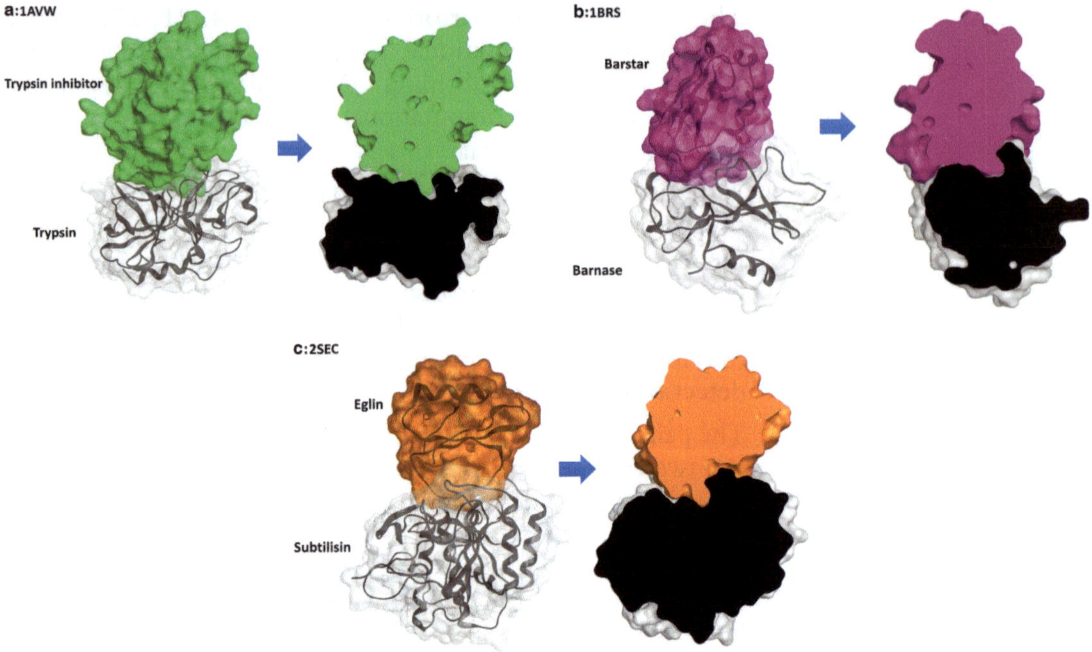

Fig. 1 Overall architecture of the three tested systems: for trypsin-soy bean inhibitor (PDB code 1AVW [30]) (**a**), barnase-barstar (PDB code 1BRS [31]) (**b**) and subtilisin/eglin-C (PDB code 2SEC [32]) (**c**). Left side: overall architecture of the protein-protein complex when structure of each protein represented as transparent complexes and its backbone as a ribbon. Right side: section through the protein's surfaces illustrating interface shape complementarity

2 Methods

2.1 Pair Interaction Energy Analysis (PIE Analysis)

1. For analysis of PPI, we used FMO as implemented [33] in the general atomic and molecular electronic structure system (GAMESS) [34]. In FMO calculations, a large biological system is partitioned into fragments [14, 21]. Each residue is a fragment, and the interaction energies can be sorted out so that they correspond to actual amino acid residues, not to residue fragments. The detailed description of the FMO methodology can be found in Chapter 4 of this book and in the other published reports [14, 21, 26]. To speed up the PPI calculations, we truncated the proteins to retain only residues located on protein-protein interface using 'Contacts' module of MOE v. 2018.01 (Computational Chemistry Group). Only those residues of one protein positioned within a radius of ≤4.5 Å from residues of the second protein were defined as interface residues and included in our calculations.
2. We used the MP2 method (second-order Møller-Plesset perturbation theory [35]) with the 6–31G* basis set as a compromise between speed and accuracy. To describe solvent, the conductor polarizable continuum solvent model (C-PCM) was used [36].

3. In PIEDA/MP2, the pair interaction energy ΔE_{IJ} or PIE between two fragments I and J is divided into the electrostatic (ES), exchange repulsion (EX), charge transfer and mix terms (CT + mix), dispersion and remainder correlation (DI + RC) and solvent screening (SOLV) [27, 37] as shown in Equation 1:

$$\Delta E_{IJ} = \Delta E_{IJ}^{\mathrm{ES}} + \Delta E_{IJ}^{\mathrm{EX}} + \Delta E_{IJ}^{\mathrm{CT+mix}} + \Delta E_{IJ}^{\mathrm{DI+RC}} + \Delta E_{IJ}^{\mathrm{SOLV}} \quad (1)$$

4. From these components, one can define the pair interaction character (PIC) describing the chemical nature of a given interaction. In this work, PIC is defined differentiating between electrostatics and dispersion as:

$$f_{IJ} = \frac{\left|\Delta E_{IJ}^{\mathrm{ES}} + \Delta E_{IJ}^{\mathrm{SOLV}}\right|}{\left|\Delta E_{IJ}^{\mathrm{ES}} + \Delta E_{IJ}^{\mathrm{SOLV}}\right| + \left|\Delta E_{IJ}^{\mathrm{DI+RC}}\right|} \quad (2)$$

5. Because $\Delta E_{IJ}^{\mathrm{DI+RC}}$ is usually dominated by dispersion, f_{IJ} is equal to 1 and 0 for purely electrostatic and dispersive interactions, respectively. The remaining two components, $\Delta E_{IJ}^{\mathrm{EX}} + \Delta E_{IJ}^{\mathrm{CT+mix}}$, describing exchange repulsion and charge transfer, typically have an opposite sign and partially cancel each other [26], although the sum may still be substantial for very short contacts, representing the important non-electrostatic physical effect of short-distance repulsion. Here, for the sake of differentiating between electrostatics and dispersion, the two terms $\Delta E_{IJ}^{\mathrm{EX}} + \Delta E_{IJ}^{\mathrm{CT+mix}}$ were not considered in the definition of PIC.
6. These PIE maps in Fig. 2 were made automatically using a Python script developed in this work. This functionality will be included into future releases of the ProGA (Protein Graph Analyser) package since 2016 [freely available upon e-mail request at sladek.vladimir@savba.sk]. This toolkit is intended for the analysis of protein residue networks based on distances or pair interaction energies. Based on previous reports [14], we considered any interaction with an absolute PIE $\geq$3.0 kcal/mol to be significant [24].

2.2 Subsystem Analysis (SA)

1. To analyse the binding directly, the subsystem analysis [38] was applied to the same truncated models as in the analysis of PIEs. It provides a complete integrated picture, in which interaction energies are added to the desolvation penalty, protein polarization energy and deformation energy. In this work, the repulsive deformation energy describing the energy cost of deforming isolated protein structures during the complex formation was not considered. The binding energies in solution ΔE in SA are divided into fragment contributions as:

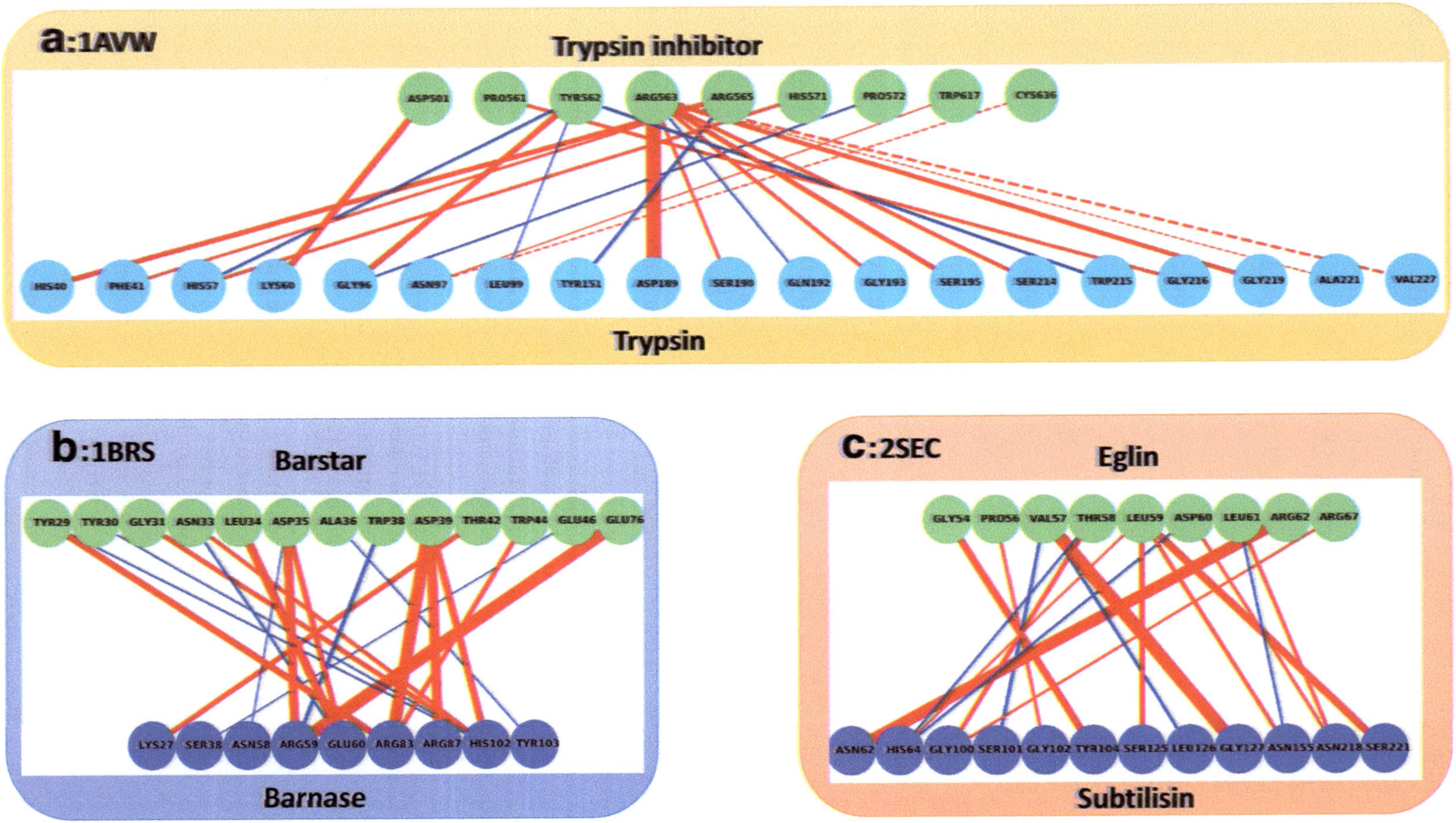

Fig. 2 PPI maps derived from FMO calculations for trypsin-soy bean inhibitor (PDB code 1AVW [30]) (**a**), barnase-barstar (PDB code 1BRS [31]) (**b**) and subtilisin/eglin-C (PDB code 2SEC [32]) (**c**). The line between a pair of circles indicates that the interaction energy $|\Delta E_{IJ}| > 3.0$ kcal/mol, and the thickness of the line is proportional to $|\Delta E_{IJ}|$. The lines are coloured by their chemical character (PIC): dark blue (hydrophobic, driven by dispersion) and red (electrostatics). The dashed line indicates repulsion $\Delta E_{IJ} \geq 3.0$ kcal/mol, and non-dashed line represents attraction $\Delta E_{IJ} \leq -3.0$ kcal/mol. The residues in the first protein in each complex are on the bottom, and those of the second residues are in the top row

$$\Delta E = E(\text{complex}) - E(\text{protein 1}) - E(\text{protein 2})$$
$$= \sum_{I=1}^{\text{protein 1}} \Delta E_I^{\text{bind}} + \sum_{I=1}^{\text{protein 2}} \Delta E_I^{\text{bind}} \quad (3)$$

2. Here, ΔE_I^{bind} includes a sum over J of fragment-fragment interaction energies ΔE_{IJ} between the two proteins (divided by 2 to avoid double counting) and partial fragment energies. The latter incorporates the effect of the mutual polarization (of one protein by another) and desolvation (the loss of the solute-solvent energy because of the complexation). The effect of the intra-protein charge transfer is included in the partial fragment energies, whereas the effect of the inter-protein charge transfer is incorporated in inter-protein ΔE_{IJ}. Further details can be found in Chapter 4 of this book.

2.3 Analysis of Protein Residue Networks (PRNs)

1. The interface picture described above is an important but simplified model of PPI. For a more comprehensive analysis, one has to consider whole proteins. In doing so, it is very useful to use the topological analysis of protein residue networks.
2. Protein-protein interactions are not limited to interfaces. In reality, proteins and protein-protein complexes are dynamic systems undergoing constant small conformational fluctuations. The formation of a protein-protein complex is usually a multistep process. It is often the case that the initial binding area is relatively small and will not be identical to the binding hotspot as identified in the final complex. In contrast, upon initial binding, some conformational, potentially long-reaching allosteric changes are likely to take place. These, together with the mutual attraction, cause a desolvation of the binding hotspots and the arrangement of the interacting proteins into the final binding pose. One can think of these conformational changes as an initially localized structural (and energetic) perturbation that is subsequently propagated throughout the protein.
3. The formation of a protein-protein complex can be facilitated if certain domains containing the final binding hotspot residues are pre-oriented to fit the surface of the binding partner. It is shown below that 'hubs' are responsible for this local stability. This process involves whole proteins and not just the final interface; hence, hub identification should also take into account all PIEs within the proteins.
4. Protein residue networks (PRNs) are a relatively new topic. In 1999, Kannan and Vishveshwara [39] developed an approach for identifying side-chain clusters in proteins by graph spectral methods. Since then the network theoretic analysis was applied to PRNs mainly for identification of important residues [40].

In the past, distance-based PRNs (D-PRN) were predominantly used, and the use of pair interaction energy (PIE) as a criterion for elucidating the residue contact is a relatively recent phenomenon [41]. The studies of PRN topology and related properties of proteins that have been conducted both for particular cases and generalized for proteins as a class of molecules [42] have focused primarily on D-PRN models.

5. Recently, a detailed analysis of PIE-based networks, PIE-PRN, was developed [43]. It was shown that the topology of these energy-based PRNs is in some regards quite different to that obtained from the distance-based D-PRN, particularly in regard to the small-world character of the network.
6. Global and/or local efficiencies do not reveal anything about the hubs in the PRN. In contrast, the efficiency centrality, a parameter quantifying the contribution of each residue to the global efficiency, has been found to be useful for identification of hubs that maintain the tertiary structure of the protein [44, 45].
7. In this chapter, PIE-PRN is used to demonstrate that not only are residues in direct contact with another protein at the interface important but also that residues embedded deeper within the protein may be crucial for quantifying PPI.
8. In PIE-PRN, one building block of the network (usually called node or vertex) corresponds to one fragment as defined in the FMO calculation. Hence, the PIEs ΔE_{IJ} from FMO define the relationship (i.e. the connection) of the residues I and J. The total interaction energy for fragment I is defined as:

$$\Delta E_I^{\mathrm{tot}} = \frac{1}{2} \sum_{J=1}^{\text{proteins 1and 2}} \Delta E_{IJ} \tag{4}$$

(where 1/2 is added to avoid double counting). The sum in $\Delta E_I^{\mathrm{tot}}$ includes both inter- and intra-protein PIEs; thus the picture offered by total PIEs does not necessarily agree with the binding hotspot picture, based on inter-protein PIEs only. In Eq. 4, index I runs over all residue fragments in both proteins (like index J).
9. As mentioned in **item 6**, Subheading 2.1, we used the criterion that the absolute PIE between residues is ≥3.0 kcal/mol in order for this interaction to be represented in the network. Within a network model of the protein, it is possible to find a connection between two residues even if they do not interact directly with each other. In such a case, the connection will include some intermediate residues forming a path.
10. If two residues I and J interact directly, then the path length $I \leftrightarrow J$ is one. However, if in order to reach residue J from I one has to pass through residue K, then the path has length

2 ($I \leftrightarrow K \leftrightarrow J$). Although I and J do not directly interact, both interact with K. The distance between two residues I and J, called the shortest path length d_{IJ} in network terms, is in the PIE-PRN model the sum of all reciprocal values of PIEs of residues included in the path (in the above example, ΔE_{IK} and ΔE_{KJ}). The FMO results predict whether residues I and J interact directly. The network model predicts whether there is possible long-range, or allosteric, interaction/influence between the protein residues I and J.

11. It is useful to have a measure of how well a network is internally connected. The definition of the global efficiency by Latora and Marchiori [46] fits this purpose.

$$E_{\mathrm{G}} = \frac{1}{N(N-1)} \sum_{I \neq J} d_{IJ}^{-1} \tag{5}$$

Herein N is the number of residues in the network. The global efficiency is usually normalized between 0 and 1, and it quantifies how efficiently messages can pass through the network. The closer the number is to 1, the more efficient is the communication across the network. In other words, the protein or PPI complex is more prone to allosteric effects.

12. Additionally, one can define the local efficiency E_{loc}. This metric reveals to what extent the system is fault tolerant or, in other words, resilient to removal of residue interactions. Consider for each residue I a subset of the network consisting only of the direct contacts J of the Ith residue (i.e. only those residues to which I has the path length equal to one; hence I is not in the subset). The efficiencies for these subsets $\widetilde{E}_I$ allow the definition of the local efficiency of the whole network E_{loc}. $\widetilde{E}_I$ is calculated via Eq. 5, where one uses N equal to the number of residues in the subset.

$$E_{\mathrm{loc}} = \frac{1}{N} \sum_{I} \widetilde{E}_I \tag{6}$$

PIE-PRN networks with larger local efficiency values are more fault tolerant to local residue removal. One can think of this as either a deletion of a residue from the sequence or its replacement (e.g., by mutation) by a residue that has different interactions.

13. A small-world network is usually used as a term describing a network in which paths between any random pair of I and J do not include many intermediate connections. In other words, the paths are short, and the distances are small. For the purpose of this work, one can anticipate that a PIE-PRN with a large E_{G} may be considered a small-world network [44].

14. Lastly, the efficiency centrality [45, 46] is defined. This metric characterizes each residue in the residue network by how much it contributes to the small-world character of the PIE-PRN. It can be viewed as an index measuring how much the global efficiency changes, when the I-th residue is removed from the network ($E'_{G,I}$), compared to the efficiency with all residues included, E_G.

$$C_I^{\text{eff}} = 1 - \frac{E'_{G,I}}{E_G} \qquad (7)$$

Larger values of C_I^{eff} indicate that the node is important with respect to the global efficiency in the PRN. This metric is used to identify both binding hotspots and hubs in PIE-PRN.

3 Results

In this chapter, we demonstrated how three different FMO-based approaches (pair interaction energy analysis (PIE analysis), subsystem analysis (SA) and analysis of protein residue networks (PRNs)) can be applied to explore PPIs between three protein complexes: trypsin/inhibitor (PDB code 1AVW [30]), barnase/barstar (PDB code 1BRS [31]) and subtilisin/eglin-C (PDB code 2SEC [32]).

3.1 Pair Interaction Energy Analysis (PIE Analysis)

1. PIE calculations (*see* Subheading 2.1) provided a quantitative breakdown of the key interactions formed between residues of two proteins, including their strength (in kcal/mol) and chemical nature (electrostatic or hydrophobic). Typical raw FMO output as calculated for trypsin/inhibitor is shown in Table 1.
2. The raw data calculated by FMO as demonstrated in Table 2 can be converted to convenient PPI maps (*see* Subheading 2.1, **item 6**) as shown in Fig. 2.
3. FMO detected that the proteins form a large number of interactions with each other (Fig. 2), with some residues forming more than one interaction with residues of the other protein. For example, residue ARG563 of trypsin inhibitor forms seven interactions with the residues of trypsin (Fig. 2a), ASP39 of barstar forms three interactions with the residues of barnase (Fig. 2b), and LEU59 of eglin forms four interactions with residues of subtilisin; *see* Figs. 2c and 3. Such information is essential for structure-based PPI drug design. Overall, the above interactions are dominated by electrostatics, although a hydrophobic (dispersion) contribution is also noticeable.
4. FMO also detected that backbone atoms play an important role in those PPIs that form a large number of backbone-backbone and backbone-side-chain interactions.

Table 1
Typical raw FMO output for PIE analysis, calculated for trypisin/inhibitor (PDB code 1AVW [30]) when F.name1 are residues of trypsin inhibitor and F.name2 are residues of trypsin

F.name1	F.name2	PIEtotal	PIEes	PIEex	PIEct	PIEdisp	PIEsolv	PIC
ASP501	LYS60	−15.16	−53.91	8.37	−3.72	−4.56	38.65	0.77
TYR562	HIS57	−6.41	−5.53	7.15	−2.36	−8.45	2.77	0.25
TYR562	GLY96	−13.39	−19.99	9.25	−2.88	−3.59	3.83	0.82
TYR562	LEU99	−3.22	−1.03	5.50	−2.73	−4.97	0.01	0.17
TYR562	TRP215	−7.16	−4.06	6.69	−4.32	−6.25	0.78	0.34
PRO561	GLY216	−8.30	−13.65	3.47	0.70	−2.40	3.58	0.81
ARG563	**ASP189**	−35.17	−134.46	28.62	−10.57	−7.49	88.73	0.86
ARG563	**SER190**	−6.71	−17.25	2.58	−1.42	−3.24	12.63	0.59
ARG563	**GLN192**	−4.42	−7.90	6.11	−2.65	−5.95	5.97	0.25
ARG563	**SER195**	−11.34	−16.36	8.23	−2.34	−7.59	6.71	0.56
ARG563	**SER214**	−9.06	−13.80	4.89	−1.87	−3.78	5.50	0.69
ARG563	**GLY219**	−10.63	−31.43	11.52	−3.29	−5.55	18.13	0.71
ARG563	**GLY193**	−8.43	−17.75	12.11	−3.19	−3.45	3.85	0.80
ARG565	HIS40	−11.96	−28.45	7.78	−3.85	−5.28	17.84	0.67
ARG565	PHE41	−6.86	−10.35	2.88	−0.52	−2.85	3.98	0.69
ARG565	TYR151	−7.29	−3.08	3.53	−2.24	−6.27	0.76	0.27
HIS571	HIS57	−7.89	−14.20	4.38	−1.64	−2.66	6.23	0.75
PRO572	GLY96	−4.90	−4.11	2.43	−1.39	−3.69	1.86	0.38
TRP617	ASN97	−3.32	−5.79	0.52	−0.90	−1.34	4.20	0.54

In **bold** we marked potential hotspots. PIEtotal (total PIE as calculated by Eq. 1), PIEes (electrostatic term), PIEex (exchange repulsion term), PIEct (charge transfer term), PIEdisp (dispersion), PIEsolv (solvation term) and PIC (pair interaction character as calculated by Eq. 2)

3.2 Subsystem Analysis (SA)

1. Subsystem analysis (SA) was used to analyse the PPI as described in Subheading 2.2. The results of this analysis are shown in Table 2. The values mean attraction, the top part of the list for each complex, indicates potential hotspots, such as ASP39 for 1BRS. There is a good correlation between hotspots identified by FMO-PIEs in Fig. 2 and potential hotspots in Table 2. However, the two are not identical; in the subsystem analysis, large PIEs for charged and polar residues are corrected by the addition of partial fragment energies, and the scores are also conveniently evaluated fragment-wise. LEU59 in eglin, for instance, has a substantial interaction energy with subtilisin; however, its binding energy is repulsive due to the addition of contributions missing in the interaction energy.

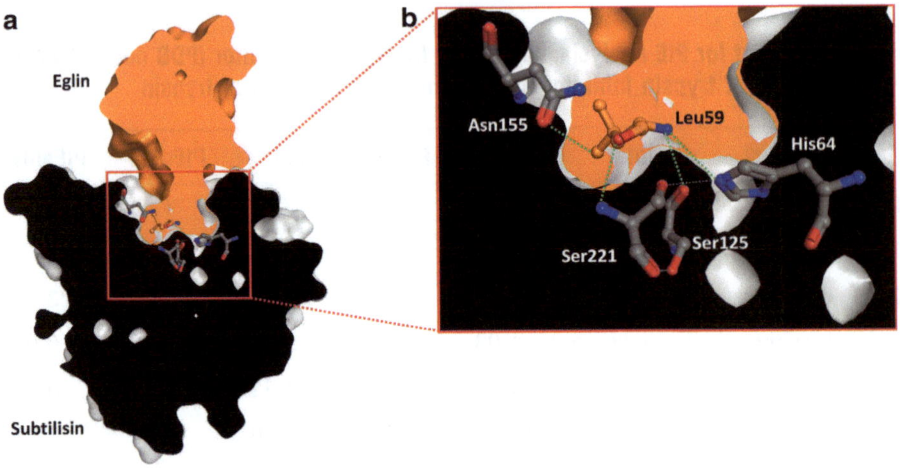

Fig. 3 Example of interaction between subtilisin and eglin (PDB code 2SEC [32]). (**a**) Section through the protein surfaces illustrating interface shape complementarity and four interactions formed between LEU59 of eglin with four residues of subtilisin. (**b**) Zoom figure when the key interactions detected by FMO shown as green dashed line

2. In addition to attractive hotspots, there are some repulsive spots, as PPI cannot please every residue in each protein, and there are points of dissent. ASN62 in 2SEC is an example of an unhappy fragment. There is more repulsion found in the binding energies than in the interaction energies due to the desolvation and mutual polarization, both of which are usually repulsive when summed over all fragments (although individual fragment contributions may be attractive). In the above discussion, the protein polarization incorporated into the partial fragment energies means the repulsive destabilization component of the polarization according to the PIEDA picture, whereas the stabilization component of the polarization is a part of PIEs.

3.3 Analysis of Protein Residue Networks (PRNs)

1. As described in Subheading 2.3, one of the interesting properties of proteins is that they are adapted to perform relatively fast and precise 'communication' across their structure, which is utilized in long-range conformational adaptations—the so-called allosteric effects. The concept of the global and local efficiency introduced by Latora and Marchiori [46] is useful in describing how well the protein is adapted to such long-range 'communications'. The global efficiency E_G is a value, usually normalized to be between 0 and 1, which tells how efficiently messages can pass through the network. The closer the number is to 1, the more efficient is the communication across the network. As an example, in the subtilisin-eglin complex (PDB code 2SEC [32]), the global efficiency is $E_G = 0.82$, which is a fairly typical value for protein residue networks [43]. High

Table 2
Potential hotspots in PPI using subsystem analysis (score is equal to ΔE_I^{bind}, in kcal/mol, and only values $|\Delta E_I^{bind}| > 3.0$ are shown)

Fragment 1BRS[a]	Protein	Score	Fragment 1AVW[b]	Protein	Score	Fragment 2SEC[c]	Protein	Score
ASP39	2	−18.2	ILE564	2	−11.8	GLY63	1	−10.2
ARG59	1	−11.5	ASP189	1	−10.6	ASP60	2	−6.3
ASN84	1	−9.4	TYR562	2	−7.6	ARG62	2	−5.4
GLU60	1	−8.8	CYS220	1	−5.7	THR58	2	−3.9
ASP35	2	−8.5	PHE41	1	−5.6	TYR104	1	−3.8
HIS102	1	−7.6	TRP215	1	−5.5	LEU126	1	−3.6
GLU76	2	−7.6	CYS58	1	−3.2	PRO56	2	4.9
ARG83	1	−7.4	VAL227	1	4.0	ASN62	1	5.9
ARG87	1	−7.2	CYS636	2	4.2			
TYR103	1	−6.1						
TYR29	2	−5.9						
THR42	2	−4.4						
ASN58	1	−3.3						
ASN33	2	−3.2						
LYS27	1	−3.0						
SER38	1	3.1						
GLY43	2	4.2						

[a]1 is barnase, 2 is barstar
[b]1 is trypsin, 2 is trypsin inhibitor
[c]1 is subtilisin, 2 is eglin

global efficiency is characteristic for so-called small-world networks. Another typical feature is the presence of highly connected nodes, often called hubs, which facilitate efficient communication. The local efficiency of the network is a value quantifying the resilience of the network to the loss of its capacity for efficient communication (its global efficiency) if one removes the nodes (i.e. the residues). The local efficiency is normally defined as one number for the whole network and should be understood as some average topological property, i.e. it is not used for identification of the hubs. Again, in the subtilisin-eglin complex, the value is about 0.11 (*see* Subheading 2.3). From this, one can deduce that potential removal of some residues could break the ability of the protein for

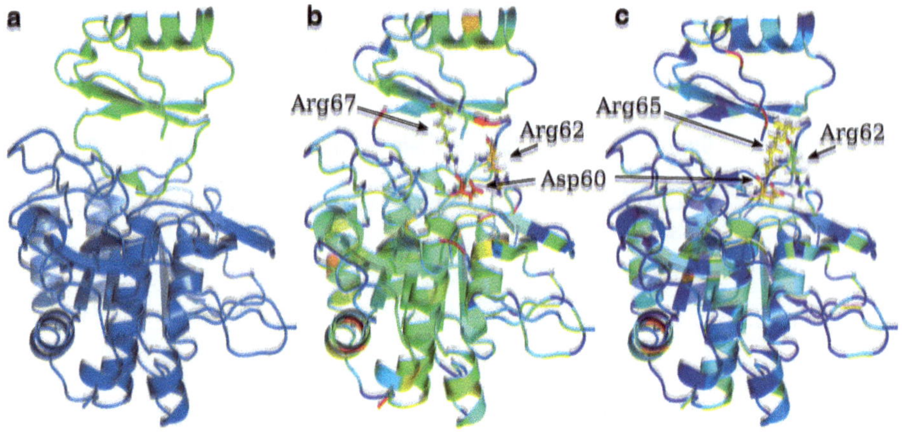

Fig. 4 Structure of the complex of subtilisin and eglin (PDB code 2SEC [32]). (**a**) Eglin is in green; subtilisin is blue. (**b**) Colour heat map based on the total interaction energy (*see* Subheading 2.3, Eq. 4) of each residue, ΔE_I^{tot} (some potential hotspots are shown in the stick representation). Red and blue colours indicate strong and weak binding, respectively. (**c**) Colour heat maps based on efficiency centrality C_I^{eff}; red and blue colours represent high and low ranking, respectively. Some domain hubs identified by the efficiency centrality in PIE-PRN are shown in the stick representation

allosteric, long-range communication and/or for formation of PPI complexes.

2. The structures of proteins have characteristic features at several levels: the primary amino acid sequence, the secondary structure, including helices, loops and sheets which are the key structural units of proteins. These secondary structures are stabilized by side-chain interactions. Moreover, the relative position of these secondary structure units is also governed by side-chain interactions, creating a tertiary structure. Residues responsible for maintaining this tertiary structure may be considered hubs in the framework of PRN (*see* Subheading 2.3).

3. The difference between a hub and a hotspot is that a hotspot, in the context of PPI, is directly responsible for binding between the two proteins. In the binding energy picture, potential hotspots are identified based on integrated fragment energies including contributions from all residues, not just those at the interface. If PIEs are used as a guide, a hotspot is a residue in one of the proteins forming the PP complex strongly interacting with, potentially multiple, residues within the other protein in the complex. Hence, hotspots describe only immediate contacts. On the other hand, the network analysis enables one to identify the hubs, the important central residues which do not necessarily mediate the binding but are indispensable in maintaining the stability of the binding hotspot neighbourhood.

4. To summarize, potential hotspot residues are typically located within a secondary structure moiety, e.g. a loop. The position and orientation of the loop can be stabilized by a hub. To

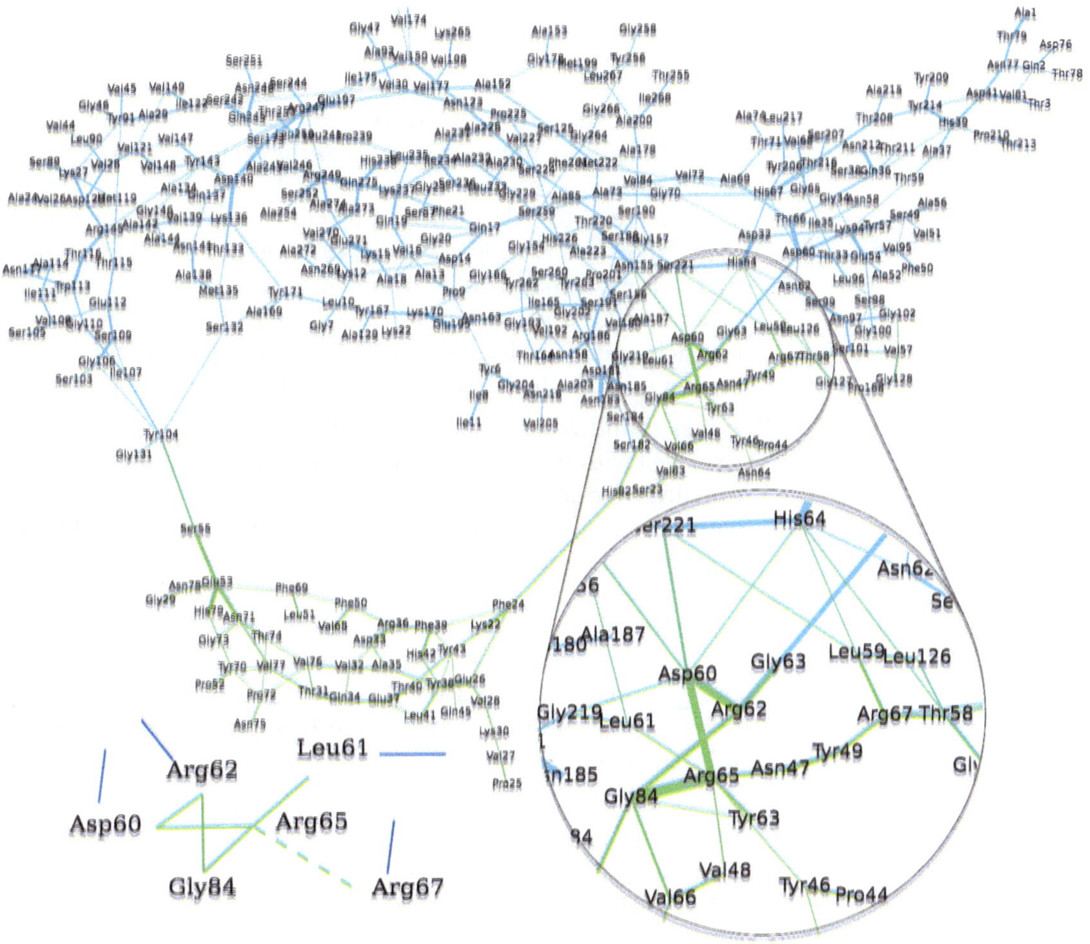

Fig. 5 The PIE-PRN of the complex of subtilisin (blue) and eglin (green) (PDB code 2SEC [32]). Nodes with $|\Delta E_{IJ}| > 3.0$ kcal/mol were connected (the sign of ΔE_{IJ} is not used in PRN). The line thickness is proportional to $|\Delta E_{IJ}|$. The bottom-left scheme highlights the cluster in which some residues are binding hotspots (ASP60, LEU61, ARG62) and ARG65 (a hub, but not a hotspot)

identify the hubs, the whole network of pair interactions both within and between proteins is considered (Fig. 4). Hubs, therefore, describe residues that can take part in the inter-protein energy transfer but are more responsible for intra-protein redistribution of the energy. The hotspot view is simpler as it is, in some sense 'one-dimensional' (*see* the row of interactions in Fig. 2 or values in Table 1). The hub view is 'two-dimensional', as demonstrated in Fig. 5.

5. ASP60 is an example of a potential hotspot with a highly attractive total interaction (denoted in Fig. 4b, in red). Other examples are ARG62 and ARG67 of eglin. These three residues are also identified as hotspots in Fig. 2c. However, ARG65 is

not a hotspot, despite having a very strong total attractive interaction—might it be a candidate for a domain hub? We answer this below.

6. Because of the extra dimensionality, topological plots can be more difficult to interpret visually, as they reflect the complexity of the protein residue network. Global network values are characteristic for the whole protein network. Examples of such values are the global and local efficiencies. These can be useful for comparisons between networks/proteins. On the other hand, 'centralities' are properties that are determined for each node (residue) and can be used to compare the contribution of individual residues to the topology of the protein.
7. It was found recently that the efficiency centrality C_I^{eff} correlates well with the total interaction energy ΔE_I^{tot} [44, 45]. Additionally, residues ranking highly in C_I^{eff} tend to mediate contact with the second protein [43]. From this, it follows that they are important in the spreading of conformational perturbations and that in the equilibrated state they often maintain favourable positions of the binding hotspots. Residues with high efficiency centrality are residues that are responsible for the inter-protein energy transfer and intra-protein redistribution of the energy—they are the hubs. Figure 5 displays the PIE-PRN of the subtilisin-eglin complex.
8. An example of a residue with a high efficiency centrality is ARG65 of eglin in the subtilisin-eglin complex. ARG65 is not a binding hotspot; it has no direct contact to subtilisin residues, yet Fig. 4 shows that it has a large ΔE_I^{tot} value. A further investigation of the PIE-PRN reveals why it scores highly in the C_I^{eff} ranking, i.e. why it is a domain hub. ARG65 has strong interactions with two key binding hotspots, ASP60 and LEU61. Furthermore, it interacts strongly with GLY84 which, in turn, interacts strongly with ARG62, another hotspot. The interaction with ARG67 is indirect, but takes place over a strongly interacting path of two additional nodes. Hence, ARG65 is in close contact with four hotspots and can be considered crucial for the stability of this small cluster of hotspots. Figure 4c shows that there are more such hubs (coloured red, orange or green depending on their ranking). Several of them are, however, quite distant from the binding hotspot and, presumably, are likely responsible for maintaining the stability of these distant secondary structure units. Most frequently, the strong stabilizing interactions are electrostatic in nature.

4 Conclusions

1. In this chapter, we described how three different FMO-based approaches (FMO-PPI) can be applied for PPI analysis of protein-protein complexes. The results produced by these methods demonstrated significant correlations in results, but differences were also observed.
2. Further improvement of the predictive power of FMO-PPI can be achieved if the conformational averaging in molecular dynamics [47] would take into account temperature and the possible multiple minima on the energy surface.
3. In this stage, the further development and optimization of these methods require experimental validation where it will be cleared when and how to apply each of these pioneering approaches in structure-based PPI drug design.

Acknowledgements

A.H. and A.T.N. would like to acknowledge the support of EU H2020 CompBioMed project (http://www.compbiomed.eu/, 675451). V.S. acknowledges the grants VEGA 2/0035/16 and 2/0031/19 by the Agency of the Ministry of Education of Slovak Republic and the Slovak Academy of Sciences.

References

1. Katchalski-Katzir E, Shariv I, Eisenstein M, Friesem AA, Aflalo C, Vakser IA (1992) Molecular surface recognition: determination of geometric fit between proteins and their ligands by correlation techniques. Proc Natl Acad Sci U S A 89(6):2195–2199
2. Snider J, Kotlyar M, Saraon P, Yao Z, Jurisica I, Stagljar I (2015) Fundamentals of protein interaction network mapping. Mol Syst Biol 11(12):848–848
3. Heifetz A, Katchalski-Katzir E, Eisenstein M (2002) Electrostatics in protein-protein docking. Protein Sci 11(3):571–587
4. Gan Z, Jessica A, Guillermo G-N (2018) Peptidomimetics targeting protein-protein interactions for therapeutic development. Protein Pept Lett 25(12):1076–1089
5. Robertson NS, Spring DR (2018) Using peptidomimetics and constrained peptides as valuable tools for inhibiting protein(−)protein interactions. Molecules 23(4):pii: E959
6. Díaz-Eufracio BI, Naveja JJ, Medina-Franco JL (2018) Protein–protein interaction modulators for epigenetic therapies. In: Donev R (ed) Advances in protein chemistry and structural biology, vol 110. Academic, New York, NY, pp 65–84
7. Stevers LM, Sijbesma E, Botta M, MacKintosh C, Obsil T, Landrieu I, Cau Y, Wilson AJ, Karawajczyk A, Eickhoff J et al (2018) Modulators of 14-3-3 protein-protein interactions. J Med Chem 61(9):3755–3778
8. Moreira IS, Fernandes PA, Ramos MJ (2007) Hot spots—a review of the protein–protein interface determinant amino-acid residues. Proteins Struct Funct Bioinformatics 68 (4):803–812
9. Bissantz C, Kuhn B, Stahl M (2010) A medicinal chemist's guide to molecular interactions. J Med Chem 53(14):5061–5084
10. Tong Y, Mei Y, Li YL, Ji CG, Zhang JZ (2010) Electrostatic polarization makes a substantial contribution to the free energy of avidin-biotin binding. J Am Chem Soc 132(14):5137–5142
11. Raha K, Peters MB, Wang B, Yu N, Wollacott AM, Westerhoff LM, Merz KM Jr (2007) The

role of quantum mechanics in structure-based drug design. Drug Discov Today 12 (17–18):725–731

12. Beratan DN, Liu C, Migliore A, Polizzi NF, Skourtis SS, Zhang P, Zhang Y (2015) Charge transfer in dynamical biosystems, or the treachery of (static) images. Acc Chem Res 48 (2):474–481
13. Ozawa T, Okazaki K, Kitaura K (2011) CH/pi hydrogen bonds play a role in ligand recognition and equilibrium between active and inactive states of the beta2 adrenergic receptor: an ab initio fragment molecular orbital (FMO) study. Bioorg Med Chem 19(17):5231–5237
14. Fedorov DG, Nagata T, Kitaura K (2012) Exploring chemistry with the fragment molecular orbital method. Phys Chem Chem Phys 14 (21):7562–7577
15. Lu Y-X, Zou J-W, Wang Y-H, Yu Q-S (2007) Substituent effects on noncovalent halogen/π interactions: theoretical study. Int J Quant Chem 107(6):1479–1486
16. Gallivan JP, Dougherty DA (1999) Cation-pi interactions in structural biology. Proc Natl Acad Sci U S A 96(17):9459–9464
17. Johnston RC, Cheong PH (2013) C-H...O non-classical hydrogen bonding in the stereomechanics of organic transformations: theory and recognition. Org Biomol Chem 11 (31):5057–5064
18. Pace CN, Fu H, Fryar KL, Landua J, Trevino SR, Shirley BA, Hendricks MM, Iimura S, Gajiwala K, Scholtz JM et al (2011) Contribution of hydrophobic interactions to protein stability. J Mol Biol 408(3):514–528
19. Popov P, Peng Y, Shen L, Stevens RC, Cherezov V, Liu ZJ, Katritch V (2018) Computational design of thermostabilizing point mutations for G protein-coupled receptors. Elife 7:e34729
20. Yu N, Li X, Cui G, Hayik SA, Merz KM II (2006) Critical assessment of quantum mechanics based energy restraints in protein crystal structure refinement. Protein Sci 15 (12):2773–2784
21. Fedorov DG, Kitaura K (2007) Extending the power of quantum chemistry to large systems with the fragment molecular orbital method. J Phys Chem A 111(30):6904–6914
22. Phipps MJ, Fox T, Tautermann CS, Skylaris CK (2015) Energy decomposition analysis approaches and their evaluation on prototypical protein-drug interaction patterns. Chem Soc Rev 44(10):3177–3211
23. Kitaura K, Ikeo E, Asada T, Nakano T, Uebayasi M (1999) Fragment molecular orbital method: an approximate computational method for large molecules. Chem Phys Lett 313(3–4):701–706
24. Heifetz A, Chudyk EI, Gleave L, Aldeghi M, Cherezov V, Fedorov DG, Biggin PC, Bodkin MJ (2016) The fragment molecular orbital method reveals new insight into the chemical nature of GPCR-ligand interactions. J Chem Inform Model 56(1):159–172
25. Alexeev Y, Mazanetz MP, Ichihara O, Fedorov DG (2012) GAMESS as a free quantum-mechanical platform for drug research. Curr Top Med Chem 12(18):2013–2033
26. Fedorov DG, Kitaura K (2007) Pair interaction energy decomposition analysis. J Comput Chem 28(1):222–237
27. Fedorov DG, Kitaura K (2012) Energy decomposition analysis in solution based on the fragment molecular orbital method. J Phys Chem A 116(1):704–719
28. Chudyk EI, Sarrat L, Aldeghi M, Fedorov DG, Bodkin MJ, James T, Southey M, Robinson R, Morao I, Heifetz A (2018) Exploring GPCR-ligand interactions with the fragment molecular orbital (FMO) method. Methods Mol Biol 1705:179–195
29. Ben-Shimon A, Eisenstein M (2010) Computational mapping of anchoring spots on protein surfaces. J Mol Biol 402(1):259–277
30. Song HK, Suh SW (1998) Kunitz-type soybean trypsin inhibitor revisited: refined structure of its complex with porcine trypsin reveals an insight into the interaction between a homologous inhibitor from Erythrina caffra and tissue-type plasminogen activator. J Mol Biol 275(2):347–363
31. Buckle AM, Schreiber G, Fersht AR (1994) Protein-protein recognition: crystal structural analysis of a barnase-barstar complex at 2.0-A resolution. Biochemistry 33(30):8878–8889
32. McPhalen CA, James MN (1988) Structural comparison of two serine proteinase-protein inhibitor complexes: eglin-c-subtilisin Carlsberg and CI-2-subtilisin Novo. Biochemistry 27(17):6582–6598
33. Fedorov DG, Kitaura K (2004) The importance of three-body terms in the fragment molecular orbital method. J Chem Phys 120 (15):6832–6840
34. Schmidt MW, Baldridge KK, Boatz JA, Elbert ST, Gordon MS, Jensen JH, Koseki S, Matsunaga N, Nguyen KA, Su S et al (1993) General atomic and molecular electronic structure system. J Comput Chem 14 (11):1347–1363
35. Fedorov DG, Kitaura K (2004) Second order Møller-Plesset perturbation theory based upon

the fragment molecular orbital method. J Chem Phys 121(6):2483–2490

36. Li H, Fedorov DG, Nagata T, Kitaura K, Jensen JH, Gordon MS (2010) Energy gradients in combined fragment molecular orbital and polarizable continuum model (FMO/PCM) calculation. J Comput Chem 31(4):778–790
37. Fedorov DG (2019) Solvent screening in zwitterions analyzed with the fragment molecular orbital method. J Chem Theor Comput 15 (10):5404–5416
38. Fedorov DG, Kitaura K (2016) Subsystem analysis for the fragment molecular orbital method and its application to protein–ligand binding in solution. J Phys Chem A 120 (14):2218–2231
39. Kannan N, Vishveshwara S (1999) Identification of side-chain clusters in protein structures by a graph spectral method. J Mol Biol 292 (2):441–464
40. Salamanca Viloria J, Allega MF, Lambrughi M, Papaleo E (2017) An optimal distance cutoff for contact-based Protein Structure Networks using side-chain centers of mass. Sci Rep 7 (1):2838
41. Vijayabaskar MS, Vishveshwara S (2010) Interaction energy based protein structure networks. Biophys J 99(11):3704–3715
42. Estrada E, Hatano N, Benzi M (2012) The physics of communicability in complex networks. Phys Rep 514(3):89–119
43. Sladek V, Tokiwa H, Shimano H, Shigeta Y (2018) Protein residue networks from energetic and geometric data: are they identical? J Chem Theory Comput 14(12):6623–6631
44. Wang S, Du Y, Deng Y (2017) A new measure of identifying influential nodes: efficiency centrality. Commun Nonlinear Sci Num Simulat 47:151–163
45. Sladek V (2018) A note on the interpretation of the efficiency centrality. Commun Nonlinear Sci Num Simulat 61:225–229
46. Latora V, Marchiori M (2001) Efficient behavior of small-world networks. Phys Rev Lett 87 (19):198701
47. Fedorov DG, Kitaura K (2018) Pair interaction energy decomposition analysis for density functional theory and density-functional tight-binding with an evaluation of energy fluctuations in molecular dynamics. J Phys Chem A 122(6):1781–1795

Chapter 14

Conformational Searching with Quantum Mechanics

Matthew Habgood, Tim James, and Alexander Heifetz

Abstract

Estimating the range of three-dimensional structures (conformations) that are available to a molecule is a key component of computer-aided drug design. Quantum mechanical simulation offers improved accuracy over forcefield methods, but at a high computational cost. The question is whether this increased cost can be justified in a context in which high-throughput analysis of large numbers of molecules is often key. This chapter discusses the application of quantum mechanics to conformational searching, with a focus on three key challenges: (1) the generation of ensembles that include a good approximation to a molecule's bioactive conformation at as prominent a ranking as possible; (2) rational analysis and modification of a pre-established bioactive conformation in terms of its energetics; and (3) approximation of real solution-phase conformational ensembles in tandem with NMR data. The impact of QM on the high-throughput application (1) is debatable, meaning that for the moment its primary application is still lower-throughput applications such as (2) and (3). The optimal choice of QM method is also discussed. Rigorous benchmarking suggests that DFT methods are only acceptable when used with large basis sets, but a trickle of papers continue to obtain useful results with relatively low-cost methods, leading to a dilemma that the literature has yet to fully resolve.

Key words Bioactive conformation, DFT, B3LYP, M06-2X, Solution-phase ensembles, NMR

1 The Goals of Conformational Searching

The range of physically feasible three-dimensional structures (conformations) available to a candidate drug molecule is essential to understanding and rational modification of its interactions with biomolecular systems, including its target protein. While X-ray crystal structures can provide a single (or, rarely, a small number of) three-dimensional structures, these will not cover the full conformational space of a flexible molecule, with multiple rotatable single bonds or unsaturated ring systems. The solution adopted both in industrial and academic settings is to use computational tools to generate a set of structures—a conformational ensemble—in silico. These can then be used in further analysis and calculations, such as docking into a crystal structure, matching to the shape of a known active compound, comparison to an established

Alexander Heifetz (ed.), *Quantum Mechanics in Drug Discovery*, Methods in Molecular Biology, vol. 2114,
https://doi.org/10.1007/978-1-0716-0282-9_14,

pharmacophore, or input in the development of a quantitative structure-activity relationship. In ligand-based drug design (LBDD), such ensembles may be the only 3D structural information used.

A very wide variety of software tools have been developed for the generation of conformational ensembles, using a number of different approaches to structure generation. These include Monte Carlo or molecular dynamics—guided exploration of a molecular potential energy surface as defined by a classical mechanical energy function 'forcefield' (as in the ConfGen module of Schrodinger's Macromodel [1, 2], the Conformational Search module of CCG's Molecular Operating Environment (MOE) [3, 4], or the free software Frog2 [5]); the mining of molecular geometry databases, typically derived from crystal structures, for bond length, angle, and torsion values (as in the CCDC's Mercury 'Conformer Generation' module [6, 7], OpenEye's Omega [8, 9], the free software ConFab [10], and the recently developed Conformator [11]); or simply the achievement of maximum geometric diversity compatible with chemically sensible constraints (as in RDKit's 'Add Conformers' module [3] or the free software ALFA [12]). A number of studies have compared the performance of these tools [13–17]. Perhaps the simplest form of conformational 'ensemble' generation is a scan of energy for a single dihedral angle, assessing the stability of a subset of conformations based around one structurally significant coordinate.

As with any modelling exercise, the central concern for in silico conformational ensemble generation is that it should be faithful to the real behaviour of the molecule. A particular concern is that the ensemble should include a good approximation to the 'bioactive conformation'—that is, the three-dimensional structure of the molecule that interacts with the desired protein target (of course, if alternative targets or off-targets are considered in the drug design process, there may be multiple bioactive conformations). Indeed, success in this task is used as a key indicator of success in generation algorithm comparison studies. At the same time, the ensemble should contain as few as possible 'non-relevant' conformations—in the broadest definition, a non-relevant conformation is one that is non-physically feasible. Historically, a convenient solution has been to assess the stability of a set of generated conformers, using a potential energy. Whether part of the initial structure generation process or not, the generated structures are ranked according to their energy, calculated for the molecule in isolation ('in the gas phase'), or sometimes with a highly simplified approximation to a condensed phase environment in the form of an implicit solvent [18, 19]. It is then assumed that all generated conformations up to some potential energy cut-off $\Delta U_{\max}$ above the lowest potential energy structure (the 'strain energy' relative to the 'global minimum' of the ensemble) are physically feasible and should be considered in further calculations. It is assumed that any conformations higher in energy than ΔU_{max} are so unstable that they cannot take part in any supermolecular complex that will have non-negligible population.

The value of ΔU_{max} must be set high enough to account for inaccuracies in the potential energy calculation method, as well as errors arising from the use of potential energy to approximate the free energy that will dictate the behaviour of the molecule in reality. The primary concern, though, is that the value should be high enough to include any candidate bioactive conformations that might have been generated. This presents a serious challenge, since the bioactive conformation will not be dictated by the stability of the molecular geometry in isolation, but by the stability in complex with the target to which it binds. For drug-sized molecule, interactions with a macromolecular target can be of the order of tens of kcal/mol [20, 21]. The value of ΔU_{max} must hence be set large enough to include conformations that could be preferentially stabilised by this amount to yield a maximally stable complex.

Literature reports on the appropriate value have typically focused on the stability (strain energy) of conformations found for drug-sized molecules bound to macromolecules in crystal structures. These are often used as convenient proxies for the bioactive conformations of the relevant systems. Unfortunately, a consensus has not emerged as to the upper bound of these strain energies. Studies in this area are complicated by the fact that the specific molecular geometry that appears in a crystal structure is usually the result of fitting to an electron density distribution and may be calculated as having a very high strain energy in terms of molecular energetics.

Perola and Charifson suggest [22] a value of 5 kcal/mol based on forcefield energy calculations. This value is questionable in so far as the same study's results also suggest that a significant proportion of bioactive conformations actually have higher strain energy than this, with 10% of studied ligands having bioactive conformations with strain energies of >9 kcal/mol and isolated cases having energies as high as 20 kcal/mol. It has also been suggested [23] by Butler et al. that the former study overestimates strain energies by failing to account for the imprecisions and inaccuracies of X-ray crystal structures, although the calculations of the latter study refer to strain energies relative to local rather than global conformational energy minima and are hence of questionable value. Sitzmann et al. have disputed the study of Butler et al. with another rigorous re-examination of crystallographic data [24] to estimate that strain energies of bioactive conformations routinely reach up to 25 kcal/mol.

A more recent study by Avgy-David and Senderowitz uses [25] improved approximations to the X-ray conformation and replaces the global energy minimum of the ensemble with an ensemble average energy to represent a 'solution-phase energy'. On this basis, it is suggested that bioactive conformations will be found in a range of 2–5 kcal/mol above the ensemble average, compared with 4–6 kcal/mol above the global minimum, when calculations are carried out with forcefields.

Indeed, enough studies have been published on this subject to warrant at least two recent review articles. Peach et al. have collected [26] calculations suggesting a wide range of strain energies (up to several tens of kcal/mol) for bioactive conformations and suggest that although in many cases very high calculated strain energies are artefacts stemming from the solution of the crystal structures, there are a small proportion of cases in which the bound ligand genuinely is highly strained compared to its isolated molecule potential energy surface. A similar conclusion was reached by Avgy-David and Senderowitz based on restricted energy minimisations aimed at resolving the same kind of crystal-structure artefacts. Foloppe and Chen report [27] on a variety of studies placing the maximum calculated strain energy of bioactive conformations at anywhere from 5 to 25 kcal/mol (with outlier figures of up to 100 kcal mol^{-1}); the authors of this study conclude that high calculated bioactive strain energies are in many cases spurious, due to the tendency of isolated-molecule calculations to assign higher stabilities to non-practically relevant 'collapsed' conformations that maximise intramolecular contacts (and which available implicit solvent models seem to only partially compensate for). This may be seen as a restatement of the idea that ΔU_{max} must be set high enough to account for the intermolecular interactions that stabilise bioactive conformations in practice.

2 The Role of Quantum Mechanics in Conformational Searching

Typically, the molecular geometry energy functions used in the ranking and (sometimes) generation of conformational ensembles are molecular mechanics forcefields—mathematical functions parameterised to reproduce the properties of a test set of molecules. These can be rapidly evaluated computationally, allowing for timely analysis of flexible molecules with many possible conformers and for analysis of large libraries of flexible molecules. However, any calculation using forcefields is essentially just an extrapolation from the relatively small test set, to some (generally distant) point in chemical space. The inevitable result is a degree of inaccuracy. Quantum mechanical (QM) calculations offer the possibility of removing or reducing this inaccuracy by recalculating the electronic structure for a specific molecule in a specific three-dimensional geometry, from first principles. Many different schemes for QM simulation have been embodied in a variety of software; each scheme represents a differing approximation to the task of exactly solving the Schrodinger equation, with a trade-off between accuracy and computational cost. All successful QM schemes offer improved accuracy over FF, albeit with greatly increased computational cost.

In the context of conformational searching and analysis, quantum mechanical simulation offers two improvements. The first is superior geometries. In many cases, conformational searching includes an optimisation of any generated conformers with respect to their internal energy. Performing a local optimisation using QM calculation rather than a forcefield should result in a structure that more closely resembles the molecule's real three-dimensional geometries, hence those found in crystal structures and other contexts. The second improvement is energetics. For a given three-dimensional geometry, quantum mechanics should give a more accurate estimate of the associated stability. As noted above, one of the principal concerns in carrying out conformational searching with forcefields is the uncertainty in the calculated energies; quantum mechanical calculation removes or reduces this uncertainty. For this reason, QM calculations are often used for dihedral angle scanning. Conversely, of course, it should be noted that such calculations cannot remove the uncertainties associated with approximating free energy by potential energy changes and with drug interactions with the environment. In particular, since bioactive conformations are determined by interactions with a target, there will still be a requirement to retain structures up to a strain energy cut-off ΔU_{max} for this fundamental application.

Perennial attempts have been made to obtain superior estimates of the strain energy (and hence appropriate values of ΔU_{max}) by quantum mechanical modelling of bound ligand poses. An early study by Allen et al. [28], supported by more recent work by Hao et al. [29], has established that the distribution of torsion angles adopted by given molecular fragments over the whole set of available crystal structures is broadly in agreement with the range of low energy torsion values suggested by these approaches. This provides the reassuring conclusion that it is indeed worthwhile searching for bound-state conformational candidates using QM methods. As previously noted, further investigations have reached sharply differing conclusions about the feasible strain energy of bound ligands. Butler et al. conclude [23] that strain energies are typically very low, whereas Sitzmann et al. concluded [24] that strain energies can reach very high values (up to 25 kcal/mol). Both conclusions were based on QM calculations of ligand geometry energies.

There are hence both unresolved questions and potentially substantial advantages in carrying out conformational searching with quantum mechanics. In spite of the latter, higher computational costs have limited their application in the literature. Broadly, there are three possible routes to incorporating QM calculations in a conformational search. One of these is to carry out all energy calculations using QM. This is rigorous but extremely costly; the author is aware of only one such study [30], and this was a demonstration of process using a cheap and low-accuracy QM scheme that is unlikely to give a sufficient level of accuracy for practical applications. A more common approach is to generate conformers using a forcefield and then carry out geometry optimisation and energy recalculation using QM methods, usually for a subset of the generated conformations. A common criterion is to pick the most stable conformers under the forcefield energy calculation, e.g. the first 20 kcal/mol of conformations. This approach is very widely used for analysis of individual molecules and is often successful in reproducing gas-phase properties associated with low energy conformers [31–33]. Unfortunately it is seldom systematically applied to generation of full ensembles for CADD purposes, though Gutten et al. have recently advocated for the desirability and feasibility of its usage [34] and Fu et al. have demonstrated the generation of QM-enhanced ensembles for comparison with bound poses of the bioactive compounds retinoic acid and donepezil [35, 36].

A potentially more efficient approach has been pioneered in the study of Juarez-Jiminez et al. [37]. Conformational ensembles are again generated using classical forcefields. These are then grouped into conformational energy 'basins' of structurally similar conformations. The structure of each basin is used to calculate a conformational free energy, while the lowest energy conformation from each basin is refined in a QM calculation to provide an exemplary QM 'correction' for the whole basin. This technique was used to provide ensembles that were consistent with available conformation-dependent NMR data for a series of phenylethylamines and for the aminoglycoside streptomycin.

Perhaps the most wide-ranging exploration of conformational searching in the context of bioactive conformation discovery has been the study [25] of Avgy-David and Senderowitz. Examining a dataset of 123 FDA-approved drugs for which target-bound crystal structures are available, the authors explored the strain energy of the bound ligands compared to a generated conformational ensemble, with energies calculated and recalculated both by several different forcefields and by a set of differing QM methods. Constrained minimisations were used in an attempt to identify the underlying 'true' or 'energetically feasible' geometries of the bound ligands. The conclusion was that on average, bioactive conformations were found at lower strain energies when energies were calculated using QM methods than with forcefields; but that strain energies were still substantial—frequently several kcal/mol—and still very high (>20 kcal/mol) in rare cases. These conclusions are consistent with the ideas that isolated-molecule QM calculation of energy eliminates many of the errors associated with isolated-molecule forcefield calculations but that isolated-molecule energies are only ever one factor among many that determine bioactive and other relevant geometries—an ever-present complication when carrying out conformational searching for drug discovery.

3 Bioactive Conformation and Rational Drug Design

The bioactive conformation is defined as the geometry adopted by the ligand when it binds to its biological target [38]. The internal energy of the bioactive conformation contributes to the total free energy of binding and is therefore reflected in the protein-ligand potency. Highly potent ligands often bind in a relatively low energy conformation; however, this parameter has to be taken into account during rational drug discovery. When the bioactive conformation of a ligand is available, the internal energy can be calculated with computational methods, which rely on forcefields or QM. Accurate characterisation of the intramolecular interactions in the bioactive conformation allows medicinal chemist to optimise ligand structures so as to reduce the repulsion and increase the number of attractions when the ligand adopts this geometry [39].

Several notable reports have been published [40–43] that emphasised the crucial role of a large number of 'underappreciated' interactions in non-covalent binding. These include CH/π [44, 45], halogen/π [46], cation/π [47], and non-classical hydrogen bonds [48], which are typically not properly parameterised in currently available forcefields (FF) [41]. Furthermore, the role of hydrophobic interactions is vital for biomolecular recognition, but there is still no reliable non-QM predictive method for its quantification [43]. The strength of these interactions can be evaluated [49] with QM methods, which are considered to be a reliable approach for the exploration of non-covalent interactions [50, 51] and an improvement over dispersion potentials in FF. Here, we outline two cases taken from real drug discovery programmes where stabilisation of the bioactive conformation was essential for project progress and where QM approaches were used to address this challenge.

Case 1: Suvorexant [52] is a medication for the treatment of insomnia made by Merck and Co [53] acting as a dual orexin (OX_1 and OX_2) receptor antagonist. Suvorexant was developed through extensive structure-activity relationship (SAR) exploration (Table 1) in conjunction with small-molecule X-ray crystallography and NMR [57, 58] on closely related analogues. This work suggested that the bioactive conformation of suvorexant is a face-to-face (F2F) 'U' shape (Fig. 1a) [58], and this was later confirmed by the crystal structure of the complex (PDB entry 4S0V [56]).

It was proposed [52] that the F2F bioactive conformation is maintained by the intramolecular interaction between the left-hand side (LHS) and right-hand side (RHS) of the ligand (Fig. 1b). To

Table 1
Raw data collected from the literature [54, 55] for the FMO SAR analysis of suvorexant analogues, complex with human orexin OX_2 receptor with (PDB entry 4SOV [56])

Structure	ID	pK_i
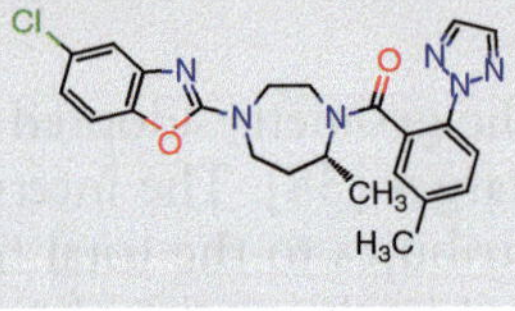	Suvorexant	9.40
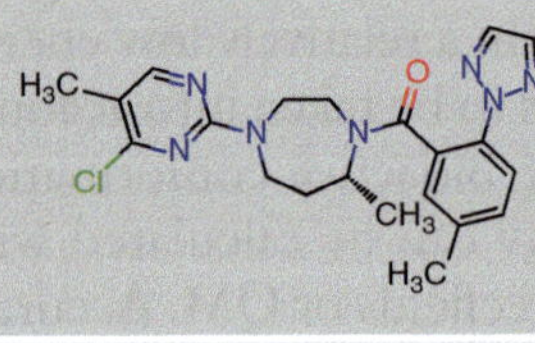	5	9.40
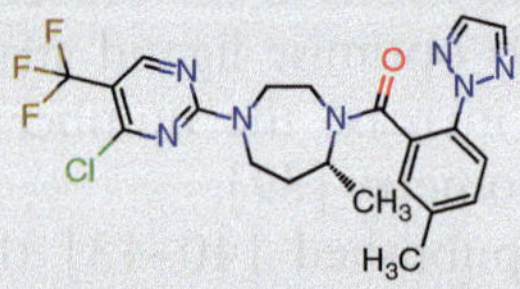	7	9.22
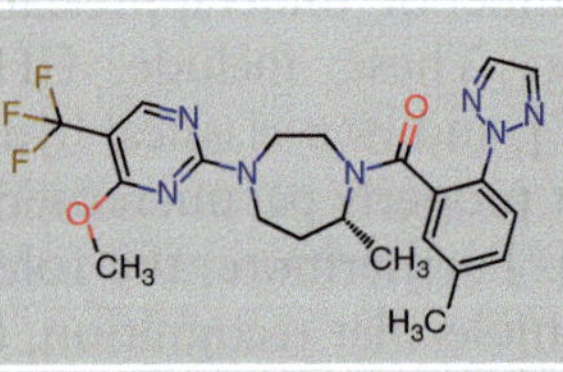	10	9.40
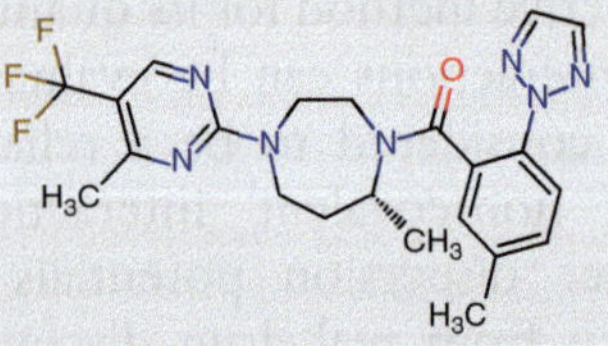	11	9.30
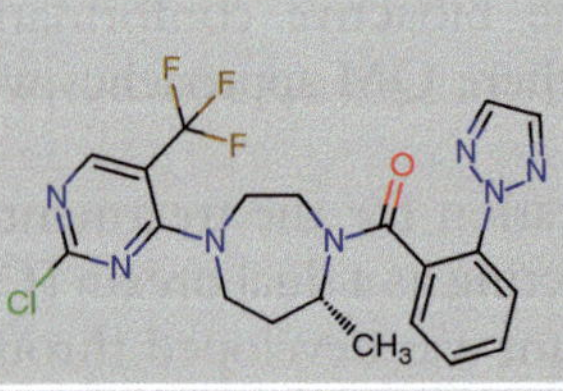	14	8.11
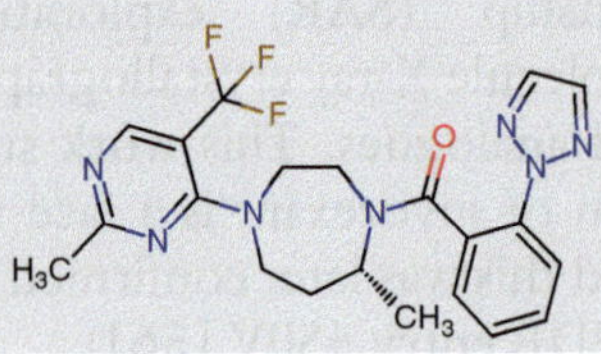	15	7.51

(continued)

Table 1
(continued)

Structure	ID	pK_i
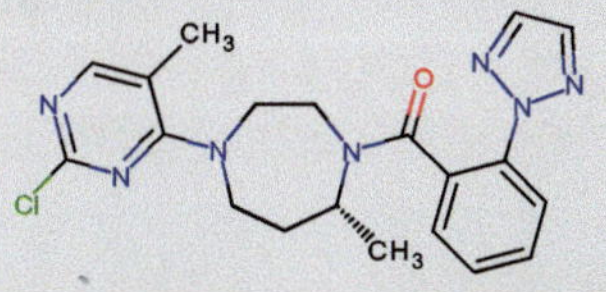	16	8.59
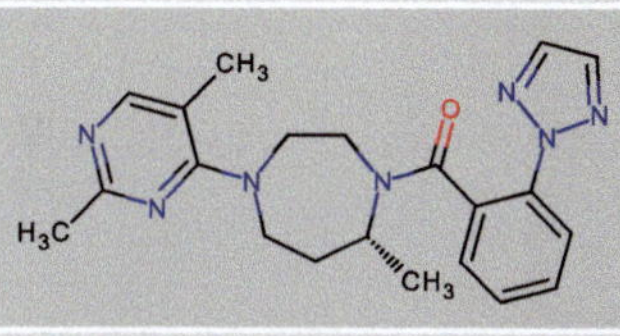	17	7.92
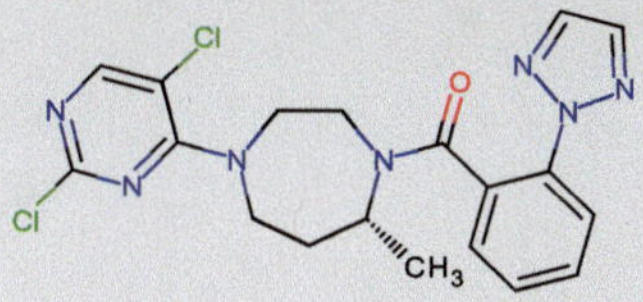	18	8.96
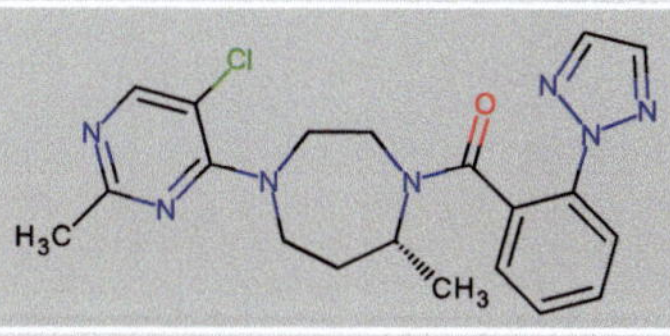	19	8.44
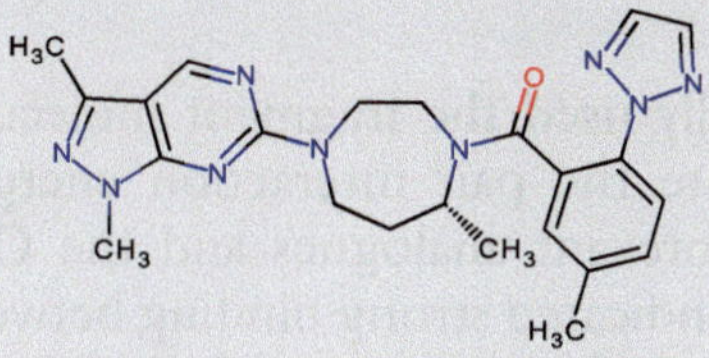	30	9.10
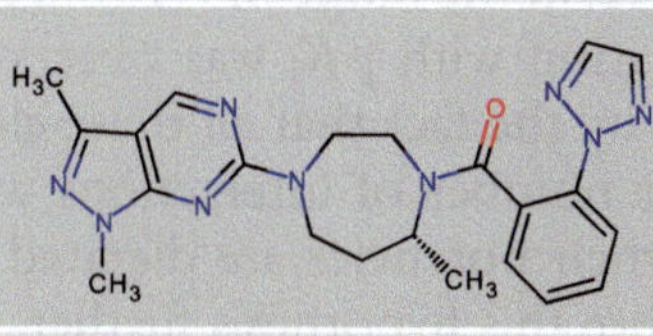	31	8.35
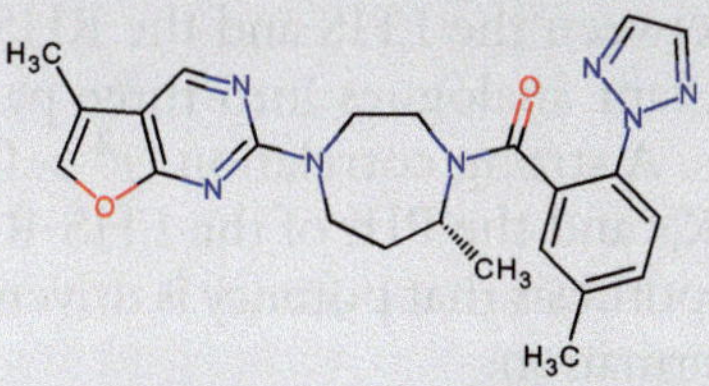	32	9.70
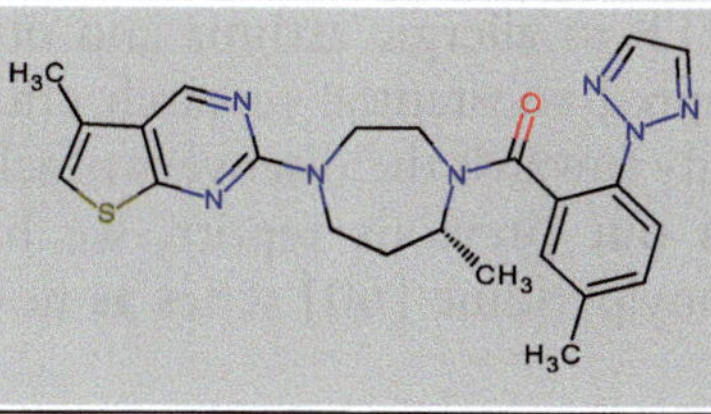	34	9.52

(continued)

Table 1 (continued)

Structure	ID	pK_i
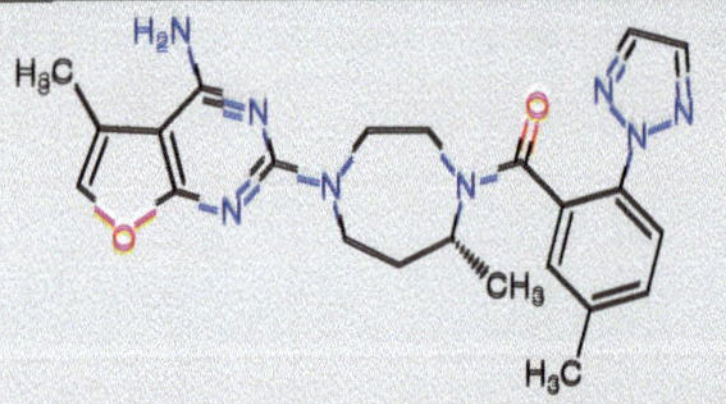	35	8.74
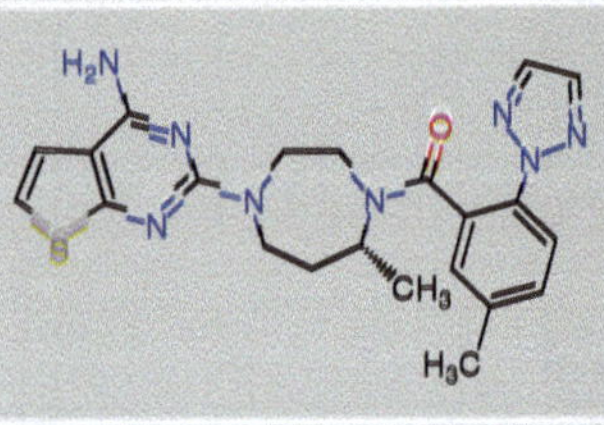	37	9.30
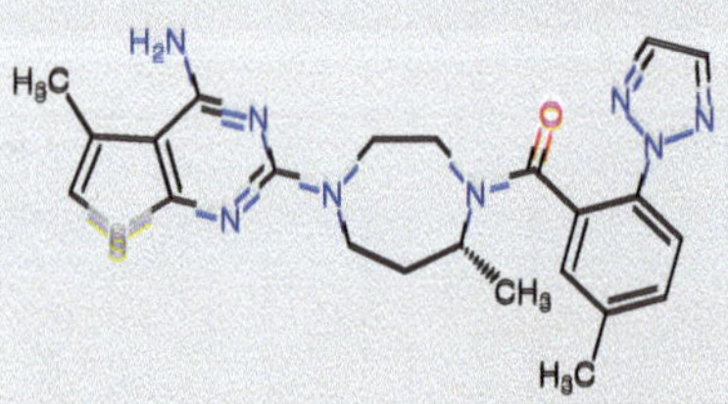	41	9.52

explore this hypothesis, we initially used the fragment molecular orbital (QM) method to calculate the pair interaction energies (PIE) between a number of suvorexant analogues and the OX_2 receptor [59]. These calculations indicated strong binding between these ligands and the receptor with PIE in the range of −113.0 and −99 kcal/mol. However, no correlation with pK_i was observed. This discrepancy has been explained by the fact that all these close suvorexant analogues share a similar number of interactions with the receptor and the differences in their potencies are dictated by intramolecular interactions. To explore this hypothesis further, we used FMO to calculate the PIE between the LHS and the RHS of the ligand by fragmenting suvorexant analogues into three parts: LHS, middle, and RHS fragments. A strong correlation ($r^2 > 0.7$, Fig. 1b) was observed between pK_is and the PIE of the LHS-RHS interaction, consistent with the hypothesis that potency is driven by stabilisation of the bioactive conformation.

Case 2: The involvement of ITK in allergic asthma and other inflammatory disorders has prompted substantial research efforts from the pharmaceutical community towards the discovery of selective inhibitors of this kinase. In our previous report, we have described the discovery of a sulfonylpyridine [60] series as novel,

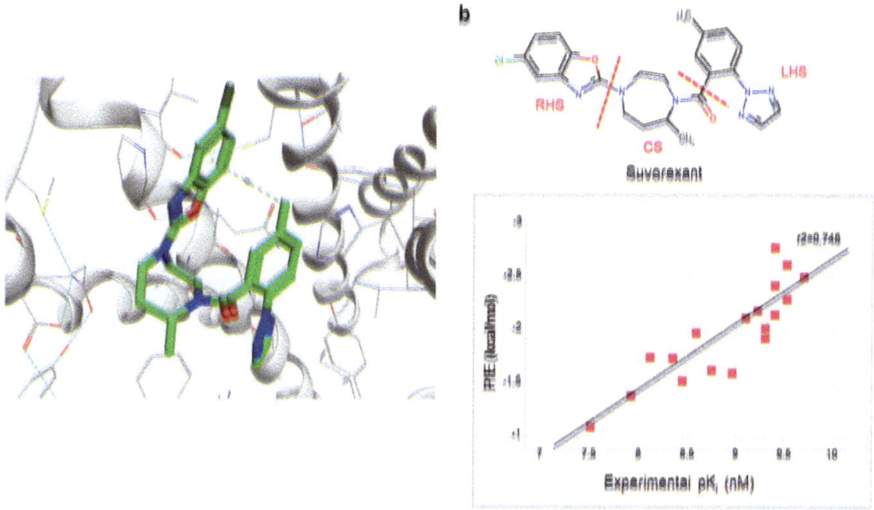

Fig. 1 (**a**) Human orexin OX_2 receptor with suvorexant (PDB entry 4SOV [56]). (**b**) Correlation plot between experimentally measured affinity (suvorexant analogue with receptor) and PIE (RHS and LHS of analogues (Table 1)). PIE is calculated using FMO

sub-nanomolar, selective ITK inhibitors. We have demonstrated how X-ray crystallography and computational tools such as QM dihedral scans, docking, and physiochemical property calculations were applied to progress the hit-to-lead and lead optimisation programmes for this series. Our target was to optimise the bioactive conformation of compound **1** (initial hit, Fig. 2a). The bioactive conformation of compound **1** was initially predicted using a homology model and later confirmed by X-ray crystallography (Fig. 2b). We predicted that the relatively low affinity of compound **1** could be the result of an electrostatic repulsion between the oxygen atoms of the sulfonyl linker and the lone pair of the proximal nitrogen of the pyrimidine core. This repulsion disfavoured the phenyl ring from adopting the desired bioactive conformation required for an interaction with Phe437 (Fig. 2b).

To validate our hypothesis, we performed a dihedral scan for compound **1** using the NWChem QM package [61]. We rotated around the S-C bond in increments of 5°, followed by constrained minimisation and ligand energy calculation. The results of the scan are shown in Fig. 2c. According to this scan, the desired bioactive conformation had a strain of approximately 4 kcal/mol above the energy minimum. Following this observation, compound **2** was designed. The dihedral scan of compound **2** suggested that it would have a much lower conformational strain of 1 kcal/mol in

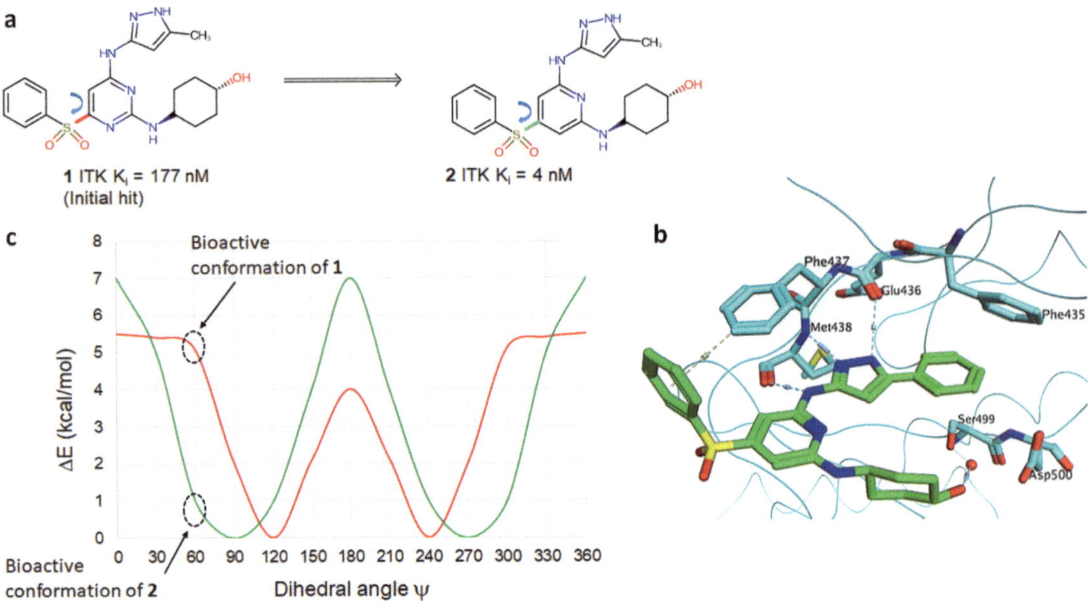

Fig. 2 Results of dihedral torsional scan [60]. (**a**) Compounds 4-methanesulfonyl pyrimidine (compound **1**) and on 4-methanesulfonyl-pyridine (compound **2**). (**b**) X-ray structure of **2** in the ITK kinase domain (resolution 2.6 Å, PDB code 4QD6). Only selected residues of ITK are shown for clarity. (**c**) Dihedral torsional scan performed with NWChem [61]

the bioactive conformation. The experimental results were consistent with this prediction. Compound **2** was very potent, showing a 44-fold increase in ITK potency (ITK $K_i = 4$ nM) compared to compound **1**.

4 Searching for Solution-Phase Conformational Ensembles

Solution-phase conformational ensembles—the subset of conformations that a molecule adopts with high probability in solution—are of interest in drug discovery, since there is evidence that drug candidates will tend to have higher affinity when their target-bound conformation matches (one component of) the solution-phase ensemble [62, 63]. This will presumably lead to lower entropy and enthalpy penalties for binding, as well as more rapid binding kinetics. Proposing a candidate solution-phase conformational ensemble is therefore a specific conformational searching problem of practical interest.

From an algorithmic point of view, this is a problem with a more complete solution than the attempt to produce a candidate bioactive conformation. In the absence of structural information about the target, all that can ever be achieved by conformational searching is a set of realistic candidate structures. By contrast, the fluidity of the solution phase ensures that the molecule will eventually explore all energetically feasible conformations; and the structure of liquid water is well-understood and straightforward to model (though not straightforward to model to high accuracy). Hence, the solution-phase conformational ensemble can in principle be completely described by a sufficiently accurate model, including accurate energetics. Geometric reoptimisation and recalculation of energies using quantum mechanical methods has a clear role to play in this. The two-level approach of Juarez-Jiminez et al. [37], which furthermore initially generates conformations through a solution-phase molecular dynamics simulation, represents an efficient approach to using QM calculations to define a solution-phase conformational ensemble.

The accessibility of the solution-phase conformational ensemble is improved still further by the availability of experimental data that is conformation-dependent. In particular, NMR spectroscopy can provide distance-dependent NOE (nuclear Overhauser effect) coupling and torsion-dependent *J* coupling parameters that provide constraints on the three-dimensional structures that could be present in a solution (the calculations of Juarez-Jiminez et al. [37] were validated against NMR data). This has been exploited by the well-established NAMFIS technique [64] (and its derivative DISCON [65]). In these methods, a conformational ensemble is generated for a molecule by any of the available computational methods; NOEs and *J* couplings are then calculated for each generated conformation by standard formulae [66] and used to optimise the population of each conformer in solution. The result is an ensemble that is consistent with the experimentally measured values, hence leading to a model solution-phase ensemble generated by simulation and experimental data. Historically the in silico calculations have been carried out using forcefield methods.

QM calculations can be used to provide more realistic conformation geometries for this process. However, there are also well-established quantum methods for estimating chemical shifts (and related properties) [67]. Unlike NOE or *J* coupling parameters, these cannot be simply estimated from molecular geometry or connectivity. The chemical shift capability has been used to determine 2D molecular structures, i.e. the atomic connectivity of a molecule, with a single 3D structure assumed [68, 69]. It has also been used in the context of conformations generated and refined entirely in silico, for verification against experimental NMR data, hence facilitating an approach to solution-phase ensembles [70–72]. Recently developed procedures, though, return to the NAMFIS-like approach of combined experimental and in silico information. Additional use is made of the chemical shifts and chemical shift-related, conformation-dependent, residual chemical shift anisotropy (RCSA) and residual dipolar couplings (RDCs), calculated by QM methods [73]. The overall process of matching solution-phase ensembles to experimental data then encompasses NOE coupling, calculated for QM-optimised conformational geometries, and QM-calculated *J* coupling, chemical shifts, and RCSA parameters. QM calculations are hence placed at the heart of solution-phase conformational ensemble simulation.

5 The Correct Choice of Quantum Approximation for Conformational Searching

As previously noted, available software embodies a variety of different approximations to the 'true' equations of quantum mechanics. These are used to generate linear combinations of basis sets, standardised atom-centred functions that describe the position of electrons within molecules. In general, the more accurate these approximations, and the larger the basis set (and hence the more complete the description of the electrons), the greater the computational cost. For any QM application, questions therefore arise of how accurate a set of quantum approximations and how large a basis set to use; and hence how costly a set of calculations will be required, to obtain results that reach the required level of physical accuracy.

In a very loose sense, the cost-accuracy ordering of commonly used quantum approximations is coupled-cluster methods (CCSD, CCSD(T)) > second order Moller-Plesset perturbation theory (MP2) > variations on density functional theory (DFT) methods > Hartree-Fock (HF). The nomenclature of commonly used basis sets of descending size is correlation consistent basis sets (aug-cc-pVTZ, cc-pVTZ > aug-cc-pVDZ, cc-pVDZ) > large 'Pople' basis sets (6-311++G**, 6-311+G**) > medium Pople basis sets (6-31G**, 6-31+G*, 6-31G*) > small Pople basis sets (3-21+G*, 3-21G*). Calculations with multiple basis set sizes can be extrapolated to give the 'complete basis set' (CBS) limit, which in principle represents maximal accuracy for a given approximation scheme. Computational chemistry professionals often choose to use DFT schemes where possible and to pair them with large or medium Pople basis sets. This facilitates execution of large numbers of calculations at purportedly reasonable accuracy.

It should be noted that in many studies, the 'geometry' and 'energy' aspects of the calculations are treated separately. Optimisation (or minimisation, or relaxation) of molecular geometry with respect to molecular energy is much more computationally costly than energy calculation. It is therefore common for generated conformations to be 'refined' with a geometry optimisation using a lower-cost QM scheme, followed by energy calculation using a higher-cost higher-accuracy scheme. Alternatively, the QM geometry optimisation may be neglected altogether, with energy calculated for a geometry defined by the forcefield initially used for conformation generation.

Studies concerning QM applied to conformational searching can be divided into two categories. In the first category, multiple QM schemes are compared with a small, highly restricted dataset in an attempt to determine the optimal scheme. In the second category, QM calculations are applied for comparison with, and explanation of, existing experimental results. Perhaps regrettably, there is little overlap. Studies in the second category usually use a single method based on perceived breadth of literature usage, without reference to conformation-specific studies in the first category. Examples of both types of study are discussed below.

In studies that have addressed the necessary level of accuracy required for conformational searching, the favoured method is comparison with very high-cost/high-accuracy QM calculations (indeed, there appear to be more of these than there are direct comparisons with experiment). Following this approach, Sellers et al. used [74] structures with MP2/6-311+G** level geometry optimisation and energies calculated at the CCSD(T) level with complete basis set (CBS) as benchmarks and hence conclude that HF/6-311+G** geometry optimisation with energies calculated at the MP2/CBS level is the preferred level of calculation, although energy calculation using the density functional theory (DFT) technique B3LYP-D3 with a 6-311+G** basis set gave almost the same accuracy at much lower computational cost. Fu et al. used [35] structures geometry-optimised at HF/6-31G* level throughout but used benchmark energy calculations at the MP2/CBS level to conclude that M06-2X [75] with an aug-cc-pVDZ or aug-cc-pVTZ basis set has an acceptable level of accuracy. Similarly, Valdes et al. used [76] CCSD(T)/CBS benchmark calculations on peptide conformations to conclude that M06-2X and TPSS-D [77] DFT methods provide a high level of accuracy, while Li et al. conformationally analysed [78] the trialanine molecule with geometry optimisation at the HF/6-31G* level and used benchmark density-fitted local CCSD(T) (DF-LCCSD(T)) with CBS calculations to conclude that the DFT methods M06-2X-D and M06-L-D, with one of a variety of low-cost basis sets, provided sufficient accuracy. Contrastingly, Rezacs et al. concluded [79] that while a wide range of DFT methods give roughly equivalent (and adequate) accuracies in reproducing CCSD(T)/CBS benchmarks when using TZVPD (equivalent to large Pople) basis sets, the best results are obtained for the PBE0 functional [80], while results from M06-2X are more disappointing. Finally, Yu et al. have used [81] benchmark calculations at the CCSD/6-311++G** level of theory to conclude that the DFT method BHandHLYP/6-311++G** [82] offers an acceptable level of accuracy. Overall, the set of studies comparing QM conformational energies with high-level (in principle maximal accuracy) QM calculations is not completely consistent, but the implication is that the M06-2X functional can provide a good match with much higher-cost and higher-accuracy methods, albeit when using a very large basis set (such as aug-cc-pVTZ).

A rare study of quantum approximations in comparison with a set of experimental conformational data (albeit for a small set of ten compounds) was carried out by Riley et al. The conclusion was reached [83] that MP2/aug-cc-pVDZ calculations provided optimal accuracy; out of the DFT methods tested, MPWB95/cc-pVTZ [84] provided optimal accuracy; and out of DFT methods coupled with Pople basis sets (and therefore having relatively low computational cost), B1B95/6-31+G* [85] provided optimal accuracy (it should be noted that this study did not test the Minnesota set of DFT methods, including M06-2X). The decrease in accuracy from MP2/aug-cc-pVDZ to MPWB95/cc-pVTZ is not dramatic (6.5% average energy error to 8% energy error), though the decrease to B1B95 with the smaller basis set 6-31+G* is more substantial (12% average energy error). This study also emphasises that the very smallest available Pople basis sets, 3-21G* and 3-21+G*, are simply unsuitable for conformational calculations, with average energy errors often in excess of 30% depending on the DFT method used.

Overall, the consensus of these conformational QM testing exercises seems to be that DFT methods can viably approximate the highest levels of theoretical accuracy, albeit with potentially large (hence computationally expensive) basis sets (Table 2).

Table 2
Comparison of 'optimal' QM methods for conformers according to various literature sources

Optimal QM method	Comparator	Reference
MP2/CBS *or* B3LYP-D3/6-311+G** (acceptable substitute)	Same ensemble; CCSD(T)/CBS	[74]
M06-2X/aug-cc-pVTZ	Same ensemble; MP2/CBS	[35]
M06-2X/6-311+G(2df,2pd) *or* TPSS-D/6-311++G(3df,3pd)	Same ensemble; CCSD(T)/CBS	[76]
M06-2X-D/aug-cc-pVDZ	Same ensemble; DF-LCCSD(T)/ CBS	[78]
PBE0/TZVPD	Same ensemble; CCSD(T)/CBS	[79]
BHandHLYP/6-311++G**	Same ensemble; CCSD(T)/6-311+ +G**	[81]
MP2/aug-cc-pVTZ *or* MPWB95/cc-pTZ (best DFT) *or* B1B95/6-31+G* (best DFT w/ Pople basis)	Experimental ensemble data	[83]
M06-2X/6-31G*	Bioactive conformer searching	[25]

'Same ensemble' indicates that optimal methods were assessed by comparing energies calculated for one set of conformers by each trial method, with the energies of the same conformer set calculated by the benchmark method

By contrast, studies in which QM conformational calculations have been applied practically for interpretation of experimental data have in general taken the (perhaps more pragmatic) approach of selecting basis sets on the basis of computational affordability. The possibility of inaccuracy arising from too low level a quantum approximation is accepted as one of the many risks surrounding any scientific endeavour. Juarez-Jiminez et al. have taken the opportunity [37] afforded by the relative small number of QM calculations inherent in their protocol to employ a relatively costly calculation method, with energies calculated at the MP2/aug-cc-pVTZ level, although geometries were optimised at the B3LYP/6-31G* level [82]. For calculating strain energies of bound ligand poses, Sitzmann et al. have taken a similar approach [24], with B3LYP/6-31G* geometry optimisation followed up by a medium-cost energy calculation at the B3LYP/6-311++G (3df,2p) level. Hao et al. have opted [29] for yet cheaper calculations, all at the B3LYP/6-311++G** level. In their older study, Butler et al. have opted [23] for the popular and cost-efficient B3LYP/6-31G* calculation level. Similarly, the QM component of Fu et al.'s QM/MM target-bound ligand calculations was carried out [35] at the HF/6-31G* level.

It is notable that many of these studies use the well-known B3LYP method either for geometry optimisation or for energy calculation, in spite of the method's almost-equally-well-known inability to model long-range interactions, which may be critical to the conformational behaviour of larger molecules. A solution to this problem is a dispersion correction, in which a straightforward $1/r^6$ energy function is added to the system calculations to crudely capture one component of these long-range interactions, but this solution does not seem to have been regularly adopted in QM conformational studies. It has been suggested that in many cases a cancellation of errors may occur for uncorrected DFT between lack of dispersion and internal basis set superposition error (BSSE), which may vindicate these studies' choice of functional to a certain extent [86].

By contrast, while the study of Avgy-David and Senderowitz uses [25] very low-cost basis sets at the 6-31G* level (presumably necessitated by the study of large molecular and conformational libraries), the B3LYP/6-31G* calculation method is contrasted with a BSSE and dispersion-corrected equivalent, B3LYP/6-31G*/gcP-D3 [86], and with the M06-2X/6-31G* method. As noted above, these display clear but not dramatic improvements over forcefield energy calculation, with M06-2X demonstrating the largest advantage. This is the only study the author is aware of in which QM methods are directly assessed for bioactive conformer searching.

Recent studies in which solution-phase conformational ensembles are generated have used the B3LYP DFT method with a medium Pople basis set for geometry optimisation and energy calculation and the MPW1PW91 method [84] with a medium or large Pople basis set (6-31G** or 6-311+G(2d,p)) for calculation of NMR parameters [70, 73]. Nazarski et al. note [70] that B3LYP performs better (with equivalent basis sets) than the methods M06-2X, ωB97X-D, or the dispersion-corrected B3LYP-GD3BJ. This is despite M06-2X performing better at reproducing accurate high-level quantum approximations in many studies. The authors note that non-B3LYP methods result in more collapsed conformations with more intramolecular interactions, so it may be that in this case B3LYP's documented lack of dispersion interactions actually provides a serendipitous counterbalance for the tendency of isolated-molecule energy calculations to favour collapsed conformations, as noted by Foloppe and Chen [27].

As with molecular mechanics isolated-molecule energy calculations, it is common for quantum mechanical energy calculations to include a crude approximation to a condensed phased environment in the form of an implicit solvent model (ISM) (sometimes in the form of a polarisable continuum model (PCM)) [87]. Unfortunately the effect of these on the accuracy of QM conformational searching has not been systematically explored, with most studies that carry out comparisons to condensed phase data simply assuming that use of an ISM will give superior results and using them for all calculations. This is probably a result of the relatively low additional computational cost associated with these models. Butler et al. [23] find that use of ISMs give slightly better geometries in reproduction of protein-bound ligand poses and much improved energy gaps relative to local structural energy minima. On the other hand, Sitzmann et al. find [24] that use of ISMs makes little difference to the overall strain energy of bound ligand poses. This is supported, over a limited dataset, by the QM/MM calculations of Fu et al. [35]. The difference may be due to Butler et al.'s focus on local energy minima rather than global energy minima for calculation of the strain energy, which can be expected to more closely resemble a given bound structure, with the application of a simulated condensed phase environment. At least one study, though, reports that full correspondence between generated conformations for Aib residue containing peptides and X-ray crystal data is only achieved with the aid of an ISM [88].

6 Conclusions and Outlook

Comprehensive molecular analysis in silico requires the generation and ranking of ensembles of three-dimensional molecular structures—conformations. Quantum mechanical calculations offer more realistic geometries and more accurate energies for generated conformations than classical forcefield methods. The much higher computational cost of QM methods relative to more conventional forcefield-, data-, or geometry-based approaches has prevented their widespread adoption by computer-aided drug design (CADD) professionals; their main value appears to be in in-depth high-accuracy studies of small numbers of molecules. This approach is supported by the finding that cheaper, more accessible QM methods (including smaller basis sets) have difficulty reaching the full accuracy QM is capable. The work of Juarez-Jiminez, in which small numbers of conformations are treated as representative of large volumes of conformational space for the purposes of QM calculations, may point the way forward to wider (if not universal) adoption. The choice of whether to use QM calculations must also be influenced by the observation that for one of the central tasks of conformational searching in drug discovery—generation of candidate bioactive conformations—QM calculation can only eliminate one source of error in conventional analyses, the energetic inaccuracy of forcefields, and cannot overcome the fundamental complication that isolated-molecule calculations (quantum or forcefield) cannot capture the interaction with a structurally unknown target.

In this context, the most immediate use for QM in conformational searching may be in the simulation of solution-phase conformational ensembles. QM-calculated conformation-dependent NMR properties such as chemical shifts, RCSAs, and RDCs can enable a computer-generated conformational ensemble to be matched to experimental NMR data for a compound and hence to provide a theoretically and observationally supported approximation to the solution-phase conformational ensemble, which can be used in further molecular design.

Apart from increased computer power and cheaper methods with higher accuracy, QM calculations in conformational searching would benefit from further and more systematic testing of differing methods for real applications (such as bioactive conformation elucidation, SAR analysis, and simulation of solution-phase ensembles) and comparison with experimental data. The author looks forward to reading such studies in years to come.

References

1. Watts KS, Dalal P, Murphy RB, Sherman W, Friesner RA, Shelley JC (2010) ConfGen: a conformational search method for efficient generation of bioactive conformers. J Chem Inf Model 50:534–546
2. Yuan X, Xu Y (2018) Recent trends and applications of molecular modeling in GPCR(−) ligand recognition and structure-based drug design. Int J Mol Sci 19:E2105
3. Sheng Y, Watanabe H, Maruyama K, Watanabe C, Okiyama Y, Honma T, Fukuzawa K, Tanaka S (2018) Towards good correlation between fragment molecular orbital interaction energies and experimental IC50 for ligand binding: a case study of p38 MAP kinase. Comput Struct Biotechnol J 16:421–434
4. Labute P (2010) LowModeMD: implicit low-mode velocity filtering applied to conformational search of macrocycles and protein loops. J Chem Inf Model 50:792–800

5. Miteva MA, Guyon F, Tuffery P (2010) Frog2: efficient 3D conformation ensemble generator for small compounds. Nucleic Acids Res 38: W622–W627
6. Macrae CF, Bruno IJ, Chisholm JA, Edgington PR, McCabe P, Pidcock E, Rodriguez-Monge-L, Taylor R, van de Streek J, Wood PA (2008) Mercury CSD 2.0 - new features for the visualization and investigation of crystal structures. J Appl Crystallogr 41:466–470
7. Cole J, Groom CR, Korb O, McCabe P, Shields GP (2016) Knowledge-based optimization of molecular geometries using crystal structures. J Chem Inf Model 56:652–661
8. Hawkins PCD, Skillman AG, Warren GL, Ellingson BA, Stahl MT (2010) Omega 2.5.1.4, OpenEye Scientific Software, Santa Fe, NM
9. Hawkins PCD, Nicholls A (2012) Conformer generation with OMEGA: learning from the data set and the analysis of failures. J Chem Inf Model 52:2919–2936
10. O'Boyle NM, Vandermeersch T, Flynn CJ, Maguire AR, Hutchison GR (2011) Confab - systematic generation of diverse low energy conformers. J Chem 3:8
11. Friedrich N-O, Flachsenberg F, Meyder A, Sommer K, Kirchmair J, Rarey M (2019) Conformator: a novel method for the generation of conformer ensembles. J Chem Inf Model 59:731–742
12. Klett J, Cortes-Cabrera A, Gil-Redondo R, Gago F, Morreale A (2014) ALFA: automatic ligand flexibility assignment. J Chem Inf Model 54:314–323
13. Kirchmair J, Wolber G, Laggner C, Langer T (2006) Comparative performance assessment of the conformational model generators omega and catalyst: a large-scale survey on the retrieval of protein-bound ligand conformations. J Chem Inf Model 46:1848–1861
14. Agrafiotis DK, Gibbs AC, Zhu F, Izrailev S, Martin E (2007) Conformational sampling of bioactive molecules: a comparative study. J Chem Inf Model 47:1067–1086
15. Ebejer J-P, Morris GM, Deane CM (2012) Freely available conformer generation methods: how good are they? J Chem Inf Model 52:1146–1158
16. Chen I-J, Foloppe N (2013) Tackling the conformational sampling of larger flexible compounds and macrocycles in pharmacology and drug discovery. Bioorg Med Chem 21:7898–7920
17. Hawkins PCD (2017) Conformation generation: the state of the art. J Chem Inf Model 57:1747–1756
18. Wojciechowski M, Lesyng B (2004) Generalized Born model: analysis, refinement and applications to proteins. J Phys Chem B 108:18368–18376
19. Habgood M (2017) Bioactive focus in conformational ensembles: a pluralistic approach. J Comput Aided Mol Des 31:1073–1083
20. Alzate-Morales JH, Contreras R, Soriano A, Tunon I, Silla E (2007) A computational study of the protein-ligand interactions in CDK2 inhibitors: using quantum mechanics/molecular mechanics interaction energy as a predictor of the biological activity. Biophys J 92:430–439
21. Moraca F, Acharya K, Melillo B, Smith ABI, Chaiken I, Abrams CF (2016) Computational evaluation of HIV-1 gp120 conformations of soluble trimeric gp140 structures as targets for de novo docking of first and second-generation small-molecule CD4 mimics. J Chem Inf Model 56:2069–2079
22. Perola E, Charifson PS (2004) Conformational analysis of drug-like molecules bound to proteins: an extensive study of ligand reorganization upon binding. J Med Chem 47:2499–2510
23. Butler KT, Luque JF, Barril X (2009) Toward accurate relative energy predictions of the bioactive conformation of drugs. J Comput Chem 30:601–610
24. Sitzmann M, Weidlich IE, Filippov IV, Liao C, Peach ML, Ihlenfeldt W-D, Karki RG, Borodina YV, Cachau RE, Nicklaus MC (2012) PDB ligand conformational energies calculated quantum-mechanically. J Chem Inf Model 52:739–756
25. Avgy-David HH, Senderowitz H (2015) Toward focusing conformational ensembles on bioactive conformations: a molecular mechanics/quantum mechanics study. J Chem Inf Model 55:2154–2167
26. Peach ML, Cachau RE, Nicklaus MC (2017) Conformational energy range of ligands in protein crystal structures: the difficult quest for accurate understanding. J Mol Recognit 30: e2618
27. Foloppe N, Chen I-J (2019) Energy windows for computed compound conformers: covering artefacts or truly large reorganization energies? Future Med Chem 11:97–118
28. Allen FH, Harris SE, Taylor R (1996) Comparison of conformer distributions in the crystalline state with conformational energies calculated by ab initio techniques. J Comput Aided Mol Des 10:247–254
29. Hao M-H, Haq O, Muegge I (2007) Torsion angle preference and energetics of small-

molecule ligands bound to proteins. J Chem Inf Model 47:2242–2252

30. Supady A, Blum V, Baldauf C (2015) First-principles molecular structure search with a genetic algorithm. J Chem Inf Model 55:2338–2348
31. Rossi M, Chutia S, Scheffler M, Blum V (2014) Validation challenge of density-functional theory for peptides-example of Ac-Phe-Ala5-LysH+. J Phys Chem A 118:7349–7359
32. Schubert F, Rossi M, Baldauf C, Pagel K, Warnke S, von Helden G, Filsinger F, Kupser P, Meijer G, Salwiczek M, Koksch B, Scheffler M, Blum V (2015) Exploring the conformational preferences of 20-residue peptides in isolation: Ac-Ala19-Lys + H + vs. Ac-Lys-Ala19 + H+ and the current reach of DFT. Phys Chem Chem Phys 17:7373–7385
33. Lukova K, Nesvadba R, Uhlikova T, Obenchain DA, Wachsmuth D, Grabow J-U, Urban S (2018) Ab initio conformational analysis of 1,2,3,4-tetrahydroquinoline and the high-resolution rotational spectrum of its lowest energy conformer. Phys Chem Chem Phys 20:14664–14670
34. Gutten O, Bim D, Rezac J, Rulisek L (2018) Macrocycle conformational sampling by DFT-D3/COSMO-RS methodology. J Chem Inf Model 58:48–60
35. Fu Z, Li X, Merz KM Jr (2012) Conformational analysis of free and bound retinoic acid. J Chem Theory Comput 8:1436–1448
36. Fu Z, Li X, Miao Y, Merz KM Jr (2013) Conformational analysis and parallel QM/MM X-ray refinement of protein bound anti-Alzheimer drug donepezil. J Chem Theory Comput 9:1686–1693
37. Juarez-Jiminez J, Barril X, Orozco M, Pouplana R, Luque JF (2015) Assessing the suitability of the multilevel strategy for the conformational analysis of small ligands. J Phys Chem B 119:1164–1172
38. Butler K, Javier Luque F, Barril X (2009) Towards accurate energy predictions of the bioactive conformation of drugs. J Comput Chem 30(4):601–610
39. Heifetz A, Southey M, Morao I, Townsend-Nicholson A, Bodkin MJ (2018) Computational methods used in hit-to-lead and lead optimization stages of structure-based drug discovery. Methods Mol Biol 1705:375–394
40. Tong Y, Mei Y, Li YL, Ji CG, Zhang JZ (2010) Electrostatic polarization makes a substantial contribution to the free energy of avidin-biotin binding. J Am Chem Soc 132:5137–5142
41. Raha K, Peters MB, Wang B, Yu N, Wollacott AM, Westerhoff LM, Merz KM Jr (2007) The role of quantum mechanics in structure-based drug design. Drug Discov Today 12:725–731
42. Beratan DN, Liu C, Migliore A, Polizzi NF, Skourtis SS, Zhang P, Zhang Y (2015) Charge transfer in dynamical biosystems, or the treachery of (static) images. Acc Chem Res 48:474–481
43. Bissantz C, Kuhn B, Stahl M (2010) A medicinal chemist's guide to molecular interactions. J Med Chem 53:5061–5084
44. Ozawa T, Okazaki K, Kitaura K (2011) CH/pi hydrogen bonds play a role in ligand recognition and equilibrium between active and inactive states of the beta2 adrenergic receptor: an ab initio fragment molecular orbital (FMO) study. Bioorg Med Chem 19:5231–5237
45. Fedorov DG, Nagata T, Kitaura K (2012) Exploring chemistry with the fragment molecular orbital method. Phys Chem Chem Phys 14:7562–7577
46. Lu Y-X, Zou J-W, Wang Y-H, Yu Q-S (2007) Substituent effects on noncovalent halogen/π interactions: theoretical study. Int J Quantum Chem 107:1479–1486
47. Gallivan JP, Dougherty DA (1999) Cation-pi interactions in structural biology. Proc Natl Acad Sci U S A 96:9459–9464
48. Johnston RC, Cheong PH (2013) C-H...O non-classical hydrogen bonding in the stereomechanics of organic transformations: theory and recognition. Org Biomol Chem 11:5057–5064
49. Crespo A, Rodriguez-Granillo A, Lim VT (2017) Quantum-mechanics methodologies in drug discovery: applications of docking and scoring in lead optimization. Curr Top Med Chem 17:2663–2680
50. Yu N, Li X, Cui G, Hayik SA, Merz KM II (2006) Critical assessment of quantum mechanics based energy restraints in protein crystal structure refinement. Protein Sci 15:2773–2784
51. Fedorov DG, Kitaura K (2007) Extending the power of quantum chemistry to large systems with the fragment molecular orbital method. J Phys Chem A 111:6904–6914
52. Wallace DJ, Mangion I, Coleman P (2016) Discovery and chemical development of suvorexant - a dual orexin antagonist for sleep disorder. In: Comprehensive accounts of pharmaceutical research and development: from discovery to late-stage process development, vol 1. American Chemical Society, Washington, DC, pp 1–36

53. Baxter CA, Cleator E, Brands KMJ, Edwards JS, Reamer RA, Sheen FJ, Stewart GW, Strotman NA, Wallace DJ (2011) The first large-scale synthesis of MK-4305: a dual orexin receptor antagonist for the treatment of sleep disorder. Org Process Res Dev 15:367–375
54. Roecker AJ, Mercer SP, Harrell CM, Garson SL, Fox SV, Gotter AL, Prueksaritanont T, Cabalu TD, Cui D, Lemaire W, Winrow CJ, Renger JJ, Coleman PJ (2014) Discovery of dual orexin receptor antagonists with rat sleep efficacy enabled by expansion of the acetonitrile-assisted/diphosgene-mediated 2,4-dichloropyrimidine synthesis. Bioorg Med Chem Lett 24:2079–2085
55. Roecker AJ, Mercer SP, Bergman JM, Gilbert KF, Kuduk SD, Harrell CM, Garson SL, Fox SV, Gotter AL, Tannenbaum PL, Prueksaritanont T, Cabalu TD, Cui D, Lemaire W, Winrow CJ, Renger JJ, Coleman PJ (2015) Discovery of diazepane amide DORAs and 2-SORAs enabled by exploration of isosteric quinazoline replacements. Bioorg Med Chem Lett 25:4992
56. Yin J, Mobarec JC, Kolb P, Rosenbaum DM (2015) Crystal structure of the human OX2 orexin receptor bound to the insomnia drug suvorexant. Nature 519:247–250
57. McGaughey G, Bayly CI, Cox CD, Schreier JD, Breslin MJ, Bogusky M, Pitzenberger S, Ball R, Coleman PJ (2014) Shaping suvorexant: application of experimental and theoretical methods for driving synthetic designs. J Comput Aided Mol Des 28:5–12
58. Cox CD, McGaughey GB, Bogusky MJ, Whitman DB, Ball RG, Winrow CJ, Renger JJ, Coleman PJ (2009) Conformational analysis of N,N-disubstituted-1,4-diazepane orexin receptor antagonists and implications for receptor binding. Bioorg Med Chem Lett 19:2997–3001
59. Heifetz A, Chudyk EI, Gleave L, Aldeghi M, Cherezov V, Fedorov DG, Biggin PC, Bodkin MJ (2016) The fragment molecular orbital method reveals new insight into the chemical nature of GPCR-ligand interactions. J Chem Inf Model 56:159–172
60. Trani G, Barker JJ, Bromidge SM, Brookfield FA, Burch JD, Chen Y, Eigenbrot C, Heifetz A, Ismaili MH, Johnson A, Krulle TM, MacKinnon CH, Maghames R, McEwan PA, Montalbetti CA, Ortwine DF, Perez-Fuertes Y, Vaidya DG, Wang X, Zarrin AA, Pei Z (2014) Design, synthesis and structure-activity relationships of a novel class of sulfonylpyridine inhibitors of interleukin-2 inducible T-cell kinase (ITK). Bioorg Med Chem Lett 24:5818–5823
61. Valiev M, Bylaska EJ, Govind N, Kowalski K, Straatsma TP, Van Dam HJJ, Wang D, Nieplocha J, Apra E, Windus TL, de Jong WA (2010) NWChem: a comprehensive and scalable open-source solution for large scale molecular simulations. Comput Phys Commun 181:1477–1489
62. Blundell CD, Packer MJ, Almond A (2013) Quantification of free ligand conformational preferences by NMR and their relationship to the bioactive conformation. Bioorg Med Chem 21:4976–4987
63. Foloppe N, Chen I-J (2016) Towards understanding the unbound state of drug compounds: implications for the intramolecular reorganization energy upon binding. Bioorg Med Chem 24:2159–2189
64. Cicero DO, Barbato G, Bazzo R (1995) NMR analysis of molecular flexibility in solution: a new method for the study of complex distributions of rapidly exchanging conformations. Application to a 13-residue peptide with an 8-residue loop. J Am Chem Soc 117:1027–1033
65. Atasoylu O, Furst G, Risatti CA, Smith ABI (2010) Solution structure of (+)-spongistatin 1 in DMSO. Org Lett 12:1788–1791
66. Karplus M (1959) Contact electron-spin coupling of nuclear magnetic moments. J Chem Phys 30:11–15
67. Schreckenbach G, Ziegler T (1998) Density functional calculations of NMR chemical shifts and ESR g-tensors. Theor Chem Accounts 99:71–82
68. Grimblat N, Sarotti AM (2016) Computational chemistry to the rescue: modern toolboxes for the assignment of complex molecules by GIAO NMR calculations. Chem Eur J 22:12246–12261
69. Navarro-Vazquez A (2016) State of the art and perspectives in the application of quantum chemical prediction of 1H and 13C chemical shifts and scalar couplings for structural elucidation of organic compounds. Magn Reson Chem 55:29–32
70. Nazarski RB, Walejko P, Witkowski S (2016) Multi-conformer molecules in solutions: an NMR-based DFT/MP2 conformational study of two glucopyranosides of a vitamin E model compound. Org Biomol Chem 14:3142–3158
71. Grimme S, Bannwarth C, Dohm S, Hansen A, Pisarek J, Pracht P, Seibert J, Neese F (2017) Fully automated quantum-chemistry-based computation of spin–spin-coupled nuclear magnetic resonance spectra. Angew Chem Int Ed Engl 56:14763–14769

72. Aliev AE, Mia ZA, Khaneja HS, King FD (2012) Structures in solutions from joint experimental-computational analysis: applications to cyclic molecules and studies of noncovalent interactions. J Phys Chem A 116:1093–1109
73. Liu Y, Sauri J, Mevers E, Peczuh MW, Hiemstra H, Clardy J, Martin GE, Thomas Williamson R (2017) Unequivocal determination of complex molecular structures using anisotropic NMR measurements. Science 356:43–49
74. Sellers BD, James NC, Gobbi A (2017) A comparison of quantum and molecular mechanical methods to estimate strain energy in druglike fragments. J Chem Inf Model 57:1265–1275
75. Zhao Y, Truhlar DG (2008) The M06 suite of density functionals for main group thermochemistry, thermochemical kinetics, noncovalent interactions, excited states, and transition elements: two new functionals and systematic testing of four M06-class functionals and 12 other functionals. Theor Chem Accounts 120:215–241
76. Valdes H, Pluhackova K, Pitonak M, Rezac J, Hobza P (2008) Benchmark database on isolated small peptides containing an aromatic side chain: comparison between wave function and density functional theory methods and empirical forcefield. Phys Chem Chem Phys 10:2747–2757
77. Tao J, Perdew JP, Staroverov VN, Scuseria GE (2003) Climbing the density functional ladder: nonempirical meta-generalized gradient approximation designed for molecules and solids. Phys Rev Lett 91:146401/146401–146401/146404
78. Li Y, Zhang S, Zhang JZH, He X (2016) Assessing the performance of popular QM methods for calculation of conformational energies of trialanine. Chem Phys Lett 652:136–141
79. Rezacs J, Bim D, Gutten O, Rulisek L (2018) Toward accurate conformational energies of smaller peptides and medium-sized macrocycles: MPCONF196 benchmark energy data set. J Chem Theory Comput 14:1254–1266
80. Carlo A, Barone V (1999) Toward reliable density functional methods without adjustable parameters: the PBE0 model. J Chem Phys 110:6158–6170
81. Yu W, Liang L, Lin Z, Ling S, Haranczyk M, Gutowski M (2009) Comparison of some representative density functional theory and wave function theory methods for the studies of amino acids. J Comput Chem 30:589–600
82. Becke AD (1993) Density-functional thermochemistry. III. The role of exact exchange. J Chem Phys 98:5648–5652
83. Riley KE, Op't Holt BT, Merz KM Jr (2007) Critical assessment of the performance of density functional methods for several atomic and molecular properties. J Chem Theory Comput 3:407–433
84. Adamo C, Barone V (1998) Exchange functionals with improved long-range behavior and adiabatic connection methods without adjustable parameters: the mPW and mPW1PW models. J Chem Phys 108:664–675
85. Becke AD (1996) Density-functional thermochemistry. IV. A new dynamic correlation functional and implications for exact-exchange mixing. J Chem Phys 104:1040–1046
86. Kruse H, Goerigk L, Grimme S (2012) Why the standard B3LYP/6-31G* model chemistry should not be used in DFT calculations of molecular thermochemistry: understanding and correcting the problem. J Org Chem 77:10824–10834
87. Tomasi J, Mennucci B, Cammi R (2005) Quantum mechanical continuum solvation models. Chem Rev 105:2999–3093
88. Walesa R, Broda MA (2017) The influence of solvent on conformational properties of peptides with Aib residue—a DFT study. J Mol Model 23:1–12

Chapter 15

User-Friendly Quantum Mechanics: Applications for Drug Discovery

Martin Kotev, Laurie Sarrat, and Constantino Diaz Gonzalez

Abstract

Quantum mechanics (QM) methods provide a fine description of receptor-ligand interactions and of chemical reactions. Their use in drug design and drug discovery is increasing, especially for complex systems including metal ions in the binding sites, for the design of highly selective inhibitors, for the optimization of bi-specific compounds, to understand enzymatic reactions, and for the study of covalent ligands and prodrugs. They are also used for generating molecular descriptors for predictive QSAR/QSPR models and for the parameterization of force fields. Thanks to the continuous increase of computational power offered by GPUs and to the development of sophisticated algorithms, QM methods are becoming part of the standard tools used in computer-aided drug design (CADD). We present the most used QM methods and software packages, and we discuss recent representative applications in drug design and drug discovery.

Key words Quantum mechanics, Quantum chemistry, DFT, SEQM, Drug discovery, Drug design, CADD, Binding affinity, Protein-ligand interaction, Reaction mechanism, Hit identification, Hit to lead, Lead optimization, QM/MM, Molecular dynamics, Virtual screening, QSAR, FMO

1 Introduction

It has almost been a century after publishing the famous quantum mechanical (QM) equation $H\psi\left(\vec{r}\right) = E\psi\left(\vec{r}\right)$ of E. Schrödinger and more than 20 years since the Nobel Prize in Chemistry was awarded "to Walter Kohn for his development of the density-functional theory and to John Pople for his development of computational methods in quantum chemistry." Today QM calculations are used routinely in almost every aspect concerning the computer-aided drug design (CADD) techniques. Initially, QM calculations were mainly and broadly applied to chemistry problems, due to smaller system size. With gradual increase of supercomputers' performance and graphics processing unit (GPU) implementations, QM calculations were performed for a variety of complex chemical, biochemical, and biomolecular systems. Applications in CADD

Alexander Heifetz (ed.), *Quantum Mechanics in Drug Discovery*, Methods in Molecular Biology, vol. 2114,
https://doi.org/10.1007/978-1-0716-0282-9_15,

include calculations at the following level of theory: semi-empirical quantum mechanical (SEQM) methods, Hartree-Fock (HF) approach, post-Hartree-Fock (pHF) methods, and density functional theory (DFT) methods. Hybrid quantum mechanics/molecular mechanics (QM/MM) methods are also widely used in CADD techniques.

2 Methods and Software Packages

2.1 Methods

Solving the Schrödinger equation, even for very simple systems, has led to the development of numerous approximations implemented in different QM methods which can describe all systems of interest today, such as organic, bioorganic, inorganic molecules and metal ions. Depending on the theory used, and the level of the simplification (the number and type of approximations), as aforementioned, we may summarize all QM methods into four major groups: SEQM, HF, pHF, and DFT. Although a specific combination, QM/MM approaches always include at least one of the four groups of QM methods. After finding the approximate solutions of the Schrödinger equation, we can obtain different measurable properties for the studied system [1–3].

The fastest and the most approximated approaches are SEQM methods. Around three orders of magnitude faster than HF calculations, they ignore some terms and replace others by empirical parameters to fit experimental heats of formation, dipole moments, and geometries. The most used SEQM are NDDO, MNDO, AM1, SAM1, RM1, PM3, PDDG/PM3, PDDG/MNDO, PM6, PM7, and OM2. Some of them were developed by John Pople, then many by Michael Dewar and lately by James J. P. Stewart group. SEQM undergo frequent improvement and revival and therefore stay quite popular on the scene [4–13].

Next in the order of increasing complexity (slow performance/high accuracy) is the HF method. The main approximation (among several others) assumes that each electron interacts with an average field of all other electrons in the system, thus reducing a many-body to a one-body problem. QM methods solve the Schrödinger equation representing the electron wave function by a set of finite number functions called basis functions (basis set). The quality of the solution is improved by a higher number of basis set functions. Usually for small drug molecules, we can get reasonable geometries with two basis functions for each valence electron and one for the core electrons. The most used are Pople basis sets, i.e., 3-21G, 6-31G, or 6-311G, where we may add also for some cases polarization functions (generally marked with *) and/or diffuse functions (denoted with +) study [1–3, 14, 15]. Polarization and diffuse-type functions could be added on heavy atoms and/or on hydrogens. Other widely used basis sets are developed by Dunning and co-workers (correlation-consistent basis sets) [16–18]. Dunning series are noted as cc-pVNZ, where N could be D, T, Q, 5, 6,... (D is double, T is triple, and so on) and "cc-p" stands for "correlation-consistent polarized" and the V for valence basis sets. Diffuse function can also be added to Dunning basis set and denoted with "aug-" prefix. We should define basis sets also for all pHF and DFT methods. The minimal basis set for each SEQM is defined by the method itself; that is why basis sets for SEQM are not explicitly specified.

Post-HF methods [1–3] were developed to overcome the main approximation in the HF by adding electron correlation. These QM methods are more accurate compared to HF methods but they are more computationally expensive by orders of magnitude, compared to HF and DFT. The simplest way to correct HF is the second order of Møller-Plesset perturbation theory method or MP2 [19], and it is the most used pHF method due to its acceptable cost. However, the accuracy of MP2 method is considered worse compared to DFT methods, despite the slower performance. Coupled-cluster pHF method presents some of the most accurate calculations for small molecules [20]. The pHF method CCSD (T) known as the gold standard in quantum chemistry [21] is an abbreviation for "coupled-cluster calculations with single, double, and perturbative treated triple excitations."

DFT method study (as it is abbreviated in the name of the theory) [1–3, 22, 23] defines the properties of a many-electron system as functions of another function (functionals); this function is the electron density, and it is dependent from the three Cartesian coordinates. In the HF methods, the wave function (Ψ in the Schrödinger equation) also depends on three coordinates, but for each electron in the system. A number of DFT functionals have been developed, e.g., BP86, PBE, M06-L, M06-2X, B3LYP [24–28], and many others. One of the most used functionals for organic molecules is the hybrid B3LYP: Becke three-parameters is the exchange functional, and Lee, Yang, and Parr is the correlation functional. The three Becke parameters are derived from a fitting to a set of atomization energies, ionization potentials, proton affinities, and total atomic energies [26].

The main question for a scientist in order to decide which QM method to use for a CADD problem is the computational cost/effectiveness ratio that has to apply to a particular system of interest. Small drug molecule partial atomic charges, geometries and/or energies of different conformations, preferred tautomeric states in solution, covalent bonds, and electron transfer issues need to be evaluated according to the desired accuracy, and each could be solved by different QM method.

Table 1
Twenty of the most used software for QM calculations

Software package	SEQM	HF	pHF	DFT	License type	GPU	Internet link
ADF	Yes	Yes	No	Yes	Comm	Yes	www.scm.com
AMPAC	Yes	No	No	No	Acad	No	www.semichem.com
CASTEP	No	Yes	No	Yes	Acad/Comm	No	www.castep.org
CP2K	Yes	Yes	Yes	Yes	FOSS	Yes	www.cp2k.org
DIRAC	No	Yes	Yes	Yes	Acad	No	www.diracprogram.org
GAMESS (UK)	Yes	Yes	Yes	Yes	Acad/Comm	Yes	www.cfs.dl.ac.uk
GAMESS (US)	Yes	Yes	Yes	Yes	Acad	Yes	www.msg.chem.iastate.edu/gamess
Gaussian	Yes	Yes	Yes	Yes	Comm	Yes	www.gaussian.com
HyperChem	Yes	Yes	Yes	Yes	Comm	No	www.hyper.com
Jaguar	No	Yes	Yes	Yes	Comm	No	www.schrodinger.com/jaguar
MOLCAS	Yes	Yes	Yes	Yes	Acad/Comm	Yes	www.molcas.org
MOLPRO	No	Yes	Yes	Yes	Comm	Yes	www.molpro.net
MOPAC	Yes	No	No	No	Acad/Comm	Yes	www.openmopac.net
MPQC	No	Yes	Yes	Yes	FOSS	No	www.mpqc.org
NWChem	No	Yes	Yes	Yes	FOSS	Yes	www.nwchem-sw.org
Q-Chem	Yes	Yes	Yes	Yes	Comm	Yes	www.q-chem.com
QSite	No	Yes	Yes	Yes	Comm	No	www.schrodinger.com/QSite
Spartan	Yes	Yes	Yes	Yes	Comm	No	www.wavefun.com
TURBOMOLE	No	Yes	Yes	Yes	Comm	No	www.turbomole.com
VASP	No	Yes	Yes	Yes	Acad/Comm	Yes	www.vasp.at

License types include academic, commercial, and FOSS (free and open-source software)

2.2 Software Packages

In the past three or more decades, numerous QM software was developed with different level of theory and basis set implementation. In general, we can sort them by included methods. Most of the QM software now includes different combinations of SEQM, HF, pHF, and DFT methods. The use of a specific software is also related to its license: academic, commercial, and free and open source. Academic software license is possible to be acquired upon request, free of charge for academic institutions. Table 1 shows general overview of 20 of the most used software packages for QM calculations. In the past 10 years, company such as NVIDIA introduced video cards for scientific calculations. Molecular modeling calculations on GPU, including QM, have been gaining interest in the recent years due to the accelerated performance compared to

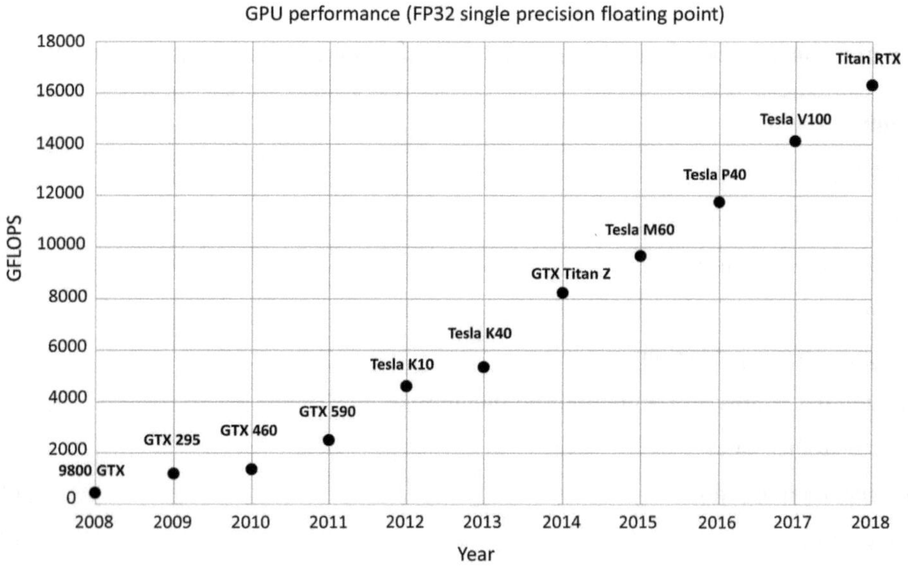

Fig. 1 Performance of GPUs, from year 2008 to 2018, in GFLOPS (10^9 floating point operations per second), for FP32 single-precision floating point calculations. The GPU models are shown above the corresponding dots. A 38-fold increase in computational power has been observed over 10 years. Data taken from https:/www.techpowerup.com/gpu-specs/

the standard CPU based. Figure 1 displays the performance of some of the best GPUs per year for the last decade, showing approximately 40-fold increase.

3 Application Domains

Computer-aided drug design (CADD) is commonly integrated in drug discovery and drug design processes [29–31]. The use of QM methods in CADD has increased during the last decade, especially due to breakthrough in computer hardware and the development of new algorithms [32–34]. QM methods provide accurate estimation of energies, electronic polarization effects, charge transfer, metal coordination, and bond formation and cleavage [35]. They allow reliable modeling and simulation of complex ligand-receptor systems, thus opening new possibilities for the study of difficult therapeutic targets and for the efficient development of optimized and selective drugs. We present and discuss diverse applications of QM in drug discovery and drug design, considering one or few examples by domain. For each application type, other examples are reported in Table 2.

3.1 QM/MM MD

Some of the most powerful approaches in the computational chemistry and in the drug design for studying protein-ligand and DNA-ligand interactions are the hybrid QM/MM methods and in particular QM/MM MD simulations [36, 37]. Together with

Table 2
Examples of various applications of QM-based methods in drug design and drug discovery

Applications	Papers
QM/MM MD	[90–92]
Parametrization of force fields	[93–96]
Protein structure refinement	[97–100]
Virtual screening	[101–103]
Protein-ligand affinity prediction	[104–109]
Identification of key drivers of protein-ligand binding	[110–113]
Protomer/tautomer states	[114–118]
Cation-π and π-π interactions	[119–122]
Bioactive conformations	[123–126]
Bi-specific inhibitors	[127–129]
Covalent compounds	[130–134]
Prodrugs	[135–139]
Reaction mechanisms	[140–146]
Drug metabolism	[147–150]
QSAR models	[151–153]
Molecular quantum similarity	[154, 155]
Enzyme design	[156–159]

the standard MD simulations, they represent two computationally expensive approaches in CADD, which recently could significantly benefit from the implementation of the GPU software and hardware.

Besides the routine use of QM methods such as calculating more accurate partial atomic ligand charges (like RESP method, [38]) for classical MD simulations in SBDD [39–41], in the past two decades, we have seen more computational chemists using QM/MM simulations [42–44]. In general, scientists treat a limited but important space of the protein (or DNA)-ligand complexes with higher accuracy at QM (but also SEQM) level. This important area could include the ligand (small peptide or small molecule) in case of a complex and the surrounding close environment—closest residues in a receptor pocket. The area can be an enzyme's active site where enzymatic cleavage reactions happen or a site with a covalent bonding possibility [45] or simply a pocket where non-covalent small drug molecules bind. The rest of the system, which is a huge part, is treated at a less accurate level using a classical force field (FF), the same as the ones in standard MD simulations

[36, 37, 42–44]. In a recent study [46], on large QM/MM calculations, authors provide a way for a rational selection of the QM region sizes for the next generation of QM/MM studies of enzymes.

Recently, hybrid methods have steadily increasing impact on the drug discovery [47–49]. In the QM/MM study of HIV-1 protease [50], the authors explored the effect of different conformations of the enzyme on its catalytic ability. Chen and co-authors [51] apply QM/MM MD simulations combining DFT and SEQM—AM1d, RM1, PM3, and PM6 methods—to study the H bonding and other interactions of inhibitors with trypsin. Their results show that the accuracy of treating the hydrogen bonding using QM/MM MD simulations of PM6 can compete with the DFT QM/MM MD simulations. Also Schirmeister et al. [52] in a study of covalent inhibitors showed that the hybrid approach was accurate enough to successfully design reversible covalent inhibitors for rhodesain. QM/MM calculations could generate electrostatic potential maps of the binding sites at QM level, thus helping to determine the protonation states of important residues and also to analyze reaction mechanisms in drug discovery targets [53]. Detailed review shows the applications of one of the most used hybrid method ONIOM including its applications for the structure-based drug design, optimization, and rescoring of docking poses or supplement for the X-ray crystallography [54].

Modern techniques add to QM/MM approach machine learning elements such as neural networks. Shen and Yang [55] put an effort to achieve the accuracy of ab initio QM/MM at the computational cost similar to SEQM/MM approaches. They developed an interesting new method and improved their previously reported QM/MM-NN (QM/MM neural network) approach with QM/MM MD simulations using an adaptive procedure.

As a future perspective and challenges, in front QM/MM MD methods in the drug design, we can point the time needed and the accuracy of these approaches. We might expect to simulate a full system at QM level in a close future. The development of both hardware and algorithms might bring quantum computers in the future to a commercial usage. Drug discovery would be a promising area of their application that will offer variety of applications [56].

3.2 Parametrization of Force Fields

Molecular mechanics force fields (FF) were developed to treat broad spectra of small organic drug-like molecules, lipids, carbohydrates, proteins, DNA, RNA, and other systems of interest. The parameters for the energy functions, in the most used FF in CADD, are derived from physical or chemical experiments and from QM calculations, usually performed at a high level of theory (pHF, DFT). The FF set of parameters includes values for different types of atoms (mass, radius, charge), chemical bonds, angles, and dihedral angles. Experimental data used to parametrize FF can be bond lengths from X-ray and neutron diffraction, dipole moments, enthalpy of vaporization, spectroscopic parameters. Some of the best parametrized FF for small molecules that are implemented in different software for CADD are OPLS3 [57], GAFF [58], CGenFF [59], and MMFF [60]—all of them rely on a number of QM-derived parameters. Some of the recently updated FF (with QM parametrizations) for proteins include AMBER ff14sb [61], OPLS3 [57], and CHARMM [62]. Significant improvements have been achieved for the simulations of DNA with Parmbsc1, parameterized from high-level QM data [63].

DNA and RNA can form a wide variety of complex tertiary structures. In nucleic acid research, the study of structural diversity and transitions between diverse types of structures is of great importance to identify druggable RNA targets and to predict small compound binding and specificity [64]. In particular, reliable structure prediction must consider the effects of ionic phosphate groups and base pairing and stacking. Zhang and co-workers developed the AMOEBA (Atomic Multipole Optimized Energetics for Biomolecular Applications) polarizable force field for DNA and RNA, based on high-level QM calculations [65]. Molecular dynamics simulations using the AMOEBA force field were performed for 20 different DNA/RNA molecules and were compared with experimental data. The conformations were reproduced with an average RMSD below or around 2.0 Å compared to NMR structures.

3.3 Protein Structure Refinement

Structures of protein-ligand complexes obtained by X-ray crystallography are key in understanding interactions and optimizing compounds. However, crystal models still contain questionable geometric coordinates, as revealed, in particular, by the clashscore, which is the number of clashes per 1000 atoms. Errors in ligand geometry and position often lead to bad perception of the interactions with the protein, which is detrimental for compound optimization. Borbulevych et al. have developed an automated refinement process using a QM/MM based on the ONIOM method implemented in the *DivCon* package [35]. They considered 80 protein-ligand structures from the Astex diverse set. After automatic refinement, they observed an improvement in the clashscore by an average factor of 4.5, as well as improvements in Ramachandran and in rotamer outlier analyses and in ligand strain energy.

3.4 Virtual Screening (VS)

VS is essentially composed of two phases: docking and scoring. Nowadays, docking is considered reliable: the poses of docked compound into target sites are relatively well reproduced when superposed to the same compounds in co-crystallized structures. Scoring is still a problem, and the results provided by diverse scoring functions are protein-dependent [66]. To overcome this problem, Zhou and Caflish [67] used small molecules, named probes, consisting of 2–10 atoms. Polar groups in the binding site of the protein were replaced by mimetic probes, and for each compound pose, the interaction energy value between the compound and the probes was calculated using PM6. They considered the target EphB4. First, a homology model was built using the structure of EphB2. Then 2.7 million compounds of the ZINC library, containing at least 1 hydrogen bond donor and 1 hydrogen bond acceptor, were docked into the ATP binding site, with AutoDock, resulting in ~100 million poses. Thirteen polar groups were replaced by QM probes in diverse parts of the binding site of EphB4, and cutoffs for filtering, according to QM values, were determined using four known inhibitors of EphB4. Finally, using the QM probe method, and additional filters, 23 compounds were selected for experimental validation. Three low micromolar inhibitors of EphB4 were identified.

In a recent mini-revue, Cavasotto et al. presented applications of QM methods in small molecule docking and scoring [33].

3.5 Protein-Ligand Affinity Prediction

Lu and co-workers [68] performed the computation of affinity between Akt kinase and eight diverse, selective, and potent inhibitors. They docked the eight compounds in a model of Akt, and they used the complexes for MD simulations. After, they tested two empirical scoring functions (DrugScore, DFIRE) and MM-PB/SA, and both methods found poor correlations with experimental affinity. Finally, they applied QM/MM method to optimize the complex structures, and they used QM/MM-PB/SA to calculate the interaction energies. A strong correlation appeared between computed and experimental affinities. Additionally, the QM/MM-PB/SA method was used to screen for analogues of apigenin, a potent anti-cancer compound. Four compounds were selected and tested in vitro against Akt kinase. They displayed nanomolar inhibitory activities.

To predict the binding affinity for a series of cyclin-dependent kinase 2 (CDK2) inhibitors, Mazanetz and co-workers [69] used the fragment molecular orbital (FMO) method. They considered 14 X-ray structures with 14 ligands. FMO calculations showed a good correlation to the experimental free energy of binding (r^2 of 0.68). Then, they combined the enthalpy contributions calculated by the FMO method, with entropy and solvation approximations, and they built a QSAR model. Calculated values displayed a strong correlation with experimental data (r^2 of 0.94).

The FMO method was also used to analyze docking poses of orexin-2 receptor agonists [70]. Sixteen analogues of non-peptidic OX_2R agonists were docked in a recent X-ray structure of the receptor. Two potential binding modes were observed. By using FMO, a strong correlation (r^2 of 0.87) was found between experimental values of EC_{50} and calculated values, for the "U"-shape poses, while no significant correlation was found for the "L"-shape poses. Such findings are important for drug design of protein modulators when no co-crystal structure is available.

To illustrate the typical results obtained using FMO, we performed calculations on the complex between galectin 1 and the inhibitor thio-digalactoside (TDG) ligand, PDB entry 3OYW, Fig. 2a, and between galectin 1 and the inhibitor 3,3′-deoxy-3,3′-bis-(4-[*m*-fluorophenyl]-1*H*-1,2,3-triazol-1-yl)-thio-digalactoside (TD139), PDB entry 4Y24, Fig. 2b. While TDG has Kd = 67.3 μM for galectin 1, TD139 has Kd = 0.22 μM for the same target, thus showing ~300-fold increase in affinity compared with TDG binding [71]. The left-hand plots show the pair interaction energies (PIE) for the residues, and those on the right show the pair interaction energy decomposition analysis (PIEDA) contributions. We consider any interaction with an absolute PIE greater than or equal to 3.0 kcal/mol to be significant [72]. FMO indicated that the two inhibitors shared significant interactions with seven residues His44, Arg48, Asp54, Asn61, Trp68, Glu71, and Arg73 and one water molecule (named HOH142 in PDB entry 3OYW and named HOH 331 in PDB entry 4Y24). For the TD139 inhibitor, the FMO calculations detected additional interactions with five residues Val31, Asn46, His52, Gly53, and Asp123 and three water molecules (HOH324, 325, 328). Difference of shared interactions between TDG and TD139 complexes is shown in Fig. 2c.

3.6 Identification of Key Drivers of Protein-Ligand Binding

The heme-containing enzyme IDO1 is an attractive immunotherapy target in cancer treatment. Zou and co-workers considered a set of 20 imidazole derivatives including 4 classes of IDO1 inhibitors. The compounds were subjected to induced fit docking into the binding site of the enzyme, followed by molecular dynamics, MM/PBSA free energy calculation, and QM/MM computation with *QSite* program in Schrödinger. They discovered that Arg231 and 7-propionate of the heme group were major contributors to ligand binding, thus suggesting possible ways to improve the binding affinity of new IDO1 inhibitors [73].

3.7 Protomer/Tautomer States

To guide drug design, the correct protomer/tautomer state of binding site residues and bound molecules must be identified. X-ray crystallography is not precise enough to detect hydrogen atoms in large systems. Neutron diffraction is a unique technique that allows experimental identification of hydrogen positions in crystal structures, but is of limited use mainly due to the necessity of deuterium exchange. In a study, Borbulevych et al. applied *XModeScore* to three cases for which X-ray diffraction model and neutron diffraction model were available in the PDB database. The method involves the QM X-ray refinement of a set of structures containing all protomer/tautomer forms followed by a statistical

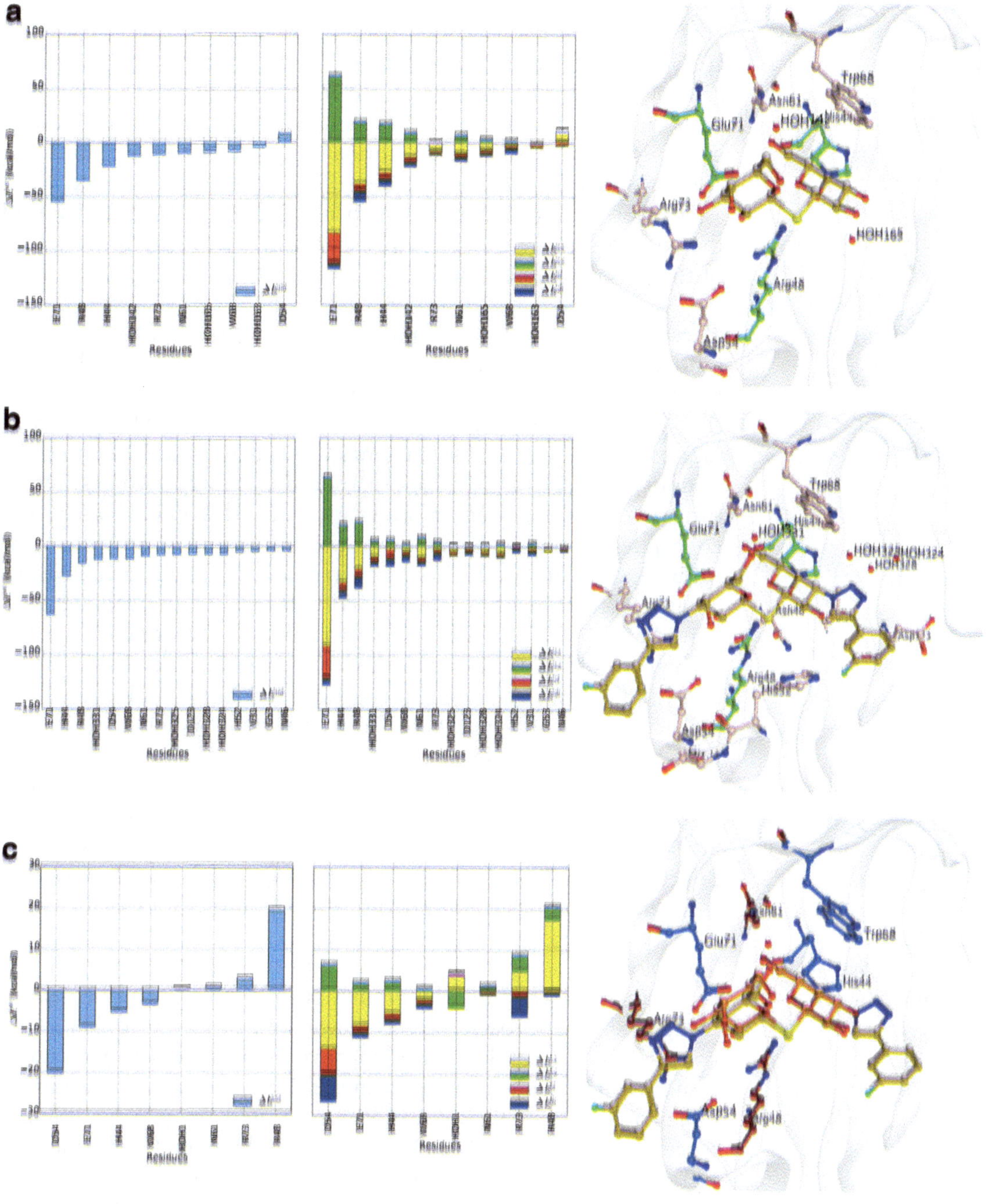

Fig. 2 FMO calculations for two galectin 1-inhibitor complexes. (**a**) Thio-digalactoside ligand (TDG), PDB entry 3OYW. (**b**) 3,3′-Deoxy-3,3′-bis-(4-[*m*-fluorophenyl]-1*H*-1,2,3-triazol-1-yl)-thio-digalactoside (TD139), PDB entry 4Y24. The left-hand plots show the total PIE for the residues, while those on the right show the PIEDA contributions. The electrostatics, exchange repulsion, charge transfer, and dispersion PIE terms are color-coded yellow, green, red, and blue, respectively. The three residues interacting more strongly with the ligands (His44, Arg48, Glu71) are shown with the carbon atoms colored in green, and the other interacting residues have their carbon atoms colored in pink. (**c**) Difference of interactions between TDG- and TD139-galectin 1 complexes. TDG is shown in orange and TD139 in gold, with the residues interacting more strongly with TD139 colored in blue and the residues interacting more strongly with TDG in dark red. The water molecules 142 and 331, in, respectively, the TDG- and the TD139-galectin 1 complexes, are at similar positions

analysis of difference electron density maps for each map candidate [74]. When applied to human carbonic anhydrase II in complex with the inhibitor acetazolamide, and to the enzyme urate oxidase in complex with uric acid monoanion, the method found the correct protomer/tautomer states of the molecules. Applied to an aspartic proteinase in complex with an inhibitor, *XModeScore* found the correct state of the catalytic aspartic acid showing protonation of the outer oxygen atom.

3.8 Cation-π and π-π Interactions

These interactions play a fundamental role in the protein-ligand binding, but they are not well described by classical force fields with fixed charge models [75]. Accurately identifying key interactions that drive the binding of β-lactam antibiotics to DD-peptidase targets, and to β-lactamase enzymes which inactivate β-lactam compounds, is essential to propose modifications to compounds in order to counter bacterial resistance. To this end, Hargis and colleagues used a combination of computational chemistry methods including quantum chemistry to analyze benzylpenicillin and a novel β-lactam peptidomimetic complexed to *Streptomyces* R61 peptidase [76]. They identified an extended π-π network for the phenyl group of benzylpenicillin with Phe120 and Trp233 that contributes into stabilizing aromatic interactions and compound efficacy for the DD-peptidase. However, structural analysis identified that this aromatic stabilization is also conserved in β-lactamases. Following their investigation, they found that interactions between the peptidomimetic tail region and the peptidase are specific when compared to interactions with class C β-lactamases. These findings suggest important modifications to β-lactam antibiotics for improving their binding and specificity.

3.9 Bioactive Conformations

The bioactive conformation of a compound is the target-bound conformation. To study the prediction of bioactive conformations using in silico approaches, Avgy-David and Senderowitz [77] have considered a set of 100 FDA-approved drugs with 1–6 rotatable bonds with available complexes in the PDB. For each compound, a large conformational ensemble was generated using three different force fields: OPLS-AA, MMFF, and CHARMm. Conformations were merged and clustered. Centroid conformations were minimized using the three force fields and four DFT methods. The probability to find the bound conformation in low-energy regions of unbound conformational ensembles was higher with QM methods and with CHARMm than with the two other force field methods.

3.10 Bi-specific Inhibitors

Restoration of p53 activity by inhibition of interaction with the two oncoproteins MDM2 and MDMX inhibits the growth of cancerous tumors in animals. Peptide and non-peptide inhibitors have been developed. They display high affinity for MDM2, but they are less efficient for inhibiting p53-MDMX interaction. To understand the origin of such difference, Chen and co-workers [78] considered two peptide inhibitors pDI6W and pDIQ and the two oncoproteins MDM2 and MDMX. They performed MD with AMBER FF02 polarizable force field during 10 ns, followed by QM/MM-GBSA calculation. The QM region was defined by the residues forming hydrogen bonds with the inhibitors and the MM region by the rest. For the two peptides, they found better binding free energies for MDM2 than for MDMX, in accordance with experimental values. A decrease in the van der Waals interaction was the main source of the weaker binding of inhibitors to MDMX than to MDM2. Considering residue-based free energy decomposition method, and residue conformations obtained during MD, they identified M53 and Y99 in MDMX as the main origin of the weaker binding of inhibitors to MDMX than to MDM2. They concluded that the structure of future bi-specific MDM2/MDMX inhibitors should be more flexible to adapt their structure to MDMX in the identified region.

3.11 Covalent Compounds

MacDonald and Boyd explored the potential of FKBP35 in *Plasmodium falciparum* as a target for novel anti-malarial [79]. Since the enzyme contains Cys106 in the active site, they investigated where to introduce a warhead into Fk506, the natural substrate of FKBP35, to form a covalently linked inactive substrate-enzyme complex. They performed MD and QM/MM (ONIOM) calculations, and they described the transition and final states representative of the bond formation between the cysteinyl sulfur in the protein and the Michael acceptor added to carbon 41 of the substrate. Interestingly, FKBP12, a human enzyme with a highly similar active site, contains His87 at the corresponding position, suggesting a potential selective inhibition of the covalent substrate for the parasite.

3.12 Prodrugs

Six-coordinate platinum(IV) prodrugs need to be reduced to four-coordinate Pt(II) drugs in order to show activity against various cancers. Despite their important role, serious side effects limit their efficacy and motivate further studies to design next-generation Pt agents. McCormick and co-workers used electrochemical experiments and QM to decipher the mechanism for the two-electron reduction of Pt(IV) prodrugs to Pt(II) drugs [80]. They considered three Pt(IV) prodrugs. By using DFT combined with continuum solvation models, they proposed the mechanism of reduction. All three prodrugs follow a stepwise mechanism, with a metastable six-coordinate Pt(III) intermediate generated upon addition of one electron. The loss of the two axial groups present in the prodrug occurs upon addition of the second electron. A strategy to design more inert Pt(IV) prodrugs with less side effects would consist to modify the two axial groups that are lost upon reduction, in such a way that lipophilicity and ease of reduction of the prodrug are appropriately tuned.

3.13 Reaction Mechanisms

Identification of reaction mechanisms for enzymes with therapeutic interest is important for the design of specific inhibitors. Type II dehydroquinase enzymes (DHQ2) are essential enzymes for the pathogenic bacterium *Mycobacterium tuberculosis* and *Helicobacter pylori*, but MtDHQ2 and HpDHQ2 show a 50-fold difference in catalytic efficiency. To understand such difference, Lence and co-workers performed QM/MM simulations for the two enzymes and for the three reaction steps which are the generation of the catalytic tyrosinate, the formation of the enolate intermediate, and the enolate dehydration [81]. They found that the second step explains the observed differences in activity between the two enzymes, with a more efficient stabilization of the enolate intermediate by the *M. tuberculosis* enzyme. Comparing the two enzymes, they found differences for a water molecule in the catalytic pocket and for the flexibility of the substrate-covering loop. The loop has limited flexibility for HpDHQ2, due to a salt bridge between two residues in this loop, while its flexibility is higher for MtDHQ2 due to an apolar residue in one of the corresponding positions. Crystal structures of DHQ2-ligand complexes exist, but they show a highly similar binding mode for both enzymes, while the inhibitors present in these structures have different inhibitory potencies. Deciphering enzymatic mechanisms and differences in enzyme reaction steps of different species is essential for the rational design of more efficient chemical entities.

3.14 Drug Metabolism

Drug metabolism has a strong impact on clearance and toxicity. Structures for main cytochrome P450 have been identified, and thus structure-based drug metabolism can be considered. In an early work, the metabolism of sirolimus, and its derivative everolimus, was investigated by Kuhn and co-workers [82]. They used a process in three steps: first, the compounds were docked into the active site of CYP3A4, then molecular dynamics calculations were performed, and finally the enzyme-substrate interactions were calculated using QM. Predictions of the regiospecificities of the hydroxylations and O-dealkylations for the two substrates were in good agreement with experimental data. They also identified substrate/CYP interactions that are important for the metabolism of the substrate, and they explained why the metabolism in everolimus is reduced compared to sirolimus. Many CYP isoforms show high levels of flexibility and can bind a wide variety of ligands. Therefore induced fit docking of compounds followed by MD and QM calculations help in incorporating the CYP flexibility in the structure-based process [83].

Recently Tyzack and Kirchmair have reviewed diverse approaches and software packages for drug metabolism predictions, including ligand-based approaches and methods for metabolite structure prediction [84].

3.15 QSAR Models

Golgi α-mannosidase II (GMII) is a Zn^{2+} co-factor-dependent glycoside hydrolase, and lysosomal α-mannosidase (LMAN) is a protein with a closely related active site. GMII inhibitors have anti-tumor activity, but inhibition of LMAN is not wanted. As all known potent inhibitors of GMII also show inhibition for LMAN, Bobovská and colleagues have built QSAR models to accurately predict the affinity of compounds for both targets [85]. They docked 57 inhibitors and 25 non-active compounds on the 2 protein sites, and they employed a combination of interaction energy descriptors computed using empirical and QM approaches. Using 5 and 7 descriptors, they achieved RMSD of 0.8 and 1.1 kcal/mol for ΔG_{bind} for, respectively, GMII and LMAN, which is enough to predict selectivity.

3.16 Molecular Quantum Similarity

Ligand-based virtual screening (LBVS) was performed by Sullivan and co-workers, using electron density attributes of chemical compounds computed at the quantum level. They described compounds by combining DFT and topological theory of atoms in molecules (AIM). Quantum similarity was calculated between known inhibitors and compounds in commercial databases, and compounds were selected without explicit reference to chemical structures. The method, applied to LBVS of anti-malarial agents, discovered new anti-malarial compounds which are chemically dissimilar, but show similarity at the quantum level [86, 87].

3.17 Enzyme Design

The design of artificial enzymes, in biotechnology and biochemistry, requires the understanding of transition states and the quantification of subtle effects due to small modifications introduced to the proteins [88]. The 34E4 catalytic antibody which catalyzes Kemp elimination of 5-nitrobenzisoxazole, as well as its Glu50Asp variant which shows a 30-fold reduction in the catalytic performance, was considered by Alexandrova and Jorgensen [89]. They used QM/MM Monte Carlo simulations and free energy perturbation theory to elucidate the mechanism of Kemp elimination catalyzed by 34E4. Glu50 is the key residue in 34E4, and substitution by Asp50 resulted in the increase of the computed activation barrier by 2.4 kcal/mol, which corresponds to a 62-fold reduction in the reaction rate at 25 °C, in good agreement with experimental data.

4 Conclusion and Outlook

QM methods provide high accuracy in the description of receptor-ligand interactions, the estimation of binding affinities, and the modeling of bond formation or breakage. They model electronic polarization, charge transfer, halogen bonding, and metal coordination, at the expense of long computing time. For complex biological systems, hybrid QM/MM approaches, which combine the accuracy of QM with the speed of calculation of MM, are used to make the most of both methods.

In early stages of drug discovery, QM methods contribute to refine X-ray structures of active sites of proteins for virtual screening, and for electronically complex binding sites, to estimate the affinity of small lists of docked compounds. In the subsequent optimization phases, QM methods are particularly important to predict the nature and the relative binding affinity of compounds to targets, mutated targets, and off-targets, in order to select the best compounds with a given selectivity profile, for synthesis and testing. QM is particularly needed for metal-containing binding site systems, for aromatic-containing compounds, for the identification of protomeric and tautomeric states, and for the study and design of covalent compounds and prodrugs. It opens the way to investigate reaction mechanisms, decipher substrate-enzyme transition states, and predict drug metabolism. QM is also key for parameterization of force fields regarding new types of targets, as DNA/RNA, and for providing QM descriptors for improved QSAR/QSPR models.

Today, QM methods are an unavoidable part of the CADD toolbox, driven by advanced algorithms and unprecedented computational power.

Acknowledgments

The authors thank Christophe Iftner, Franck Taillez, and Emilie Pihan (Evotec (France) SAS, Toulouse, France) for valuable suggestions to the manuscript.

References

1. Szabo A, Ostlund NS (1996) Modern quantum chemistry. Dover Publishing, Mineola, NY
2. Cramer CJ (2006) Essentials of computational chemistry: theories and models. Wiley, Chichester, UK
3. Jensen F (2007) Introduction to computational chemistry. Wiley, Chichester, UK
4. Pople JA, Beveridge DL (1970) Approximate molecular orbital theory. McGraw-Hill, New York
5. Dewar MJS, Zoebisch EG, Healy EF, Stewart JJP (1985) AM1: a new general purpose quantum mechanical molecular model. J Am Chem Soc 107:3902–3909. https://doi.org/10.1021/ja00299a024
6. Stewart JJP (1989) Optimization of parameters for semiempirical methods I. Method. J Comput Chem 10:209–220. https://doi.org/10.1002/jcc.540100208
7. Dewar MJS, Jie C, Yu J (1993) SAM1; the first of a new series of general purpose quantum mechanical molecular models. Tetrahedron 49:5003–5038. https://doi.org/10.1016/S0040-4020(01)81868-8
8. Repasky MP, Chandrasekhar J, Jorgensen WL (2002) PDDG/PM3 and PDDG/MNDO: improved semiempirical methods. J Comput Chem 23:1601–1622. https://doi.org/10.1002/jcc.10162
9. Stewart JJ (2007) Optimization of parameters for semiempirical methods V: modification of NDDO approximations and application to 70 elements. J Mol Model 13:1173–1213. https://doi.org/10.1007/s00894-007-0233-4
10. Řezáč J, Fanfrlík J, Salahub D, Hobza P (2009) Semiempirical quantum chemical PM6 method augmented by dispersion and H-bonding correction terms reliably describes

various types of noncovalent complexes. J Chem Theory Comput 5:1749–1760. https://doi.org/10.1021/ct9000922

11. Dral PO, Wu X, Spörkel L, Koslowski A, Weber W, Steiger R, Scholten M, Thiel W (2016) Semiempirical quantum-chemical orthogonalization-corrected methods: theory, implementation, and parameters. J Chem Theory Comput 12:1082–1096. https://doi.org/10.1021/acs.jctc.5b01046
12. Dral PO, Wu X, Thiel W (2019) Semiempirical quantum-chemical methods with orthogonalization and dispersion corrections. J Chem Theory Comput 15:1743–1760. https://doi.org/10.1021/acs.jctc.8b01265
13. Kříž K, Řezáč J (2019) Reparametrization of the COSMO solvent model for semiempirical methods PM6 and PM7. J Chem Inf Model 59:229–235. https://doi.org/10.1021/acs.jcim.8b00681
14. Hehre WJ, Ditchfield R, Pople JA (1972) Self-consistent molecular orbital methods. 12. Further extensions of Gaussian-type basis sets for use in molecular-orbital studies of organic-molecules. J Chem Phys 56:2257. https://doi.org/10.1063/1.1677527
15. Jensen F (2013) Atomic orbital basis sets. WIREs Comput Mol Sci 3:273–295. https://doi.org/10.1002/wcms.1123
16. Dunning TH (1989) Gaussian basis sets for use in correlated molecular calculations. I. The atoms boron through neon and hydrogen. J Chem Phys 90:1007–1023. https://doi.org/10.1063/1.456153
17. Kendall RA, Dunning TH, Harrison RJ (1992) Electron affinities of the first-row atoms revisited. Systematic basis sets and wave functions. J Chem Phys 96:6796–6806. https://doi.org/10.1063/1.462569
18. Weigend F, Ahlrichs R (2005) Balanced basis sets of split valence, triple zeta valence and quadruple zeta valence quality for H to Rn: design and assessment of accuracy. Phys Chem Chem Phys 7:3297–3305. https://doi.org/10.1039/b508541a
19. Møller C, Plesset MS (1934) Note on an approximation treatment for many-electron systems. Phys Rev 46:618–622. https://doi.org/10.1103/PhysRev.46.618
20. Shavitt I, Bartlett RJ (2009) Many-body methods in chemistry and physics: MBPT and coupled-cluster theory. Cambridge University Press, Cambridge
21. Ramabhadran RO, Raghavachari K (2013) Extrapolation to the gold-standard in quantum chemistry: computationally efficient and accurate CCSD(T) energies for large molecules using an automated thermochemical hierarchy. J Chem Theory Comput 9:3986–3994. https://doi.org/10.1021/ct400465q
22. Hohenberg P, Kohn W (1964) Inhomogeneous electron gas. Phys Rev 136:B864–B871
23. Kohn W, Sham LJ (1965) Self-consistent equations including exchange and correlation effects. Phys Rev 140:A1133–A1138. https://doi.org/10.1103/PhysRev.140.A1133
24. Perdew JP (1986) Density-functional approximation for the correlation energy of the inhomogeneous electron gas. Phys Rev B Condens Matter 33:8822–8824. https://doi.org/10.1103/physrevb.33.8822
25. Lee CT, Yang WT, Parr RG (1988) Development of the Colle-Salvetti correlation-energy formula into a functional of the electron density. Phys Rev B Condens Matter Mater Phys 37:785–789. https://doi.org/10.1103/PhysRevB.37.785
26. Becke AD (1993) Density-functional thermochemistry. III. The role of exact exchange. J Chem Phys 98:5648–5652. https://doi.org/10.1063/1.464913
27. Perdew JP, Burke K, Ernzerhof M (1996) Generalized gradient approximation made simple. Phys Rev Lett 77:3865–3868. https://doi.org/10.1103/PhysRevLett.77.3865
28. Zhao Y, Truhlar DG (2008) The M06 suite of density functionals for main group thermochemistry, thermochemical kinetics, noncovalent interactions, excited states, and transition elements: two new functionals and systematic testing of four M06-class functionals and 12 other functionals. Theor Chem Accounts 120:215–241. https://doi.org/10.1007/s00214-007-0310-x
29. Liao C, Sitzmann M, Pugliese A, Nicklaus MC (2011) Software and resources for computational medicinal chemistry. Future Med Chem 3(8):1057–1085. https://doi.org/10.4155/fmc.11.63
30. Kitchen DB (2017) Computer-aided drug discovery research at a global contract research organization. J Comput Aided Mol Des 31(3):309–318. https://doi.org/10.1007/s10822-016-9991-3
31. Muegge I, Bergner A, Kriegl JM (2017) Computer-aided drug design at Boehringer Ingelheim. J Comput Aided Mol Des 31(3):275–285. https://doi.org/10.1007/s10822-016-9975-3

32. Nitsche MA, Ferreria M, Mocskos EE, González Lebrero MC (2014) GPU accelerated implementation of density functional theory for hybrid QM/MM simulations. J Chem Theory Comput 10(3):959–967. https://doi.org/10.1021/ct400308n
33. Cavasotto CN, Adler NS, Aucar MG (2018) Quantum chemical approaches in structure-based virtual screening and lead optimization. Front Chem 6:188. https://doi.org/10.3389/fchem.2018.00188
34. Shi M, Xu D, Zeng J (2018) GPU accelerated quantum virtual screening: application for the natural inhibitors of New Delhi metalloprotein (NDM-1). Front Chem 6:564. https://doi.org/10.3389/fchem.2018.00564
35. Borbulevych O, Martin RI, Westerhoff LM (2018) High-throughput quantum-mechanics/molecular-mechanics (ONIOM) macromolecular crystallographic refinement with PHENIX/DivCon: the impact of mixed Hamiltonian methods on ligand and protein structure. Acta Cryst D74:1063–1077. https://doi.org/10.1107/S2059798318012913
36. Rode BM, Hofer TS, Randolf BR, Schwenk CF, Xenides D, Vchirawongkwin V (2006) Ab initio quantum mechanical charge field (QMCF) molecular dynamics: a QM/MM – MD procedure for accurate simulations of ions and complexes. Theor Chem Accounts 115:77–85. https://doi.org/10.1007/s00214-005-0049-1
37. Senn HM, Thiel W (2009) QM/MM methods for biomolecular systems. Angew Chem Int Ed Engl 48(7):1198–1229. https://doi.org/10.1002/anie.200802019
38. Cornell WD, Cieplak P, Bayly CI, Kollman PA (1993) Application of RESP charges to calculate conformational energies, hydrogen bond energies, and free energies of solvation. J Am Chem Soc 115:9620–9631. https://doi.org/10.1021/ja00074a030
39. Kotev M, Pascual R, Almansa C, Guallar V, Soliva R (2018) Pushing the limits of computational structure-based drug design with a cryo-EM structure: the Ca^{2+} channel α2δ-1 subunit as a test case. J Chem Inf Model 58(8):1707–1715. https://doi.org/10.1021/acs.jcim.8b00347
40. Bekker GJ, Araki M, Oshima K, Okuno Y, Kamiya N (2019) Dynamic docking of a medium-sized molecule to its receptor by multicanonical MD simulations. J Phys Chem B 123(11):2479–2490. https://doi.org/10.1021/acs.jpcb.8b12419
41. Wang J, Zhao C, Tu J, Yang H, Zhang X, Lv W, Zhai H (2019) Design of novel quinoline-aminopiperidine derivatives as Mycobacterium tuberculosis (MTB) GyrB inhibitors: an in silico study. J Biomol Struct Dyn 37(11):2913–2925. https://doi.org/10.1080/07391102.2018.1498806
42. Ryde U, Söderhjelm P (2016) Ligand-binding affinity estimates supported by quantum-mechanical methods. Chem Rev 116(9):5520–5566. https://doi.org/10.1021/acs.chemrev.5b00630
43. Arodola OA, Soliman ME (2017) Quantum mechanics implementation in drug-design workflows: does it really help? Drug Des Devel Ther 11:2551–2564. https://doi.org/10.2147/DDDT.S126344
44. Ganesan A, Coote ML, Barakat K (2017) Molecular dynamics-driven drug discovery: leaping forward with confidence. Drug Discov Today 22(2):249–269. https://doi.org/10.1016/j.drudis.2016.11.001
45. Awoonor-Williams E, Walsh AG, Rowley CN (2017) Modeling covalent-modifier drugs. Biochim Biophys Acta Proteins Proteom 1865(11 Pt B):1664–1675. https://doi.org/10.1016/j.bbapap.2017.05.009
46. Kulik HJ, Zhang J, Klinman JP, Martínez TJ (2016) How large should the QM region be in QM/MM calculations? The case of catechol O-methyltransferase. J Phys Chem B 120(44):11381–11394. https://doi.org/10.1021/acs.jpcb.6b07814
47. Lodola A, De Vivo M (2012) The increasing role of QM/MM in drug discovery. Adv Protein Chem Struct Biol 87:337–362. https://doi.org/10.1016/B978-0-12-398312-1.00011-1
48. Barbault F, Maurel F (2015) Simulation with quantum mechanics/molecular mechanics for drug discovery. Expert Opin Drug Discov 10(10):1047–1057. https://doi.org/10.1517/17460441.2015.1076389
49. Nascimento ÉCM, Oliva M, Świderek K, Martins JBL, Andrés J (2017) Binding analysis of some classical acetylcholinesterase inhibitors: insights for a rational design using free energy perturbation method calculations with QM/MM MD simulations. J Chem Inf Model 57(4):958–976. https://doi.org/10.1021/acs.jcim.7b00037
50. Ribeiro AJM, Santos-Martins D, Russo N, Ramos MJ, Fernandes PA (2015) Enzymatic flexibility and reaction rate: a QM/MM study of HIV-1 protease. ACS Catal 5:5617–5626. https://doi.org/10.1021/acscatal.5b00759
51. Chen J, Wang J, Zhang Q, Chen K, Zhu W (2015) A comparative study of trypsin specificity based on QM/MM molecular dynamics simulation and QM/MM GBSA calculation. J Biomol Struct Dyn 33(12):2606–2618.

https://doi.org/10.1080/07391102.2014.1003146

52. Schirmeister T, Kesselring J, Jung S, Schneider TH, Weickert A, Becker J, Lee W, Bamberger D, Wich PR, Distler U, Tenzer S, Johé P, Hellmich UA, Engels B (2016) Quantum chemical-based protocol for the rational design of covalent inhibitors. J Am Chem Soc 138(27):8332–8335. https://doi.org/10.1021/jacs.6b03052
53. Cavalli A, Carloni P, Recanatini M (2006) Target-related applications of first principles quantum chemical methods in drug design. Chem Rev 106(9):3497–3519. https://doi.org/10.1021/cr050579p
54. Chung LW, Sameera WM, Ramozzi R, Page AJ, Hatanaka M, Petrova GP, Harris TV, Li X, Ke Z, Liu F, Li HB, Ding L, Morokuma K (2015) The ONIOM method and its applications. Chem Rev 115(12):5678–5796. https://doi.org/10.1021/cr5004419
55. Shen L, Yang W (2018) Molecular dynamics simulations with quantum mechanics/molecular mechanics and adaptive neural networks. J Chem Theory Comput 14(3):1442–1455. https://doi.org/10.1021/acs.jctc.7b01195
56. Cao Y, Romero J, Aspuru-Guzik A (2018) Potential of quantum computing for drug discovery. IBM J Res Dev 62(6):1–6. https://doi.org/10.1147/JRD.2018.2888987
57. Harder E, Damm W, Maple J, Wu C, Reboul M, Xiang JY, Wang L, Lupyan D, Dahlgren MK, Knight JL, Kaus JW, Cerutti DS, Krilov G, Jorgensen WL, Abel R, Friesner RA (2016) OPLS3: a force field providing broad coverage of drug-like small molecules and proteins. J Chem Theory Comput 12:281–296. https://doi.org/10.1021/acs.jctc.5b00864
58. Wang J, Wolf RM, Caldwell JW, Kollman PA, Case DA (2004) Development and testing of a general AMBER force field. J Comput Chem 25:1157–1174. https://doi.org/10.1002/jcc.20035
59. Vanommeslaeghe K, Hatcher E, Acharya C, Kundu S, Zhong S, Shim J, Darian E, Guvench O, Lopes P, Vorobyov I, Mackerell AD Jr (2010) CHARMM general force field: a force field for drug-like molecules compatible with the CHARMM all-atom additive biological force fields. J Comput Chem 31:671–690. https://doi.org/10.1002/jcc.21367
60. Halgren TA (1999) MMFF VI. MMFF94s option for energy minimization studies. J Comput Chem 20:720–729. https://doi.org/10.1002/(SICI)1096-987X(199905)20:73.0.CO;2-X
61. Maier JA, Martinez C, Kasavajhala K, Wickstrom L, Hauser KE, Simmerling C (2015) ff14SB: improving the accuracy of protein side chain and backbone parameters from ff99SB. J Chem Theory Comput 11:3696–3713. https://doi.org/10.1021/acs.jctc.5b00255
62. Brooks BR, Brooks CL III, Mackerell AD Jr, Nilsson L, Petrella RJ, Roux B, Won Y, Archontis G, Bartels C, Boresch S, Caflisch A, Caves L, Cui Q, Dinner AR, Feig M, Fischer S, Gao J, Hodoscek M, Im W, Kuczera K, Lazaridis T, Ma J, Ovchinnikov V, Paci E, Pastor RW, Post CB, Pu JZ, Schaefer M, Tidor B, Venable RM, Woodcock HL, Wu X, Yang W, York DM, Karplus M (2009) CHARMM: the biomolecular simulation program. J Comput Chem 30:1545–1614. https://doi.org/10.1002/jcc.21287
63. Ivani I, Dans PD, Noy A, Pérez A, Faustino I, Hospital A, Walther J, Andrio P, Goñi R, Balaceanu A, Portella G, Battistini F, Gelpí JL, González C, Vendruscolo M, Laughton CA, Harris SA, Case DA, Orozco M (2016) Parmbsc1: a refined force field for DNA simulations. Nat Methods 13:55–58. https://doi.org/10.1038/nmeth.3658
64. Sztuba-Solinska J, Chavez-Calvillo G, Cline SE (2019) Unveiling the druggable RNA targets and small molecule therapeutics. Bioorg Med Chem 27(10):2149–2165. https://doi.org/10.1016/j.bmc.2019.03.057
65. Zhang C, Lu C, Jing Z, Wu C, Piquemal JP, Ponder JW, Ren P (2018) AMOEBA polarizable atomic multipole force field for nucleic acids. J Chem Theory Comput 14(4):2084–2108. https://doi.org/10.1021/acs.jctc.7b01169
66. Li Y, Liu Z, Li J, Han L, Liu J, Zhao Z, Wang R (2014) Comparative assessment of scoring functions on an updated benchmark: 2. Evaluation methods and general results. J Chem Inf Model 54(6):1700–1716. https://doi.org/10.1021/ci500080q
67. Zhou T, Caflish A (2010) High-throughput virtual screening using quantum mechanical probes: discovery of selective kinase inhibitors. ChemMedChem 5(7):1007–1014. https://doi.org/10.1002/cmdc.201000085
68. Lu J, Zhang Z, Ni Z, Shen H, Tu Z, Liu H, Lu R (2014) QM/MM-PB/SA scoring of the interaction strength between Akt kinase and apigenin analogues. Comput Biol Chem 52:25–33. https://doi.org/10.1016/j.compbiolchem.2014.07.002
69. Mazanetz MP, Ichihara O, Law RJ, Whittaker M (2011) Prediction of cyclin-dependent kinase 2 inhibitor potency using the fragment

molecular orbital method. J Chem 3:2. https://doi.org/10.1186/1758-2946-3-2

70. Heifetz A, Aldeghi M, Chudyk EI, Fedorov DG, Bodkin MJ, Biggin PC (2016) Using the fragment molecular orbital method to investigate agonist-orexin-2 receptor interactions. Biochem Soc Trans 44(2):574–581. https://doi.org/10.1042/BST20150250
71. Hsieh TJ, Lin HY, Tu Z, Lin TC, Wu SC, Tseng YY, Liu FT, Hsu ST, Lin CH (2016) Dual thio-digalactoside-binding modes of human galectins as the structural basis for the design of potent and selective inhibitors. Scientific Rep 6:29457. https://doi.org/10.1038/srep29457
72. Chudyk EI, Sarrat L, Aldeghi M, Fedorov DG, Bodkin MJ, James T, Southey M, Robinson R, Morao I, Heifetz A (2018) Exploring GPCR-ligand interactions with the Fragment Molecular Orbital (FMO) method. Methods Mol Biol 1705:179–195. https://doi.org/10.1007/978-1-4939-7465-8_8
73. Zou Y, Wang F, Wang Y, Guo W, Zhang Y, Xu Q, Lai Y (2017) Systematic study of imidazoles inhibiting IDO1 via the integration of molecular mechanics and quantum mechanics calculations. Eur J Med Chem 131:152–170. https://doi.org/10.1016/j.ejmech.2017.03.021
74. Borbulevych O, Martin RI, Tickle IJ, Westerhoff LM (2016) XModeScore: a novel method for accurate protonation/tautomer-state determination using quantum-mechanically driven macromolecular X-ray crystallographic refinement. Acta Cryst D72:586–598. https://doi.org/10.1107/S2059798316002837
75. Zhou T, Huang D, Caflisch A (2010) Quantum mechanical methods for drug design. Curr Top Med Chem 10(1):33–45. https://doi.org/10.2174/156802610790232242
76. Hargis JC, Vankayala SL, White JK, Woodcock HL (2014) Identification and characterization of noncovalent interactions that drive binding and specificity in DD-peptidases and β-lactamases. J Chem Theory Comput 10(2):855–864. https://doi.org/10.1021/ct400968v
77. Avgy-David HH, Senderowitz H (2015) Toward focusing conformational ensembles on bioactive conformations: a molecular mechanics/quantum mechanics study. J Chem Inf Model 55:2154–2167. https://doi.org/10.1021/acs.jcim.5b00259
78. Chen J, Wang J, Zhang Q, Chen K, Zhu W (2015) Probing origin of binding difference of inhibitors to MDM2 and MDMX by polarizable molecular dynamics simulation and QM/MM-GBSA calculation. Sci Rep 5:17421. https://doi.org/10.1038/srep17421
79. MacDonald CA, Boyd RJ (2015) Computational insights into the suicide inhibition of Plasmodium falciparum Fk506-binding protein 35. Bioorg Med Chem Lett 25(16):3221–3225. https://doi.org/10.1016/j.bmcl.2015.05.079
80. McCormick MC, Keijzer K, Polavarapu A, Schultz FA, Baik MH (2014) Understanding intrinsically irreversible, non-Nernstian, two-electron redox processes: a combined experimental and computational study of the electrochemical activation of platinum (IV) antitumor prodrugs. J Am Chem Soc 136(25):8992–9000. https://doi.org/10.1021/ja5029765
81. Lence E, van der Kamp MW, González-Bello C, Mulholland AJ (2018) QM/MM simulations identify the determinants of catalytic activity differences between type II dehydroquinase enzymes. Org Biomol Chem 16(24):4443–4455. https://doi.org/10.1039/c8ob00066b
82. Kuhn B, Jacobsen W, Christians U, Benet LZ, Kollman PA (2001) Metabolism of sirolimus and its derivative everolimus by cytochrome P450 3A4: insights from docking, molecular dynamics, and quantum chemical calculations. J Med Chem 44(12):2027–2034. https://doi.org/10.1021/jm010079y
83. Sun H, Scott DO (2010) Structure-based drug metabolism predictions for drug design. Chem Biol Drug Des 75:3–17. https://doi.org/10.1111/j.1747-0285.2009.00899.x
84. Tyzack JD, Kirchmair J (2019) Computational methods and tools to predict cytochrome P450 metabolism for drug discovery. Chem Biol Drug Des 93(4):377–386. https://doi.org/10.1111/cbdd.13445
85. Bobovská A, Tvaroška I, Kóňa J (2016) Using DFT methodology for more reliable predictive models: design of inhibitors of Golgi α-mannosidase II. J Mol Graph Model 66:47–57. https://doi.org/10.1016/j.jmgm.2016.03.004
86. Sullivan DJ Jr, Kaludov N, Martinov MN (2011) Discovery of potent, novel, non-toxic anti-malarial compounds via quantum modelling, virtual screening and in vitro experimental validation. Malar J 10:274. https://doi.org/10.1186/1475-2875-10-274
87. Sullivan DJ, Liu Y, Mott BT, Kaludov N, Martinov MN (2015) Discovery of novel liver-stage antimalarials through quantum

similarity. PLoS One 10(5):e0125593. https://doi.org/10.1371/journal.pone.0125593

88. Świderek K, Tuñón I, Moliner V, Bertran J (2015) Computational strategies for the design of new enzymatic functions. Arch Biochem Biophys 582:68–79. https://doi.org/10.1016/j.abb.2015.03.013
89. Alexandrova AN, Jorgensen WL (2009) Origin of the activity drop with the E50D variant of catalytic antibody 34E4 for Kemp elimination. J Phys Chem B 113(2):497–504. https://doi.org/10.1021/jp8076084
90. Gong W, Wu R, Zhang Y (2015) Thiol versus hydroxamate as zinc binding group in HDAC inhibition: an ab initio QM/MM molecular dynamics study. J Comput Chem 36:2228–2235. https://doi.org/10.1002/jcc.24203
91. Hitzenberger M, Schuster D, Hofer TS (2017) The binding mode of the sonic hedgehog inhibitor robotnikinin, a combined docking and QM/MM MD study. Front Chem 5:76. https://doi.org/10.3389/fchem.2017.00076
92. Steinmann C, Olsson MA, Ryde U (2018) Relative ligand-binding free energies calculated from multiple short QM/MM MD simulations. J Chem Theory Comput 14:3228–3237. https://doi.org/10.1021/acs.jctc.8b00081
93. Zhu T, Xiao X, Ji C, Zhang JZ (2013) A new quantum calibrated force field for zinc-protein complex. J Chem Theory Comput 9(3):1788–1798. https://doi.org/10.1021/ct301091z
94. Xiong X, Chen Z, Cossins BP, Xu Z, Shao Q, Ding K, Zhu W, Shi J (2015) Force fields and scoring functions for carbohydrate simulation. Carbohydr Res 401:73–81. https://doi.org/10.1016/j.carres.2014.10.028
95. Cole DJ, Vilseck JZ, Tirado-Rives J, Payne MC, Jorgensen WL (2016) Biomolecular force field parameterization via atoms-in-molecule electron density partitioning. J Chem Theory Comput 12(5):2312–2323. https://doi.org/10.1021%2Facs.jctc.6b00027
96. Visscher KM, Geerke DP (2019) Deriving force-field parameters from first principles using a polarizable and higher order dispersion model. J Chem Theory Comput 15(3):1875–1883. https://doi.org/10.1021/acs.jctc.8b01105
97. Hsiao YW, Sanchez-Garcia E, Doerr M, Thiel W (2010) Quantum refinement of protein structures: implementation and application to the red fluorescent protein DsRed.M1. J Phys Chem B 114(46):15413–15423. https://doi.org/10.1021/jp108095n
98. Li X, Hayik SA, Merz KM Jr (2010) QM/MM X-ray refinement of zinc metalloenzymes. J Inorg Biochem 104(5):512–522. https://doi.org/10.1016/j.jinorgbio.2009.12.022
99. Fu Z, Li X, Miao Y, Merz KM Jr (2013) Conformational analysis and parallel QM/MM X-ray refinement of protein bound anti-Alzheimer drug donepezil. J Chem Theory Comput 9(3):1686–1693. https://doi.org/10.1021/ct300957x
100. Dittrich B, Lübben J, Mebs S, Wagner A, Luger P, Flaig R (2017) Accurate bond lengths to hydrogen atoms from single-crystal X-ray diffraction by including estimated hydrogen ADPs and comparison to neutron and QM/MM benchmarks. Chemistry 23(19):4605–4614. https://doi.org/10.1002/chem.201604705
101. Chaskar P, Zoete V, Röhrig UF (2014) Toward on-the-fly quantum mechanical/molecular mechanical (QM/MM) docking: development and benchmark of a scoring function. J Chem Inf Model 54(11):3137–3152. https://doi.org/10.1021/ci5004152
102. Wichapong K, Rohe A, Platzer C, Slynko I, Erdmann F, Schmidt M, Sippl W (2014) Application of docking and QM/MM-GBSA rescoring to screen for novel Myt1 kinase inhibitors. J Chem Inf Model 54(3):881–893. https://doi.org/10.1021/ci4007326
103. Zang P, Gong A, Zhang P, Yu J (2016) Targeting druggable enzymome by exploiting natural medicines: an in silico-in vitro integrated approach to combating multidrug resistance in bacterial infection. Pharm Biol 54(4):604–618. https://doi.org/10.3109/13880209.2015.1068338
104. Khandelwal A, Balaz S (2007) QM/MM linear response method distinguishes ligand affinities for closely related metalloproteins. Proteins 69(2):326–339. https://doi.org/10.1002/prot.21500
105. Sawada T, Fedorov DG, Kitaura K (2010) Role of the key mutation in the selective binding of avian and human influenza hemagglutinin to sialosides revealed by quantum-mechanical calculations. J Am Chem Soc 132(47):16862–16872. https://doi.org/10.1021/ja105051e
106. Rathore RS, Sumakanth M, Reddy MS, Reddanna P, Rao AA, Erion MD, Reddy MR (2013) Advances in binding free energies calculations: QM/MM-based free energy perturbation method for drug design. Curr Pharm Des 19(26):4674–4686. https://doi.org/10.2174/13816128113199260002

107. Otsuka T, Okimoto N, Taiji M (2015) Assessment and acceleration of binding energy calculations for protein-ligand complexes by the fragment molecular orbital method. J Comput Chem 36(30):2209–2218. https://doi.org/10.1002/jcc.24055
108. Su PC, Tsai CC, Mehboob S, Hevener KE, Johnson ME (2015) Comparison of radii sets, entropy, QM methods, and sampling on MM-PBSA, MM-GBSA, and QM/MM-GBSA ligand binding energies of *F. tularensis* Enoyl-ACP reductase (FabI). J Comput Chem 36(25):1859–1873. https://doi.org/10.1002/jcc.24011
109. Ehrlich S, Göller AH, Grimme S (2017) Towards full quantum-mechanics-based protein-ligand binding affinities. ChemPhysChem 18(8):898–905. https://doi.org/10.1002/cphc.201700082
110. Caballero J, Alzate-Morales JH, Vergara-Jaque A (2011) Investigation of the differences in activity between hydroxycycloalkyl N1 substituted pyrazole derivatives as inhibitors of B-Raf kinase by using docking, molecular dynamics, QM/MM, and fragment-based de novo design: study of binding mode of diastereomer compounds. J Chem Inf Model 51(11):2920–2931. https://doi.org/10.1021/ci200306w
111. Sun TY, Wang Q, Zhang J, Wu T, Zhang F (2013) Trastuzumab-Peptide interactions: mechanism and application in structure-based ligand design. Int J Mol Sci 14 (8):16836–16850. https://doi.org/10.3390/ijms140816836
112. De Colibus L, Wang X, Spyrou JAB, Kelly J, Ren J, Grimes J, Puerstinger G, Stonehouse N, Walter TS, Hu Z, Wang J, Li X, Peng W, Rowlands D, Fry EE, Rao Z, Stuart DI (2014) More powerful virus inhibitors from structure-based analysis of HEV71 capsid-binding molecules. Nat Struct Mol Biol 21(3):282–288. https://doi.org/10.1038/nsmb.2769
113. Zanatta G, Nunes G, Bezerra EM, da Costa RF, Martins A, Caetano EW, Freire VN, Gottfried C (2014) Antipsychotic haloperidol binding to the human dopamine D3 receptor: beyond docking through QM/MM refinement toward the design of improved schizophrenia medicines. ACS Chem Neurosci 5 (10):1041–1054. https://doi.org/10.1021/cn500111e
114. Yu N, Hayik SA, Wang B, Liao N, Reynolds CH, Merz KM Jr (2006) Assigning the protonation states of the key aspartates in β-Secretase using QM/MM X-ray structure refinement. J Chem Theory Comput 2 (4):1057–1069. https://doi.org/10.1021/ct0600060
115. Lee W, Luckner SR, Kisker C, Tonge PJ, Engels B (2011) Elucidation of the protonation states of the catalytic residues in *mt*KasA - implications for inhibitor design. Biochemistry 50(25):5743–5756. https://doi.org/10.1021/bi200006t
116. Vega-Teijido MA, El Chamy Maluf S, Bonturi CR, Sambrano JR, Ventura ON (2014) Theoretical insight into the mechanism for the inhibition of the cysteine protease cathepsin B by 1,2,4-thiadiazole derivatives. J Mol Model 20(6):2254. https://doi.org/10.1007/s00894-014-2254-0
117. Ma S, Vogt KA, Petrillo N, Ruhoff AJ (2015) Characterizing the protonation states of the catalytic residues in apo and substrate-bound human T-cell leukemia virus type 1 protease. Comput Biol Chem 56:61–70. https://doi.org/10.1016/j.compbiolchem.2015.04.002
118. Kocak A, Erol I, Yildiz M, Can H (2016) Computational insights into the protonation states of catalytic dyad in BACE1-acyl guanidine based inhibitor complex. J Mol Graph Model 70:226–235. https://doi.org/10.1016/j.jmgm.2016.10.013
119. Chakravorty DK, Wang B, Ucisik MN, Merz KM Jr (2011) Insight into the cation-π interaction at the metal binding site of the copper metallochaperone CusF. J Am Chem Soc 133 (48):19330–19333. https://doi.org/10.1021/ja208662z
120. Zhou J, Xie H, Liu Z, Luo HB, Wu R (2014) Structure-function analysis of the conserved tyrosine and diverse π-stacking among class I histone deacetylases: a QM (DFT)/MM MD study. J Chem Inf Model 54(11):3162–3171. https://doi.org/10.1021/ci500513n
121. Lenz SAP, Wetmore SD (2017) QM/MM study of the reaction catalyzed by alkyladenine DNA glycosylase: examination of the substrate specificity of a DNA repair enzyme. J Phys Chem B 121(49):11096–11108. https://doi.org/10.1021/acs.jpcb.7b09646
122. Zhou J, Wang YS (2017) Rational redesign of a cation···π···π stacking at cardiovascular Fbw7-Skp1 complex interface and its application for deriving self-inhibitory peptides to disrupt the complex interaction. J Mol Model 23(10):296. https://doi.org/10.1007/s00894-017-3456-z
123. Zhang L, Hao GF, Tan Y, Xi Z, Huang MZ, Yang GF (2009) Bioactive conformation analysis of cyclic imides as protoporphyrinogen oxidase inhibitor by combining DFT calculations, QSAR and molecular dynamic simulations. Bioorg Med Chem 17

(14):4935–4942. https://doi.org/10.1016/j.bmc.2009.06.003

124. Mandal M, Zhu Z, Cumming JN, Liu X, Strickland C, Mazzola RD, Caldwell JP, Leach P, Grzelak M, Hyde L, Zhang Q, Terracina G, Zhang L, Chen X, Kuvelkar R, Kennedy ME, Favreau L, Cox K, Orth P, Buevich A, Voigt J, Wang H, Kazakevich I, McKittrick BA, Greenlee W, Parker EM, Stamford AW (2012) Design and validation of bicyclic iminopyrimidinones as beta amyloid cleaving enzyme-1 (BACE1) inhibitors: conformational constraint to favor a bioactive conformation. J Med Chem 55 (21):9331–9345. https://doi.org/10.1021/jm301039c

125. Arooj M, Sakkiah S, Kim S, Arulalapperumal V, Lee KW (2013) A combination of receptor-based pharmacophore modeling & QM techniques for identification of human chymase inhibitors. PLoS One 8 (4):e63030. https://doi.org/10.1371/journal.pone.0063030

126. Pasha FA, Neaz MM (2013) Molecular dynamics and QM/MM-based 3D interaction analyses of cyclin-E inhibitors. J Mol Model 19(2):879–891. https://doi.org/10.1007/s00894-012-1620-z

127. Bembenek SD, Keith JM, Letavic MA, Apodaca R, Barbier AJ, Dvorak L, Aluisio L, Miller KL, Lovenberg TW, Carruthers NI (2008) Lead identification of acetylcholinesterase inhibitors-histamine H3 receptor antagonists from molecular modeling. Bioorg Med Chem 16(6):2968–2973. https://doi.org/10.1016/j.bmc.2007.12.048

128. Remko M, Remková A, Broer R (2016) Theoretical study of molecular structure and physicochemical properties of novel factor Xa inhibitors and dual factor Xa and factor IIa inhibitors. Molecules 21(2):185. https://doi.org/10.3390/molecules21020185

129. Pardhi T, Vasu K (2018) Identification of dual kinase inhibitors of CK2 and GSK3β: combined qualitative and quantitative pharmacophore modeling approach. J Biomol Struct Dyn 36(1):177–194. https://doi.org/10.1080/07391102.2016.1270856

130. Spiegel K, Magistrato A (2006) Modeling anticancer drug-DNA interactions via mixed QM/MM molecular dynamics simulations. Org Biomol Chem 4(13):2507–2517. https://doi.org/10.1039/b604263p

131. Lodola A, Capoferri L, Rivara S, Chudyk E, Sirirak J, Dyguda-Kazimierowicz E, Andrzej Sokalski W, Mileni M, Tarzia G, Piomelli D, Mor M, Mulholland AJ (2011) Understanding the role of carbamate reactivity in fatty acid amide hydrolase inhibition by QM/MM mechanistic modelling. Chem Commun (Camb) 47(9):2517–2519. https://doi.org/10.1039/c0cc04937a

132. Schmidt TC, Welker A, Rieger M, Sahu PK, Sotriffer CA, Schirmeister T, Engels B (2014) Protocol for rational design of covalently interacting inhibitors. ChemPhysChem 15 (15):3226–3235. https://doi.org/10.1002/cphc.201402542

133. Capoferri L, Lodola A, Rivara S, Mor M (2015) Quantum mechanics/molecular mechanics modeling of covalent addition between EGFR-cysteine 797 and N-(4-anilinoquinazolin-6-yl) acrylamide. J Chem Inf Model 55(3):589–599. https://doi.org/10.1021/ci500720e

134. Ren W, Pengelly R, Farren-Dai M, Shamsi Kazem Abadi S, Oehler V, Akintola O, Draper J, Meanwell M, Chakladar S, Świderek K, Moliner V, Britton R, Gloster TM, Bennet AJ (2018) Revealing the mechanism for covalent inhibition of glycoside hydrolases by carbasugars at an atomic level. Nat Commun 9(1):3243. https://doi.org/10.1038/s41467-018-05702-7

135. James C, Pettit GR, Nielsen OF, Jayakumar VS, Joe IH (2008) Vibrational spectra and ab initio molecular orbital calculations of the novel anti-cancer drug combretastatin A-4 prodrug. Spectrochim Acta A Mol Biomol Spectrosc 70(5):1208–1216. https://doi.org/10.1016/j.saa.2007.10.052

136. Karaman R (2011) Computational-aided design for dopamine prodrugs based on novel chemical approach. Chem Biol Drug Des 78(5):853–863. https://doi.org/10.1111/j.1747-0285.2011.01208.x

137. Karaman R, Fattash B, Qtait A (2013) The future of prodrugs - design by quantum mechanics methods. Expert Opin Drug Deliv 10(5):713–729. https://doi.org/10.1517/17425247.2013.786699

138. Arfeen M, Patel DS, Abbat S, Taxak N, Bharatam PV (2014) Importance of cytochromes in cyclization reactions: quantum chemical study on a model reaction of proguanil to cycloguanil. J Comput Chem 35 (28):2047–2055. https://doi.org/10.1002/jcc.23719

139. Ponte F, Russo N, Sicilia E (2018) Insights from computations on the mechanism of reduction by ascorbic acid of Pt^{IV} prodrugs with asplatin and its chlorido and bromido analogues as model systems. Chemistry 24 (38):9572–9580. https://doi.org/10.1002/chem.201800488

140. Van der Kamp MW, Chaudret R, Mulholland AJ (2013) QM/MM modelling of ketosteroid isomerase reactivity indicates that active

site closure is integral to catalysis. FEBS J 280 (13):3120–3131. https://doi.org/10.1111/febs.12158

141. Kaiyawet N, Lonsdale R, Rungrotmongkol T, Mulholland AJ, Hannongbua S (2015) High-level QM/MM calculations support the concerted mechanism for Michael addition and covalent complex formation in thymidylate synthase. J Chem Theory Comput 11 (2):713–722. https://doi.org/10.1021/ct5005033
142. Kumari M, Kozmon S, Kulhánek P, Štepán J, Tvaroška I, Koča J (2015) Exploring reaction pathways for O-GlcNAc transferase catalysis. A string method study. J Phys Chem B 119 (12):4371–4381. https://doi.org/10.1021/jp511235f
143. Fernandes HS, Ramos MJ, Cerqueira NMFSA (2017) The catalytic mechanism of the pyridoxal-5′-phosphate-dependent enzyme, histidine decarboxylase: a computational study. Chemistry 23(38):9162–9173. https://doi.org/10.1002/chem.201701375
144. Elsässer B, Zauner FB, Messner J, Soh WT, Dall E, Brandstetter H (2017) Distinct roles of catalytic cysteine and histidine in the protease and ligase mechanisms of human legumain as revealed by DFT-based QM/MM simulations. ACS Catal 7(9):5585–5593. https://doi.org/10.1021/acscatal.7b01505
145. Roy S, Kästner J (2017) Catalytic mechanism of salicylate dioxygenase: QM/MM simulations reveal the origin of unexpected regioselectivity of the ring cleavage. Chemistry 23 (37):8949–8962. https://doi.org/10.1002/chem.201701286
146. Brás NF, Fernandes PA, Ramos MJ (2018) Understanding the rate-limiting step of glycogenolysis by using QM/MM calculations on human glycogen phosphorylase. ChemMedChem 13(15):1608–1616. https://doi.org/10.1002/cmdc.201800218
147. Lonsdale R, Houghton KT, Żurek J, Bathelt CM, Foloppe N, de Groot MJ, Harvey JN, Mulholland AJ (2013) Quantum mechanics/molecular mechanics modeling of regioselectivity of drug metabolism in cytochrome P450 2C9. J Am Chem Soc 135 (21):8001–8015. https://doi.org/10.1021/ja402016p
148. Tyzack JD, Williamson MJ, Torella R, Glen RC (2013) Prediction of cytochrome P450 xenobiotic metabolism: tethered docking and reactivity derived from ligand molecular orbital analysis. J Chem Inf Model 53 (6):1294–1305. https://doi.org/10.1021/ci400058s
149. Lonsdale R, Rouse SL, Sansom MS, Mulholland AJ (2014) A multiscale approach to modelling drug metabolism by membrane-bound cytochrome P450 enzymes. PLoS Comput Biol 10(7):e1003714. https://doi.org/10.1371/journal.pcbi.1003714
150. Putkaradze N, Kiss FM, Schmitz D, Zapp J, Hutter MC, Bernhardt R (2017) Biotransformation of prednisone and dexamethasone by cytochrome P450 based systems - identification of new potential drug candidates. J Biotechnol 242:101–110. https://doi.org/10.1016/j.jbiotec.2016.12.011
151. Ferreira AM, Krishnamurthy M, Moore BM II, Finkelstein D, Bashford D (2009) Quantitative structure-activity relationship (QSAR) for a series of novel cannabinoid derivatives using descriptors derived from semi-empirical quantum-chemical calculations. Bioorg Med Chem 7(6):2598–2606. https://doi.org/10.1016/j.bmc.2008.11.059
152. Güssregen S, Matter H, Hessler G, Müller M, Schmidt F, Clark T (2012) 3D-QSAR based on quantum-chemical molecular fields: toward an improved description of halogen interactions. J Chem Inf Model 52 (9):2441–2453. https://doi.org/10.1021/ci300253z
153. Ginex T, Muñoz-Muriedas J, Herrero E, Gibert E, Cozzini P, Luque FJ (2016) Application of the quantum mechanical IEF/PCM-MST hydrophobic descriptors to selectivity in ligand binding. J Mol Model 22 (6):136. https://doi.org/10.1007/s00894-016-2991-3
154. Hudson BD, Whitley DC, Ford MG, Swain M, Essex JW (2008) Pattern recognition based on color-coded quantum mechanical surfaces for molecular alignment. J Mol Model 14(1):49–57. https://doi.org/10.1007/s00894-007-0251-2
155. Vázquez J, Deplano A, Herrero A, Ginex T, Gibert E, Rabal O, Oyarzabal J, Herrero E, Luque FJ (2018) Development and validation of molecular overlays derived from three-dimensional hydrophobic similarity with PharmScreen. J Chem Inf Model 58 (8):1596–1609. https://doi.org/10.1021/acs.jcim.8b00216
156. Cannizzaro CE, Ashley JA, Janda KD, Houk KN (2003) Experimental determination of the absolute enantioselectivity of an antibody-catalyzed Diels-Alder reaction and theoretical explorations of the origins of stereoselectivity. J Am Chem Soc 125 (9):2489–2506. https://doi.org/10.1021/ja020879d

157. Smith AJT, Müller R, Toscano MD, Kast P, Hellinga HW, Hilvert D, Houk KN (2008) Structural reorganization and preorganization in enzyme active sites: comparisons of experimental and theoretically ideal active site geometries in the multistep serine esterase reaction cycle. J Am Chem Soc 130 (46):15361–15373. https://doi.org/10.1021/ja803213p
158. Frushicheva MP, Cao J, Chu ZT, Warshel A (2010) Exploring challenges in rational enzyme design by simulating the catalysis in artificial kemp eliminase. Proc Natl Acad Sci U S A 107(39):16869–16874. https://doi.org/10.1073/pnas.1010381107
159. Singh MK, Chu ZT, Warshel A (2014) Simulating the catalytic effect of a designed mononuclear zinc metalloenzyme that catalyzes the hydrolysis of phosphate triesters. J Phys Chem B 118(42):12146–12152. https://doi.org/10.1021/jp507592g

Chapter 16

Binding Free Energy Calculation Using Quantum Mechanics Aimed for Drug Lead Optimization

Claudio N. Cavasotto

Abstract

The routine use of in silico tools is already established in drug lead design. Besides the use of molecular docking methods to screen large chemical libraries and thus prioritize compounds for purchase or synthesis, more accurate calculations of protein–ligand binding free energy has shown the potential to guide lead optimization, thus saving time and resources. Theoretical developments and advances in computing power have allowed quantum mechanical-based methods applied to calculations on biomacromolecules to be increasingly explored and used, with the purpose of providing a more accurate description of protein–ligand interactions and an enhanced level of accuracy in the calculation of binding affinities. It should be noted that the quantum mechanical formulation includes, in principle, all contributions to the energy, considering terms usually neglected in molecular mechanics force fields, such as electronic polarization, metal coordination, and covalent binding; moreover, quantum mechanical approaches are systematically improvable. By treating all elements and interactions on equal footing, and avoiding the need of system-dependent parameterizations, they provide a greater degree of transferability. In this review, we illustrate the increasing relevance of quantum mechanical methods for binding free energy calculation in the context of structure-based drug lead optimization, showing representative applications of the different approaches available.

Key words Quantum mechanics, Structure-based drug design, Lead optimization, Computer-aided drug discovery, Binding free energy calculation

1 Introduction

The drug lead design process relied for until not so long on the experimental high-throughput screening of compound libraries; in spite of an array of efforts to improve its efficiency and performance, high-throughput screening is still an expensive and time-consuming technique [1]. Since about 40 years ago, protein–ligand three-dimensional (3D) structures have been used to guide the optimization of drug lead molecules to improve both potency and selectivity [2], thus paving the way to a more rational approach in drug design. This was accompanied by theoretical developments, improved computational algorithms, and faster computing resources, which allowed the use of computational methods to in

Alexander Heifetz (ed.), *Quantum Mechanics in Drug Discovery*, Methods in Molecular Biology, vol. 2114, https://doi.org/10.1007/978-1-0716-0282-9_16,

silico characterize protein–ligand interaction, obtain approximate values of small-molecule binding free energy, and screen large chemical libraries using structure-based approaches. It can be stated that today, computational chemistry is a consolidated and valuable tool in the drug discovery process [3–6].

The central quantity in protein-small-molecule association is the binding free energy (ΔG_{bind}), a property of the utmost importance in pharmaceutical chemistry and thus of immense relevance for the pharma industry. No effort is big enough to calculate it in an accurate yet computationally efficient way. Reliable calculation of receptor-small-molecule affinities in the early stages of the drug discovery pipeline would be instrumental to rationally design new, more potent, and safer drugs, thus saving precious effort, time, and cost. The accurate calculation of ΔG_{bind} in the context of protein-small-molecule association depends on (1) the potential energy surface of the system, (2) the consideration of protein flexibility, (3) the accounting for the presence of water molecules within the binding site, and (4) the solvation model used. The last two issues have been discussed recently (Refs. 7–11, and Ref. 12, respectively). During the last 25 years, there has been a remarkable progress in theoretical formulation and algorithmic developments for the calculation of binding free energy [13–15], ranging from fast methods, to be used in high-throughput docking and scoring [16, 17], to slower—yet in principle more accurate—calculations using free energy perturbation (FEP) or thermodynamic integration [18, 19], aimed to guide chemical synthesis for hit-to-lead optimization. While most of these applications are rooted in classical mechanics-based force fields, the advent of more powerful computers allowed, in recent years, an increasing development and application of quantum mechanical tools to model macromolecular systems in the aid of drug lead discovery and design. The blind challenges ran by the Drug Design Data Resource (D3R) in 2015 [20], 2016 [21], and 2018 [22] have shown the importance of method development and benchmarking for pose assessment and binding affinity ranking of ligands.

It should be noted that the quantum mechanical formulation, being in principle theoretically exact, includes all contributions to the energy, accounting for interactions or effects usually neglected in molecular mechanics force fields: electronic polarization, charge transfer, halogen bonding, and covalent bond formation; moreover, it offers the possibility of treating all elements and interactions on equal footing, avoiding the need of system-dependent parameterizations, thus being general across the chemical space, with the corresponding practical advantages. Very recently, Grimme and Schreiner published a very interesting essay on the changes they envisage for the next 25 years in computational chemistry, being one of them how low-cost quantum mechanical methods will replace force fields [23].

It should be highlighted that quantum mechanics has had an important role since the early days of in silico drug design (cf. the seminal work of W.G. Richards on quantum pharmacology [24]). In fact, quantum mechanical calculations are routinely used to derive force field torsional potentials from high-level ab initio data and partial atomic charges by fitting to electrostatic surface potentials [25], in QSAR methods [26], to study reaction mechanisms [27], to study small-molecule strain [28, 29], and in molecular docking [30, 31], among other applications. The goal of this chapter is to illustrate the increasing importance of quantum mechanical methods to calculate binding affinities in the context of structure-based drug lead optimization, highlighting also representative applications of each of the different approaches available. Should the reader be interested in the development rather than the application of quantum mechanical tools for binding affinity calculations, an excellent recent review is available [32]; a deep coverage on the development of semi-empirical quantum mechanical and density functional methods expanded by hydrogen bonding and dispersion corrections can be found in a nice work by Yilmazer and Korth [33].

2 Calculation of Ligand Binding Affinities Using Quantum Mechanics

The pioneering works of Raha and Merz were among the first to use quantum mechanical methods to calculate binding free energies. They introduced QMScore, where the entire molecular system is treated at a quantum mechanical level for binding free energy calculation. The method was initially applied to 18 inhibitors of carbonic anhydrase and 5 of carboxypeptidase [34]. Gas-phase energy calculations were performed at the semi-empirical level using the Austin Model 1 (AM1) Hamiltonian [35] and DivCon [36], a quantum mechanical linear scaling approach. To describe non-polar interactions, the AMBER 96 force field was used, and the solvation energy contribution was calculated using the Poisson-Boltzmann (PB) implicit solvation model. In a follow-up work, QMScore was used for the discrimination of native from decoy poses on a set of 165 protein–ligand complexes, what also allowed to capture key ligand affinity trends [37].

Between the computationally demanding free energy calculations and the fast though rather inaccurate phenomenological models, such as docking scoring, end point methods using a continuous solvent model represent a reasonable alternative to calculate binding free energy. These methods use only the initial and final states of the system, and they could be regarded as efficient to capture the essential physical chemistry of binding. In this case, the binding free energy can be derived from first-principle statistical thermodynamics as [13, 38].

$$\Delta G^0_{PL} = \langle U_{PL} \rangle - \langle U_P \rangle - \langle U_L \rangle + \langle W_{PL} \rangle - \langle W_P \rangle - \langle W_L \rangle - T\Delta S^0_{config} \quad (1)$$

where the subindices refer to ligand, protein, and ligand-protein complex; $\langle U_i \rangle$ and $\langle W_i \rangle$ are the Boltzmann-averaged potential and solvation energies, respectively; the last term accounts for the change in standard entropy upon binding; and average values are calculated using molecular dynamics trajectories. Usually, only a subset of snapshots are re-evaluated using an implicit solvent model, what improves convergence without sacrificing accuracy and saves calculation time [39]. The entropy is usually calculated using a rigid rotor/harmonic oscillator approach [13]. Two very popular methods based on Eq.(1) are the molecular mechanics-Poisson-Boltzmann surface area (MM-PBSA) [40] and the MM-generalized Born surface area (MM-GBSA) [41] methods, in which molecular dynamics simulations with explicit solvent are re-scored using the PB or GB implicit solvent model, respectively. To incorporate a quantum mechanical description of the system, a new method named molecular mechanics-quantum mechanics surface area (MM-QMSA) [42] was presented, in which the binding free energy is calculated in an analogue way to Eq. (1) as

$$\Delta G^0_{PL} = \Delta \langle H^{QM} \rangle - T\Delta S^{RB} - T\Delta S^{int} \quad (2)$$

where H^{QM} represents the quantum mechanical energy (accounting for both the vacuum and solvation contributions), but evaluated on a molecular dynamics trajectory; the last two terms represent the change related to rigid body association entropy and internal degrees of freedom (vibrational) entropy. The solvation energy was calculated using the conductor-like screening model (COSMO) continuum solvent model [43], using re-optimized parameters (atomic radii and surface tension) [44] (cf. Refs. 42, 45 for a thorough explanation of this methodology). The MM-QMSA method was used to calculate ΔG_{bind} on a series of phospho-tetrapeptides (Ac-pYEEI, Ac-pYAEI, Ac-pYEAI, Ac-pYEEG, Ac-pYEEA) binding to the SH2 domain of the human protein kinase Lck. The quantum mechanical calculations were performed at the semi-empirical level using the PM3 Hamiltonian [46] and a linear scaling approach [47]. First, classical mechanics methods MM-PBSA (using standard [48] and PB-optimized radii [49]), MM-GBSA, and MM-GBSA with GB energy minimization were used to calculate the binding free energies, but they failed by almost an order of magnitude to reproduce experimental values, in agreement with an earlier work [50]. The MM-QMSA method showed a remarkable agreement for both the absolute (MAE = 0.7 kcal mol^{-1}) and relative ΔG_{bind} calculations.

The authors stated that the failure to reproduce ΔG_{bind} using FF-based methods could be attributed to the lack of force-field charge transferability within end point methods, while a quantum mechanical approach might provide a better balance to circumvent this problem. In a follow-up contribution, the binding of pSXXF phospho-peptides to the C-terminal domain of BRCA1 (BRCT) was assessed experimentally and the binding free energies calculated using three different end point approaches [51]. The MM-PBSA and the solvated interaction energy (SIE) [52] calculations were in reasonable agreement with experimental affinities, exhibiting a Pearlman predictive index (PI) [53] of 0.71 and 0.78, respectively. The MM-QMSA showed improved results, with a PI of 0.78. The correlation coefficients between experimental and calculated energies were 0.59, 0.61, and 0.69, respectively.

The binding of five acetylcholinesterase (AChE) inhibitors was analyzed using a classical mechanics approach (free energy perturbation, FEP) and a quantum mechanical-based approach, namely, quantum mechanics/molecular mechanics (QM/MM) molecular dynamics simulations [54]. The quantum mechanical calculations were performed at the semi-empirical level, using the AM1 Hamiltonian. The ΔG_{bind} was calculated as the sum of two terms, introducing two weights for the electrostatic and van der Waals contributions, within QM/MM calculations [55]. The correlation between the experimental and calculated values was very good ($R^2 = 0.96$ for 100 ps). Moreover, there was a qualitative agreement in the calculated and experimental ranking order. It was concluded that the use of quantum mechanics to describe the molecular system was very important due to the polar nature of the ligands and the high aromaticity of the enzyme binding site.

In a recent work, Ryde and Olsson evaluated the binding of nine small carboxylate ligands to the octa-acid deep-cavity host [56], using both a classical mechanics approach (reference-potential FEP calculations [57]) and full QM/MM FEP simulations. The ligand was described at the semi-empirical level using the PM6 Hamiltonian [58], augmented by the DH+ empirical dispersion and hydrogen bond corrections [59]. The authors found that the reference-potential approach is almost three times more effective than the direct approach. Regarding the quantum mechanical simulations, the convergence of the MM → QM/MM perturbations is enhanced by the addition of QM/MM MD simulations using several coupling parameter values between the MM and QM/MM energies.

A full quantum mechanical approach was presented recently where the absolute ligand binding free energy was expressed as the sum of three terms: the interaction energy, the solvation contribution, and the entropic term [60]. A truncated system was used, defined by the ligand and neighboring binding site atoms (~1000 atoms in total). For the interaction energy contribution, two

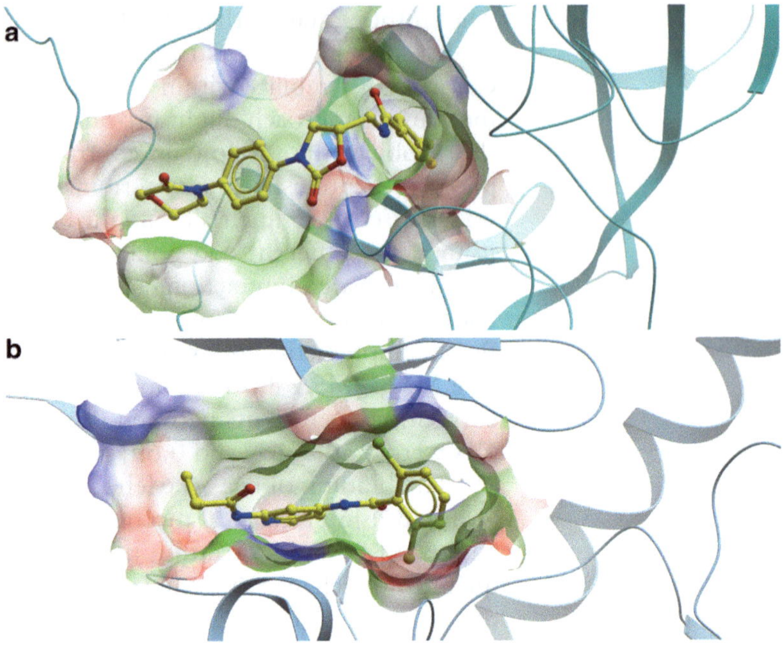

Fig. 1 Receptor-ligand complexes. (**a**) Serine protease factor X (FXa) in complex with rivaroxaban (PDB 2W26); (**b**) tyrosine-protein kinase 2 (TYK2) in complex with a cyclopropylamide (PDB 2GIH). The ligands are displayed in stick representation with yellow carbons, blue nitrogens, and red oxygens; the receptor is displayed in ribbon representation; the color code of the molecular surface of the binding site is green, hydrophobic; red, hydrogen bond acceptor; and blue, hydrogen bond donor. (Figure prepared with the ICM program (Molsoft LLC, San Diego, CA))

different approaches were used: the minimal basis Hartree-Fock HF-3c [61] including a D3 dispersion correction [62] and the composite hybrid PBEh-3c DFT, which exhibits a lower computational cost [63]; the entropic contribution was calculated using semi-empirical DFTB3-D3 hessian calculations [64, 65]; and the solvation term was calculated using the COSMO-RS method [66, 67]. The systems studied were the activated serine protease factor X (FXa) with 25 ligands and the non-receptor tyrosine-protein kinase 2 (TYK2) with 16 ligands (Fig. 1). The mean absolute deviation (MAD) values of ΔG_{bind} at the HF-3c level for both molecules were 2.8 and 2.7 kcal/mol, the Pearson correlation coefficients 0.47 and 0.51, respectively; a MAD of 2.1 kcal/mol was obtained for the FXa system using the PBEh-3c method (Pearson coefficient = 0.53). Although these results are clearly promising in terms of the use of quantum mechanical calculations to study small-molecule binding, it is also clear that this approach cannot be yet used in an industrial setting. Furthermore, errors stemming from the lack of conformational sampling and the solvation contribution would need further revision.

A QM/MM-based calculation of ΔG_{bind} was recently performed on four diverse protein targets of pharmaceutical relevance: beta-secretase 1 (BACE1), TYK2, heat shock protein 90 α (HSP90α), and protein kinase R (PKR)-like endoplasmic reticulum kinase (PERK); the number of ligands was 22, 16, 70, and 32, respectively [68]. Binding affinities were calculated using the linear interaction energy (LIE) protocol [69], where the α and β LIE coefficients were similar to those already reported [70], but re-optimized to fit the experimental binding free energies within the TYK2 set. The ligands were described at the semi-empirical level using the AM1 Hamiltonian and the proteins using molecular mechanics force fields. The ensemble averages were obtained from QM/MM calculations on the MD trajectories. The MAE values were 0.86, 0.42, 0.86, and 1.11 kcal/mol, for each of the systems, respectively, with correlation values of 0.73, 0.71, 0.60, and 0.86, respectively. The authors stated that their approach offered a reasonable balance between accuracy and computational efficiency.

The semi-empirical quantum mechanics/COSMO (SQM/COSMO) approach [71] at the DFTB3-D3H4X level of theory was used to generate a reliable affinity ranking on a challenging set of ten inhibitors binding to carbonic anhydrase II (CAII), a zinc metalloenzyme [72]. To improve the performance of the SQM/COSMO scheme, calculations were performed on a truncated system defined by the ligand and a region of 12 Å around of it, since an earlier benchmark showed that there were not significant differences between calculations on truncated and full systems [71]. Four of the most commonly used scoring functions and energy calculations based on AMBER force field [73] were also used as comparison. The best correlation between calculated and experimental binding affinities was achieved by the SQM/COSMO method on the crystal structures, but considering explicit waters within the active site ($R^2 = 0.69$; PI = 0.81); other scoring functions gave very low correlation. The correlation value was not heavily impacted when crystallographic waters were not explicitly considered ($R^2 = 0.58$; PI = 0.76), thus showing that the COSMO continuum solvent model was adequate enough to model the effects of water interactions. The ability to distinguish between strong and weak ligands was evaluated using one protein structure conformation, and the correlation with experimental values was similar ($R^2 = 0.56$; PI = 0.64), probably because the binding site is fairly rigid. On re-scoring poses from the docking experiments, SQM/COSMO calculations assigned the most negative scores to native structures, showing that it could be a valid tool to identify the experimental pose. In fact, in this case, an even better correlation with experimental binding data was found ($R^2 = 0.77$; PI = 0.92).

Recently, Pu et al. evaluated different approaches for the calculation of the binding free energy of 20 congeneric compounds with known experimental affinities toward the Polo-like kinase 1 (PLK1) structure, a serine/threonine kinase [74]. The tested methods were FEP, MM-GBSA, and QM/MM-GBSA. The ligand poses were generated with the GOLD docking program [75]. FEP calculations were performed starting with the ligand docked conformations as the initial poses and the MD software Desmond version 3.9 [76], obtaining a Spearman ranking coefficient $r_s = 0.854$, while the correlation with experimental binding data was $R^2 = 0.852$. The ranking performance of FEP was the best, followed by the MM-GBSA calculations ($r_s = 0.767$), but with a significantly lower computational cost. Interesting, treating the ligand using quantum mechanical methods improved the ranking performance.

Within their participation in the D3R grand challenge blind test competition [20], Ryde and co-workers analyzed four different approaches for predicting the binding free energies of three sets of ligands of the HSP90 [77] (Fig. 2): (1) ligand poses generated using induced-fit docking [78], followed by calculations with three energy functions; (2) MM/GBSA on minimized docked structures; (3) optimization of docked protein–ligand structures with QM/MM calculations, followed by quantum mechanical energy calculations on a truncated system of ~1000 atoms, and using a continuous solvent model; and (4) calculations of relative binding free energies using free energy simulations. Overall, the results were poor, but the authors succeeded in identifying the sources of the errors: (1) in one case, the ligand would displace water molecules (what could be found a posteriori, after the experimental data had been released), and (2) for other two, either ligands might exhibit alternative binding modes than those in the crystal or accounting for conformational changes might be critical.

3 Conclusions

During the last 10 years, we have seen an increasing interest in the application of quantum mechanical methods in drug lead discovery and optimization. This was motivated by the desire to describe interactions within biomolecular systems in a more accurate way and enabled by the steady growth of computing power. The quantum mechanical formulation is, in principle, theoretically exact; it captures the underlying physics of the system and includes all contributions to the energy. Several types of interactions or phenomena, which are usually missing in force fields, such as electronic polarization, charge transfer, halogen bonding, and covalent bond formation, are de facto accounted for in the quantum mechanical framework, which also avoids the need of system-dependent parameterizations, thus being general across the chemical space.

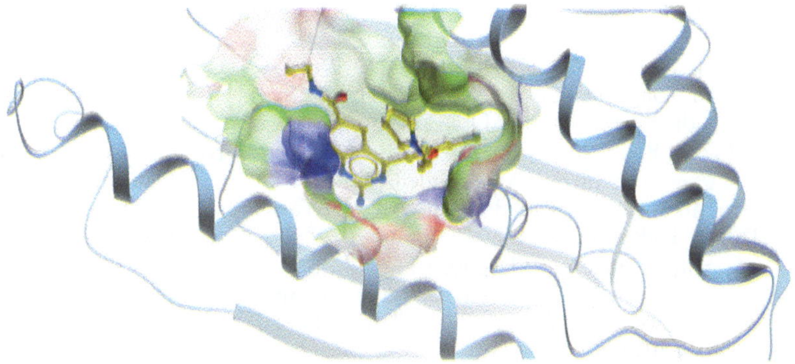

Fig. 2 Heat shock protein 90 (HSP90) complexed with a 2-aminothieno[2,3-*d*] pyrimidine ATP competitive inhibitor (PDB 2WI7). The ligand is displayed in stick representation with yellow carbons, blue nitrogens, red oxygens, and green sulfur; the receptor is displayed in ribbon representation; the color code of the molecular surface of the binding site is green, hydrophobic; red, hydrogen bond acceptor; and blue, hydrogen bond donor. (Figure prepared with the ICM program (Molsoft LLC, San Diego, CA)

We presented in this contribution the incorporation of quantum mechanical approaches to calculate binding free energy. It can be said that the results are, in general, very encouraging. However, improvements in the following fronts should be developed to enhance the accuracy and predictability of these methods: (2) further validation and benchmarking to expand their applicability to several systems; (3) optimized codes (especially those aimed for graphics processing units (GPUs)), so that quantum mechanics-based calculations could be applied in industrial settings; (3) more reliable methods to calculate the solvation contribution; and (4) an enhanced entropic contribution, which is usually omitted in this type of calculations. If these limitations could be addressed, at least partially, the use of quantum mechanical-based methods aimed for macromolecular systems would be a tremendous leap forward toward accurate and efficient binding free energy calculations.

Acknowledgments

This work was supported by the National Agency for the Promotion of Science and Technology (ANPCyT) (PICT-2014-3599 and PICT-2017-3767). CNC thanks Molsoft LLC (San Diego, CA) for providing an academic license for the ICM program. The author thanks the National System of High Performance Computing (Sistemas Nacionales de Computación de Alto Rendimiento, SNCAD), the Centro de Computación de Alto Rendimiento (Computational Centre of High Performance Computing, CeCAR), and the Centro de Cálculo de Alto Desempeño (Universidad Nacional de Córdoba) for granting use of their computational resources.

References

1. Phatak SS, Stephan CC, Cavasotto CN (2009) High-throughput and in silico screenings in drug discovery. Expert Opin Drug Discov 4:947–959
2. Congreve M, Murray CW, Blundell TL (2005) Structural biology and drug discovery. Drug Discov Today 10:895–907
3. Cavasotto CN, Orry AJ (2007) Ligand docking and structure-based virtual screening in drug discovery. Curr Top Med Chem 7:1006–1014
4. Jorgensen WL (2009) Efficient drug lead discovery and optimization. Acc Chem Res 42:724–733
5. Jorgensen WL (2012) Challenges for academic drug discovery. Angew Chem Int Ed Engl 51:11680–11684
6. Shoichet BK (2004) Virtual screening of chemical libraries. Nature 432:862–865
7. Cavasotto CN (2011) Handling protein flexibility in docking and high-throughput docking. In: Sotriffer C (ed) Virtual screening. Principles, challenges and practical guidelines. Wiley-VCH Verlag, Weinheim, Germany, pp 245–262
8. Cavasotto CN (2012) Normal mode-based approaches in receptor ensemble docking. Methods Mol Biol 819:157–168
9. Spyrakis F, Bidon-Chanal A, Barril X et al (2011) Protein flexibility and ligand recognition: challenges for molecular modeling. Curr Top Med Chem 11:192–210
10. Cozzini P, Dellafiora L, Ginex T et al (2015) How protein flexibility can influence docking/scoring simulations. In: Cavasotto CN (ed) In silico drug discovery and design: theory, methods, challenges, and applications. CRC Press, Taylor & Francis Group, Boca Raton, FL, pp 411–440
11. Campbell AJ, Lamb ML, Joseph-McCarthy D (2014) Ensemble-based docking using biased molecular dynamics. J Chem Inf Model 54:2127–2138
12. Spyrakis F, Cavasotto CN (2015) Open challenges in structure-based virtual screening: Receptor modeling, target flexibility consideration and active site water molecules description. Arch Biochem Biophys 583:105–119
13. Gilson MK, Zhou HX (2007) Calculation of protein-ligand binding affinities. Annu Rev Biophys Biomol Struct 36:21–42
14. Gohlke H, Klebe G (2002) Approaches to the description and prediction of the binding affinity of small-molecule ligands to macromolecular receptors. Angew Chem Int Ed Engl 41:2644–2676
15. Mobley DL, Gilson MK (2017) Predicting binding free energies: frontiers and benchmarks. Annu Rev Biophys 46:531–558
16. Cavasotto CN (2012) Binding free energy calculations and scoring in small-molecule docking. In: Luque FJ, Barril X (eds) Physicochemical and computational approaches to drug discovery. Royal Society of Chemistry, London, pp 195–222
17. Cross JB, Thompson DC, Rai BK et al (2009) Comparison of several molecular docking programs: pose prediction and virtual screening accuracy. J Chem Inf Model 49:1455–1474
18. Hansen N, van Gunsteren WF (2014) Practical aspects of free-energy calculations: a review. J Chem Theory Comput 10:2632–2647
19. Mobley DL, Klimovich PV (2012) Perspective: alchemical free energy calculations for drug discovery. J Chem Phys 137:230901
20. Gathiaka S, Liu S, Chiu M et al (2016) D3R grand challenge 2015: evaluation of protein-ligand pose and affinity predictions. J Comput Aided Mol Des 30:651–668
21. Gaieb Z, Liu S, Gathiaka S et al (2018) D3R grand challenge 2: blind prediction of protein-ligand poses, affinity rankings, and relative binding free energies. J Comput Aided Mol Des 32:1–20
22. Gaieb Z, Parks CD, Chiu M et al (2019) D3R grand challenge 3: blind prediction of protein-ligand poses and affinity rankings. J Comput Aided Mol Des 33:1–18
23. Grimme S, Schreiner PR (2018) Computational chemistry: the fate of current methods and future challenges. Angew Chem Int Ed Engl 57:4170–4176
24. Richards WG (1977) Quantum pharmacology. Butterworths, London; Boston
25. Mucs D, Bryce RA (2013) The application of quantum mechanics in structure-based drug design. Expert Opin Drug Discov 8:263–276
26. De Benedetti PG, Fanelli F (2014) Multiscale quantum chemical approaches to QSAR modeling and drug design. Drug Discov Today 19:1921–1927
27. Blomberg MR, Borowski T, Himo F et al (2014) Quantum chemical studies of mechanisms for metalloenzymes. Chem Rev 114:3601–3658
28. Forti F, Cavasotto CN, Orozco M et al (2012) A multilevel strategy for the exploration of the conformational flexibility of small molecules. J Chem Theory Comput 8:1808–1819
29. Juarez-Jimenez J, Barril X, Orozco M et al (2015) Assessing the suitability of the

multilevel strategy for the conformational analysis of small ligands. J Phys Chem B 119:1164–1172
30. Cavasotto CN, Adler NS, Aucar MG (2018) Quantum chemical approaches in structure-based virtual screening and lead optimization. Front Chem 6:188
31. Aucar MG, Cavasotto CN (2019) Molecular docking using quantum mechanical-based methods. Methods Mol Biol 2114:269–284
32. Ryde U, Söderhjelm P (2016) Ligand-binding affinity estimates supported by quantum-mechanical methods. Chem Rev 116:5520–5566
33. Yilmazer ND, Korth M (2016) Prospects of applying enhanced semi-empirical QM methods for 2101 virtual drug design. Curr Med Chem 23:2101–2111
34. Raha K, Merz KM Jr (2004) A quantum mechanics-based scoring function: study of zinc ion-mediated ligand binding. J Am Chem Soc 126:1020–1021
35. Dewar MJS, Zoebisch EG, Healy EF et al (1985) Development and use of quantum mechanical molecular models. 76. AM1: a new general purpose quantum mechanical molecular model. J Am Chem Soc 107:3902–3909
36. Dixon SL, Merz KM Jr (1996) Semiempirical molecular orbital calculations with linear system size scaling. J Chem Phys 104:6643–6649
37. Raha K, Merz KM Jr (2005) Large-scale validation of a quantum mechanics based scoring function: predicting the binding affinity and the binding mode of a diverse set of protein-ligand complexes. J Med Chem 48:4558–4575
38. Gilson MK, Given JA, Bush BL et al (1997) The statistical-thermodynamic basis for computation of binding affinities: a critical review. Biophys J 72:1047–1069
39. Simonson T, Archontis G, Karplus M (1997) Continuum treatment of long-range interactions in free energy calculations. J Phys Chem B 101:8349–8362
40. Srinivasan J, Cheatham TE, Cieplak P et al (1998) Continuum solvent studies of the stability of DNA, RNA, and phosphoramidate-DNA helices. J Am Chem Soc 120:9401–9409
41. Qiu D, Shenkin PS, Hollinger FP et al (1997) The GB/SA continuum model for solvation. A fast analytical method for the calculation of approximate born radii. J Phys Chem A 101:3005–3014
42. Anisimov VM, Cavasotto CN (2011) Quantum mechanical binding free energy calculation for phosphopeptide inhibitors of the Lck SH2 domain. J Comput Chem 32:2254–2263
43. Klamt A, Schüürmann G (1993) COSMO: a new approach to dielectric screening in solvents with explicit expressions for the screening energy and its gradient. J Chem Soc Perkin Trans 2:799–805
44. Anisimov VM, Cavasotto CN (2011) Hydration free energies using semiempirical quantum mechanical Hamiltonians and a continuum solvent model with multiple atomic-type parameters. J Phys Chem B 115:7896–7905
45. Cavasotto CN, Aucar MG, Adler NS (2019) Computational chemistry in drug lead discovery and design. Int J Quantum Chem 119: e25678
46. Stewart JJP (1989) Optimization of parameters for semiempirical methods I. J Comput Chem 10:209–220
47. Anikin NA, Anisimov VM, Bugaenko VL et al (2004) Local SCF method for semiempirical quantum-chemical calculation of ultralarge biomolecules. J Chem Phys 121:1266–1270
48. Bondi A (1964) van der Waals volumes and radii. J Phys Chem 68:441–451
49. Swanson JMJ, Adcock SA, McCammon JA (2005) Optimized radii for Poisson–Boltzmann calculations with the AMBER force field. J Chem Theory Comput 1:484–493
50. Woo HJ, Roux B (2005) Calculation of absolute protein-ligand binding free energy from computer simulations. Proc Natl Acad Sci U S A 102:6825–6830
51. Anisimov VM, Ziemys A, Kizhake S et al (2011) Computational and experimental studies of the interaction between phosphopeptides and the C-terminal domain of BRCA1. J Comput Aided Mol Des 25:1071–1084
52. Naïm M, Bhat S, Rankin KN et al (2007) Solvated interaction energy (SIE) for scoring protein-ligand binding affinities. 1. Exploring the parameter space. J Chem Inf Model 47:122–133
53. Pearlman DA, Charifson PS (2001) Are free energy calculations useful in practice? A comparison with rapid scoring functions for the p38 MAP kinase protein system. J Med Chem 44:3417–3423
54. Nascimento ÉCM, Oliva M, Świderek K et al (2017) Binding analysis of some classical acetylcholinesterase inhibitors: insights for a rational design using free energy perturbation method calculations with QM/MM MD simulations. J Chem Inf Model 57:958–976
55. Swiderek K, Ruiz-Pernia JJ, Moliner V et al (2014) Heavy enzymes—experimental and

computational insights in enzyme dynamics. Curr Opin Chem Biol 21:11–18
56. Olsson MA, Ryde U (2017) Comparison of QM/MM Methods to obtain ligand-binding free energies. J Chem Theory Comput 13:2245–2253
57. Rod TH, Ryde U (2005) Quantum mechanical free energy barrier for an enzymatic reaction. Phys Rev Lett 94:138302
58. Stewart JJP (2009) Application of the PM6 method to modeling proteins. J Mol Model 15:765–805
59. Korth M (2010) Third-generation hydrogen-bonding corrections for semiempirical QM methods and force fields. J Chem Theory Comput 6:3808–3816
60. Ehrlich S, Goller AH, Grimme S (2017) Towards full quantum-mechanics-based protein-ligand binding affinities. ChemPhysChem 18:898–905
61. Sure R, Grimme S (2013) Corrected small basis set Hartree-Fock method for large systems. J Comput Chem 34:1672–1685
62. Grimme S, Antony J, Ehrlich S et al (2010) A consistent and accurate ab initio parametrization of density functional dispersion correction (DFT-D) for the 94 elements H-Pu. J Chem Phys 132:154104
63. Grimme S, Brandenburg JG, Bannwarth C et al (2015) Consistent structures and interactions by density functional theory with small atomic orbital basis sets. J Chem Phys 143:054107
64. Brandenburg JG, Grimme S (2014) Accurate modeling of organic molecular crystals by dispersion-corrected density functional tight binding (DFTB). J Phys Chem Lett 5:1785–1789
65. Gaus M, Cui Q, Elstner M (2011) DFTB3: extension of the self-consistent-charge density-functional tight-binding method (SCC-DFTB). J Chem Theory Comput 7:931–948
66. Klamt A (1995) Conductor-like screening model for real solvents: a new approach to the quantitative calculation of solvation phenomena. J Phys Chem 99:2224–2235
67. Klamt A (2011) The COSMO and COSMO-RS solvation models. Wiley Interdiscip Rev Comput Mol Sci 1:699–709
68. Frush EH, Sekharan S, Keinan S (2017) In silico prediction of ligand binding energies in multiple therapeutic targets and diverse ligand sets-A case study on BACE1, TYK2, HSP90, and PERK proteins. J Phys Chem B 121:8142–8148
69. Åqvist J, Medina C, Samuelsson JE (1994) A new method for predicting binding affinity in computer-aided drug design. Protein Eng 7:385–391
70. Su Y, Gallicchio E, Das K et al (2007) Linear interaction energy (LIE) models for ligand binding in implicit solvent: theory and application to the binding of NNRTIs to HIV-1 reverse transcriptase. J Chem Theory Comput 3:256–277
71. Pecina A, Meier R, Fanfrlik J et al (2016) The SQM/COSMO filter: reliable native pose identification based on the quantum-mechanical description of protein-ligand interactions and implicit COSMO solvation. Chem Commun 52:3312–3315
72. Pecina A, Brynda J, Vrzal L et al (2018) Ranking power of the SQM/COSMO scoring function on carbonic anhydrase II-inhibitor complexes. ChemPhysChem 19(7):873–879
73. Maier JA, Martinez C, Kasavajhala K et al (2015) ff14SB: improving the accuracy of protein side chain and backbone parameters from ff99SB. J Chem Theory Comput 11:3696–3713
74. Pu C, Yan G, Shi J et al (2017) Assessing the performance of docking scoring function, FEP, MM-GBSA, and QM/MM-GBSA approaches on a series of PLK1 inhibitors. Medchemcomm 8:1452–1458
75. Jones G, Willett P, Glen RC (1995) A genetic algorithm for flexible molecular overlay and pharmacophore elucidation. J Comput Aided Mol Des 9:532–549
76. Bowers KJ, Chow E, Xu H et al (2006) Scalable algorithms for molecular dynamics simulations on commodity clusters. In: Proceedings of the 2006 ACM/IEEE conference on Supercomputing. ACM, Tampa, Florida, p 84
77. Misini Ignjatovic M, Caldararu O, Dong G et al (2016) Binding-affinity predictions of HSP90 in the D3R grand challenge 2015 with docking, MM/GBSA, QM/MM, and free-energy simulations. J Comput Aided Mol Des 30:707–730
78. Sherman W, Day T, Jacobson MP et al (2006) Novel procedure for modeling ligand/receptor induced fit effects. J Med Chem 49:534–553

Chapter 17

Molecular Docking Using Quantum Mechanical-Based Methods

M. Gabriela Aucar and Claudio N. Cavasotto

Abstract

Computational methods are a powerful and consolidated tool in the early stage of the drug lead discovery process. Among these techniques, high-throughput molecular docking has proved to be extremely useful in identifying novel bioactive compounds within large chemical libraries. In the docking procedure, the predominant binding mode of each small molecule within a target binding site is assessed, and a docking score reflective of the likelihood of binding is assigned to them. These methods also shed light on how a given hit could be modified in order to improve protein–ligand interactions and are thus able to guide lead optimization. The possibility of reducing time and cost compared to experimental approaches made this technology highly appealing. Due to methodological developments and the increase of computational power, the application of quantum mechanical methods to study macromolecular systems has gained substantial attention in the last decade. A quantum mechanical description of the interactions involved in molecular association of biomolecules may lead to better accuracy compared to molecular mechanics, since there are many physical phenomena that cannot be correctly described within a classical framework, such as covalent bond formation, polarization effects, charge transfer, bond rearrangements, halogen bonding, and others, that require electrons to be explicitly accounted for. Considering the fact that quantum mechanics-based approaches in biomolecular simulation constitute an active and important field of research, we highlight in this work the recent developments of quantum mechanical-based molecular docking and high-throughput docking.

Key words Quantum mechanics, Structure-based drug design, Molecular docking, High-throughput docking, Computer-aided drug discovery, Protein–ligand interaction

1 Introduction

It has been estimated that to bring a new drug to the market may cost up to 2.6 billion US dollars [1]. Historically, the hit identification process have been accomplished experimentally through high-throughput screening (HTS). In spite of the numerous efforts to improve its efficiency, HTS is still a time-consuming technique, often exhibiting stagnant success rates [2], has poor quality of hits [3], and is coupled with both false positives and negatives [4]. It is known that molecular recognition, including non-covalent binding

Alexander Heifetz (ed.), *Quantum Mechanics in Drug Discovery*, Methods in Molecular Biology, vol. 2114,
https://doi.org/10.1007/978-1-0716-0282-9_17,

of a ligand to its biological receptor, is a key aspect of many biological processes [5]. Thus, the availability of high-quality three-dimensional (3D) structures of protein–ligand (PL) complexes, mainly obtained using nuclear magnetic resonance or X-ray crystallography techniques, has helped toward a more rational development of drug design through the use of structure-based approaches. At the same time, computational methods began to play an increasingly major role in modeling and analyzing PL interactions, favored by theoretical developments, faster computing resources, and availability of macromolecular structures [6–11].

Understanding the structural determinants of PL interaction and knowledge of ligand binding free energy (ΔG_{bind}) are of the utmost relevance in the pharmaceutical industry, since this is instrumental for a more rational approach in the early stages of the drug discovery process, helping to save time, effort, and costs [12]. During the last years, there has been an active theoretical and computational development of the calculation of binding energy (the reader is referred to excellent reviews on that field [13–16]). Molecular docking is one of the most commonly used structure-based methods to model PL interaction and assess the binding pose of small molecules within the binding site of a receptor. Moreover, docking can be implemented in a high-throughput fashion (high-throughput docking (HTD)), where in silico chemical libraries are screened against a molecular target in order to identify potential modulators from a large pool of small molecules by exploring their binding mode and assigning a score to each molecule as a measure of the likelihood of binding; the top-scoring molecules are then advanced for chemical synthesis (or purchase from a chemical vendor), followed by biochemical and biological evaluation [17]. The docking score is usually—though not always—calculated as a fast approximation to ΔG_{bind} and is clearly influenced by the energy model used to describe the system, the solvation model and the treatment of explicit water molecules within the active site [18, 19], and the degree of consideration of receptor flexibility [9], which is usually neglected in HTD.

It should be highlighted that the most commonly used applications in the context of drug discovery and design are based on classical mechanics-based force fields (FFs). However, to properly describe protein–ligand interactions in the context of drug design, the use of quantum mechanical (QM) methods might seem to be necessary, at least in some cases. The QM formulation constitutes the most accurate level of theory, accounting for all physical phenomena involved in biomolecular systems like electronic polarization, covalent bond formation, and halogen bonding. In addition, unlike molecular mechanics (MM) methods, they do not require parameterization, since each interaction has the corresponding weight, and they are thus system-independent. It should be noted that QM has been used early in computer-aided drug design [20] and is currently used to parameterize torsional potentials from high-level ab initio data and to derive partial atomic charges by fitting to electrostatic surface potentials [21]. Furthermore, QM is used in quantitative structure–activity relationship (QSAR) methods [22], in the study of reaction mechanisms [23], and to assess small-molecule strain [24, 25].

Recent years have seen the development and application of QM methods to biomacromolecular systems in the context of drug lead discovery and design [21, 26–30]. Because of the size, and consequently high dimensionality that has to be dealt with in biological systems, a routine use of a full QM treatment at an industrial scale is still beyond reach, but it constitutes a growing area of active research.

In this work, we focused on the applications of quantum mechanical methods to docking-based structure-based drug design. It is not our aim to address the development of QM methods for neither protein–ligand binding affinity calculations nor the advances and challenges of theoretical QM methods. The reader is referred to recent reviews covering those topics (Refs. 31, 32).

2 Quantum Chemical Approaches in Protein-Small-Molecule Docking

Molecular docking is one of the most commonly used in silico methods which has had great development in terms of research and applications in the last two decades. Several reviews have been published that demonstrate the great interest of the scientific community in order to obtain more accurate strategies and extend its application to diverse and challenging biological systems [9, 17, 33, 34]. The main aim of molecular docking is to determine the optimal binding modes (poses) of small molecules within the binding site of a receptor and an estimation of their affinities. When used in a high-throughput fashion in HTD, the goal is to generate a sub-library enriched with potential ligands, which will be further experimentally evaluated. Two different stages can be distinguished in HTD: the prediction of the best binding mode(s) of each molecule (which can be called the "docking stage") and the estimation of the affinity of each molecule toward the target using a scoring function ("scoring stage") [35]. It should be noted the conceptual difference between docking energy and docking score: the former is used to select the best pose(s) from a large amount of conformations generated for each one of the molecules, while the latter is usually an estimation of the binding free energy of the molecule. Thus, docking energy discriminates among poses of the same molecule, while the docking score is used to rank the different molecules of the docked chemical library; in practice, many docking programs use the same function for both purposes. Scoring functions can be classified as force field-based, empirical, and knowledge-based [17, 36]. Recently, a new classification scheme has been published for the up-to-date scoring functions applied in molecular docking tool, adding machine learning to the list [37]. Typical interactions such as van de Waals, electrostatic and hydrogen bonding, and entropy changes and solvation/desolvation effects are usually taken into account in the calculation of the docking score for each molecule of the chemical library.

It is regarded that docking programs are usually successful to accurately identify the correct pose [38]. Indeed, a recent update of the Comparative Assessment of Scoring Functions (CASF) benchmark on 285 PL complexes and 25 scoring functions [39] confirmed that scoring functions display a better performance in terms of docking accuracy than in any of these three scoring-related metrics: (1) the ability to correlate with experimental binding data (scoring power); (2) the ability to rank ligands from the same target by their binding affinity data when the correct poses are provided (ranking power); and (3) the ability to identify actual ligands from a pool of randomly selected small molecules (screening power). These findings are in agreement with older [40] and recent [41] works. In spite of a great effort over the years to develop fast yet accurate enough scoring functions, this still constitutes an area of active research.

To properly describe PL interactions, a quantum mechanical formulation might be convenient, since by considering electrons explicitly, covalent bonding, charge transfer, halogen bonding, and polarization are included in a natural way. Cho et al. have improved docking accuracy replacing FF-based static charges of ligands with charges derived from quantum mechanics/molecular mechanics (QM/MM) calculations [42]. More recently, a QM-polarized protein docking scheme was applied on 50 protein–ligand complexes with known crystal structures; there the scoring function is enhanced using a polarized protein-specific charge model to better predict ligand poses at the binding site of the receptor [43].

Despite its great advantage, the description of PL interactions explicitly using QM theory level remains a challenging task for computational chemists, mainly because of the computational costs involved. The remarkable works of Raha and Merz were among the first aimed at this goal. They developed QMScore, a QM-based scoring function to adequately describe the protein–ligand interactions by allowing the entire system to be treated at a quantum theory level for binding free energy prediction. The method was applied to a set of an 18 carbonic anhydrase (CA) and 5 carboxypeptidase (CPA) inhibitors [44]. The Austin Model 1 (AM1) semi-empirical Hamiltonian [45] was used to compute the interaction gas phase energy using DivCon [46], a QM linear scaling approach. The AMBER 96 force field was used to describe the non-polar interactions, and the Poisson-Boltzmann (PB) implicit solvation model was used to compute solvation free energies. In a follow-up work, QMScore was used to discriminate native from decoy poses in a set of 165 PL complexes, also capturing key ligand affinity trends [47]. Fong et al. presented a study on QM/MM scoring functions aimed at the discrimination of native from decoy poses in six HIV-1 proteases, where it was found some degree of improvement over MM empirical potentials [48].

Alvarez and Burton used molecular dynamics (MD) simulations and a QM approach to study the preferred docking poses of the natural compound salpichrolide A and a synthetic analog with an aromatic D ring within the estrogen receptor α (ERα) binding site (Fig. 1) [49]. Energy calculations were performed with the MM/quantum mechanical surface area (MM/QMSA) method [50–52], using the semi-empirical PM6 Hamiltonian and the conductor-like screening *model* (COSMO) solvation model [53]. MD simulations followed by QM energy calculations for several binding modes confirmed the preferred inverted orientation of the steroids within the ERα binding site, where the aromatic ring D overlaps with the corresponding position of the A ring of estradiol. G protein-coupled receptors (GPCRs) represent an interesting case to evaluate docking

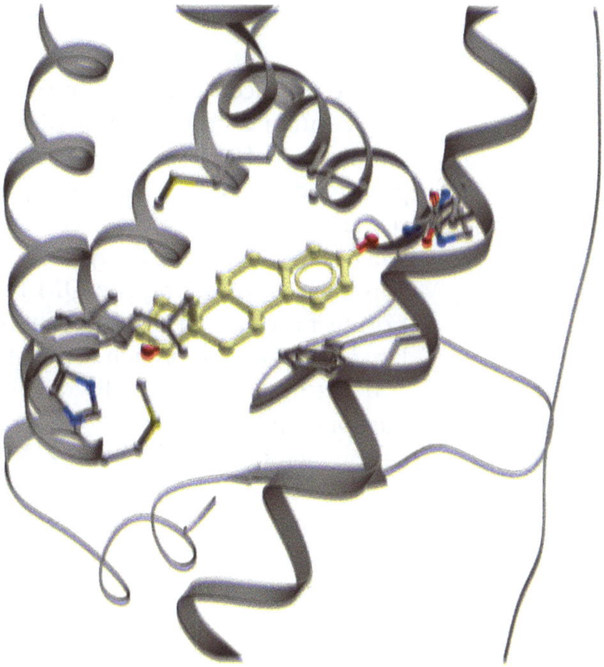

Fig. 1 Binding site of ERα complexed with estradiol (PDB 1QKU; Helix 12 not shown). Neighboring residues are displayed in stick representation. Color code: ERα gray ribbon; estradiol, light yellow. (Figure prepared with the ICM program (Molsoft LLC, San Diego, CA))

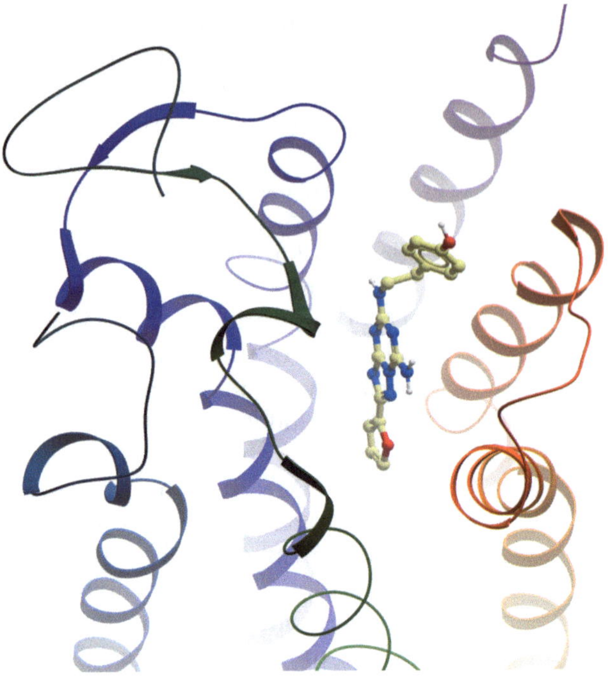

Fig. 2 Ligand ZM241835 (light yellow carbons) within the exposed binding site of GPCR A_{2A} adenosine receptor (PDB 3EML). Ribbons are colored from blue (Helix 1) to red (Helix 7). (Figure prepared with the ICM program (Molsoft LLC, San Diego, CA))

protocols since their binding sites are semi-exposed to the solvent and contain a combination of hydrophobic, polar, and aromatic amino acids [54] (Fig. 2). A new docking protocol was recently presented where QM/MM calculations coupled with a continuous solvation model were used to rescore docked ligand poses [55]. The vacuum energy was calculated using a QM/MM approach, considering the QM region as the ligand and its neighboring residues within 5 Å. The solvation energy term was calculated using a PB approach using atomic partial charges derived from ESP fitting. On 40 GPCR complexes from classes A, B, and F, the authors reported an average docking RMSD of 0.78 Å, with 100% of the ligands with RMSD <2 Å.

Recently, Chaskar et al. developed an on-the-fly approach [56] combining the EADock DSS docking algorithm [57] with QM/MM calculations, coupling the self-consistent charge density functional tight-binding (SCC-DFTB) [58] model with the CHARMM FF [59]; it was evaluated on a data set of high-quality crystallographic structures of zinc metalloproteins, and it was found to significantly improve the performance compared to classical docking programs in terms of ligand pose RMSD. In a later development, a similar approach using the Attracting Cavities docking algorithm [60] was evaluated on three docking sets [61]:

(1) the Astex Diverse data set, consisting of 85 common non-covalent drug/target complexes; (2) a zinc metalloprotein data set (281 complexes); and (3) a heme protein data set (72 complexes), where protein–ligand interactions are dominated by covalent ligand/iron binding. While on the first set the performance was comparable to standard scoring functions, on the other two, the QM/MM methodology displayed a better performance, especially on the third set.

In order to improve docking accuracy, Pecina et al. have developed a new useful methodology to be applied for medium-throughput refinement in later stages of virtual screening, aiming to discriminate native from decoy ligand docking poses [62]. Based on a simplified version of the general binding free energy function [47, 63],

$$\Delta G_{\text{bind}} = \Delta E_{\text{int}} + \Delta\Delta G_{\text{solv}} + \Delta G_{\text{conf}} - T\Delta S \tag{1}$$

(where ΔE_{int} is the gas-phase interaction energy, $\Delta\Delta G_{\text{solv}}$ the solvation energy change upon complex formation, ΔG_{conf} the change of conformational free energy, and $-T\Delta S$ the entropy change upon binding), the semi-empirical quantum mechanical/COSMO (SQM/COSMO) filter energy was defined by considering the first two dominant terms in Eq. (1) only and avoiding SQM optimization stages, in order to make it usable in virtual screening campaigns. It has been proved that this SQM-based scoring function (SF) can be useful in the description of covalent and non-covalent inhibitor bindings [63]. A later development of the SQM approach, called PM6-D3H4X, derived from the PM6 method [64] coupled with corrections for the description of dispersion and hydrogen and halogen bonding, was used to compute the ΔE_{int} values [65, 66]. For the $\Delta\Delta G_{\text{solv}}$ term, QM-based implicit solvent models (COSMO and Solvent Model D (SMD) [67]) were used based on the evidence that they are more accurate than MM-based ones [68]. Four challenging systems [acetylcholinesterase (AChE), TNF-α converting enzyme (TACE), aldose reductase (AR), and HIV-1 protease (HIV PR); cf. Fig. 3] were investigated with the main goal of identifying within a set of non-redundant poses created by docking with four different programs, the known native X-ray ligand pose within the binding site of the corresponding receptor. A superior performance over eight widely used SFs, seven empirical and one physics-based AMBER/GB (generalized Born), was reached with the SQM/COSMO filter in terms of two different metrics: the number of false positives obtained and the ligand RMSD of all poses within a given range of a normalized score. The hardest case was the TACE metalloprotein, which has a metal cation in the protein. This shows the strength of a QM description of PL interaction, since charge-transfer effects between the ligand and the cation are implicitly

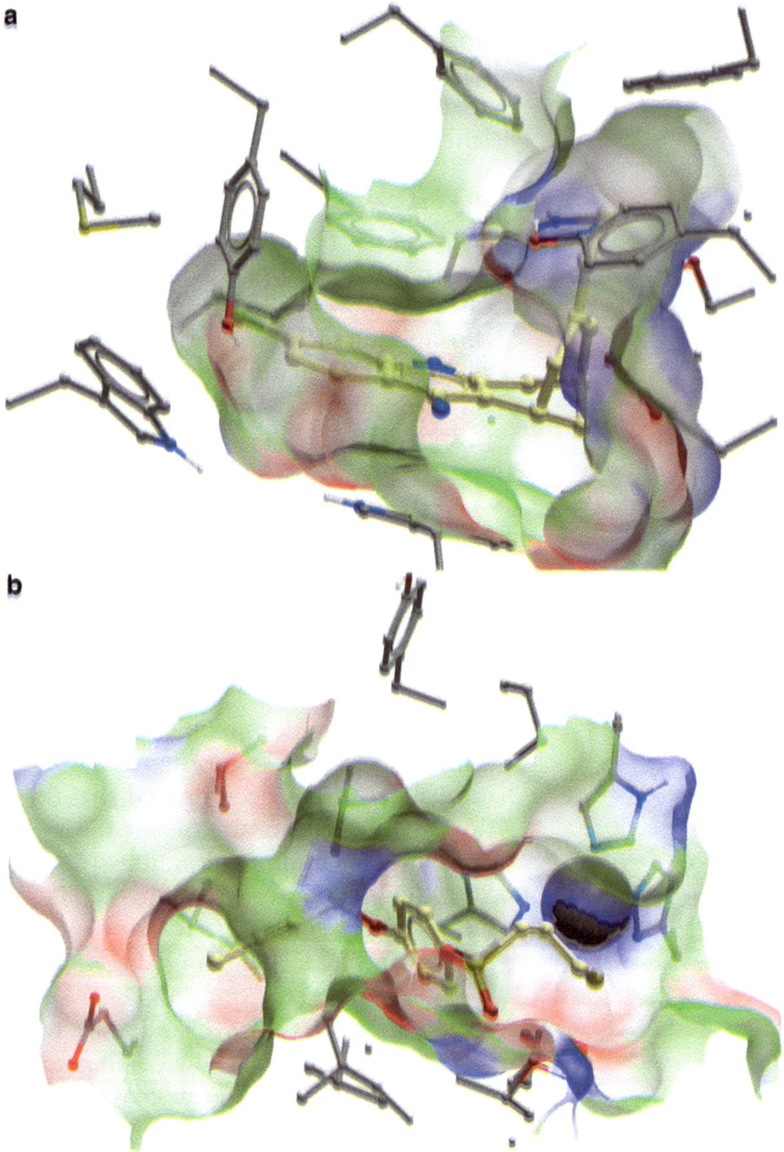

Fig. 3 Receptor–ligand complexes. (**a**) AchE with (−)-huprine X (PDB 1E66); (**b**) TACE with a thiol-containing aryl sulfone compound (PDB 3B92); the Zn atom is displayed as a blue sphere; (**c**) AR with compound IDD393 (PDB 2IKJ), NADP not shown; (**d**) HIV protease with a phenylnorstatine inhibitor (PDB 1NHO). Color code for carbons: ligands, light yellow; neighboring residues, gray. (Figure prepared with the ICM program (Molsoft LLC, San Diego, CA))

c

d

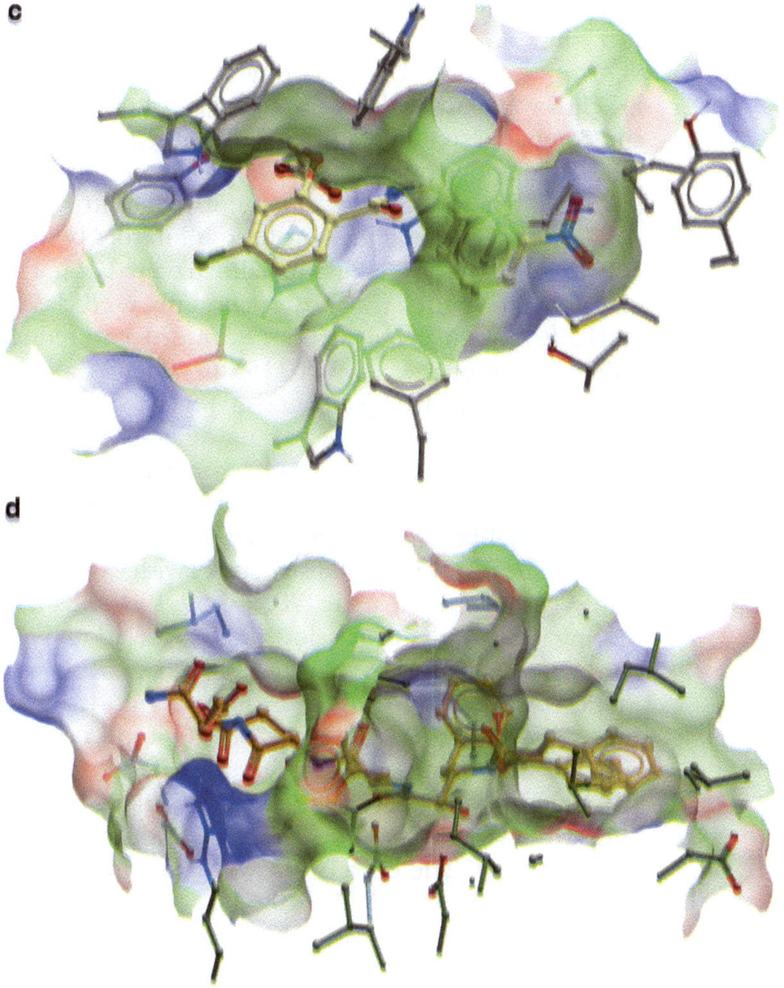

Fig. 3 (continued)

considered in this formulation. The SQM/COSMO filter performs best for this PL complex, followed by ChemScore SF, although several false positives were found. It was shown that this method is more accurate than classical force fields or statistical potentials, but at a higher computational cost. Compared to classical scoring functions, it is approximately 100 times slower and approximately ten times slower than the AMBER/GB scoring function. However it is almost 100 times faster than the standard SQM-based scoring function.

In a follow-up contribution [69], the SQM/COSMO filter was re-evaluated in the same four systems (AChE, TACE, AR, HIV PR), but using the SCC-DFTB method coupled with D3H4 corrections for dispersion and hydrogen-bond interactions [66]. This new filter (named DFTB3-D3H4) conserved its successful performance on the AChE, AR, and HIV PR systems and displayed an enhanced performance on the TACE system, however, at a (reasonably) higher computational cost. To further explore these two SQM approaches, 17 PL complexes were studied (the DFTB3-D3H4X was extended in this case to account for halogen bonding) and compared to four traditional docking programs [70]. Recently, the ranking power of SQM/COSMO at the DFTB3-D3H4X level of theory was tested by evaluating its ability to generate a reliable ranking on a challenging set of ten inhibitors binding to carbonic anhydrase II (CAII), a zinc metalloenzyme [71]. The results were compared to four of the most commonly used classical SFs and to AMBER molecular mechanics FF. In order to reduce the computational time of the SQM/COSMO scheme, a truncated system was defined by the ligand and a region of 12 Å around of it, representing the active site, having demonstrated that this truncation does not deteriorate results comparing with the full system [62]. The best correlation value between scores and the determined experimental binding affinity was achieved by the SQM/COSMO filter on the resolved crystal structures with explicit crystallographic waters within the active site ($R^2 = 0.69$; PI = 0.81), while the other scoring functions gave very low or no correlation. Furthermore, the correlation was not affected to a large extent when the crystallographic waters were not explicitly considered ($R^2 = 0.58$; PI = 0.76) showing that the COSMO implicit solvent model can simulate a real scenario capturing the effects of water interactions. Also the ability to rank the inhibitors distinguishing between strong and weak binders was tested using only one protein structure conformation. The correlation with experimental binding affinity was similar ($R^2 = 0.56$; PI = 0.64), perhaps because the active site is fairly conserved. Finally the poses generated from the docking experiment were rescored, which confirmed the capacity of the SQM/COSMO SF to assign the most negative scores to crystal structures, which means that it can be able to identify the X-ray pose. Moreover, a better correlation with experimental binding data was found ($R^2 = 0.77$; PI = 0.92). Even though QM-based scoring functions are clearly more expensive in terms of computational requirements than standard scoring functions, they have interesting benefits, such as the accuracy related to a low number of false positives obtained and the correct X-ray pose identification.

Pu et al. have evaluated the ranking power through the Spearman ranking coefficient (r_s) of a traditional docking scoring function and different binding free energy calculations: free energy perturbation (FEP), MM-generalized Born surface area (GBSA), and QM/MM-GBSA. They have used the Polo-like kinase 1 (PLK1) structure (Fig. 4), which belongs to the serine/threonine kinase PLK family, with 20 congeneric compounds of known affinity. Ligand binding modes were obtained using the GOLD docking program [72]. Starting from docking outputs, binding affinities were calculated with two different kinds of scoring functions,

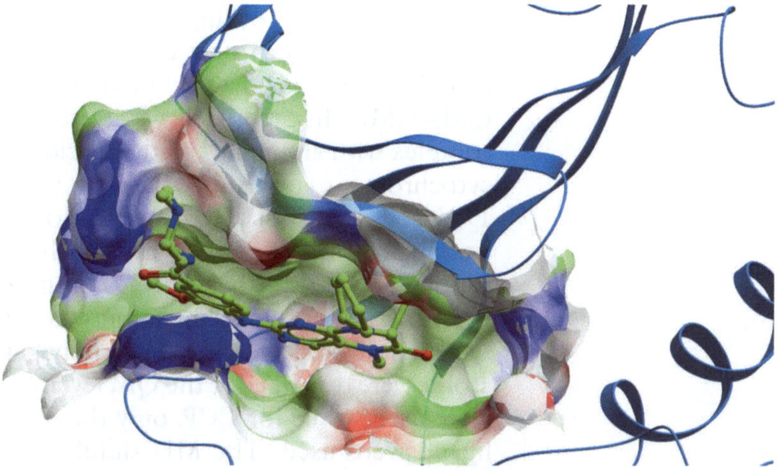

Fig. 4 Polo-like kinase 1 complexed with nanomolar inhibitor BI2536 (PDB 2RKU). Ligand is displayed in stick representation. (Figure prepared with the ICM program (Molsoft LLC, San Diego, CA))

empirical and force field-based. FEP calculations were performed using the docking conformations as the initial poses and MD software Desmond [73] version 3.9. In this case, $r_s = 0.854$, and the correlation with experimental binding affinity was $R^2 = 0.852$. The r_s of the ChemPLP empirical scoring function was 0.217. Furthermore AutoDock Vina [74] scoring function, a combination of empirical and physics-based scoring function, was also employed obtaining a very poor correlation value ($r_s = 0.171$). Then, the FF-based AMBER and AutoDock [75, 76] SFs were used, outperforming the empirical method ChemPLP and Vina score in ranking the compounds, with r_s-AMBER = 0.347 and r_s-AutoDock = 0.378. Finally the r_s was evaluated for MM-(P)GBSA and QM/MM-GBSA methods. The ranking performance of FEP was the best with $r_s = 0.854$, followed by MM-GBSA score which has reached 0.767 with much lower computational cost. Interesting, treating the ligand by QM improved the ranking performance [77].

In a recent work, Ajani et al. reported on the synthesis and activity testing of imidazo[1,2-*c*]pyrimidin-5(6*H*)-one analogs for the cyclin-dependent kinase 2 (CDK2) ATP-competitive inhibition [78]. The series consisted of aromatic-substituted scaffolds at position 8, which inhibited CDK2 in in vitro enzyme assays. Two-dimensional structure–activity relationships (SAR) revealed that small substituents usually lead to low micromolar IC_{50} values, while larger substituents decreased inhibition. Docking with Glide [79, 80] revealed that compounds made up to two hydrogen bonds with the hinge regions, differing in the orientation of the core and the 8-substituents. Semi-empirical QM-based calculations were used to identify probable binding modes, which can be used for further structure-based design and lead optimization.

With the aim of improving docked poses to obtain a better correlation between predictions and experimental values, Burger et al. have developed a new methodology that combines MD with QM/MM. This approach was tested and validated on trypsin in complex with an (amino methyl)cyclohexane (AMH) inhibitor and cytochrome c peroxidase (CCP) with 43 known ligands and 17 decoys [81]. Starting from the docked poses obtained with the docking algorithm implemented in the Molecular Operating Environment (MOE; Chemical Computing Group Inc., Montreal, Canada), a short MD simulation was performed to equilibrate the system, followed by QM/MM minimization, where only the ligand atoms were described at the QM level. For trypsin, the top 20 poses were used, while for CCP, only the 5 top-scoring poses from each ligand were used. The MD simulations coupled with QM/MM energy minimization for each one of the generated docking poses improved docking results, showing a better ranking of the generated poses (binding modes), lower RMSD values with respect to the crystal structure, and better correlation with the measured binding affinities. These results highlight the necessity of incorporating a QM-based description, with the inclusion of a van der Waals term and desolvation energy contributions.

3 Conclusions

We present quantum mechanical-based molecular docking applications in the context of drug lead discovery, limiting ourselves to those cases where QM is explicitly taken into account to calculate at least one of the energy terms. It should be highlighted that due to theoretical developments in molecular simulation and the never-ending growth of computing power, these last 10 years have seen a growing interest in the development of QM-based methods for drug design. QM methods are appealing since they capture the underlying physics of the system and, in principle, account for all energetic contributions. The QM framework de facto incorporates electronic polarization, charge transfer, covalent bond formation, and halogen bonding, terms usually missing in traditional FFs. Moreover, QM methods are consistently valid across the chemical space, thus offering greater freedom to deal with non-standard molecules and avoiding FF parameterizations.

The performance of QM-based docking approaches is, so far, very encouraging, but still the following sources of error should be taken into consideration to improve pose accuracy and ranking power:

1. Protein flexibility: it has been already addressed the impact of of neglecting protein flexibility in docking and high-throughput docking [12, 82, 83].

2. The solvation energy contribution, especially in charged systems.
3. Entropic considerations: although this is a common problem in MM-based docking algorithms, a reliable entropy term will clearly enhance ligand discrimination in HTD.

It should be added that, in spite of the developments in computing power, most QM-based applications to molecular docking cannot still be used in large industrial settings; this shows the urgent need for code development, especially those using GPUs; mainly due to computational time, in most docking protocols used in a high-throughput fashion, at most only local energy minimization is performed. In this scenario, the possibility of system cutout [30] should be considered and the optimal combinations of these types of strategies thoroughly validated. In spite of the above remarks, it should be acknowledged that improvements in QM-based docking will be an enormous step forward toward predictive methods for identification and optimization of small-molecule modulators of pharmaceutically relevant targets.

Acknowledgments

This work was supported by the National Agency for the Promotion of Science and Technology (ANPCyT) (PICT-2014-3599 and PICT-2017-3767). CNC thanks Molsoft LLC (San Diego, CA) for providing an academic license for the ICM program. The authors thank the National System of High Performance Computing (Sistemas Nacionales de Computación de Alto Rendimiento (SNCAD)), the Centro de Computación de Alto Rendimiento (Computational Centre of High Performance Computing (CeCAR)), and the Centro de Cálculo de Alto Desempeño (Universidad Nacional de Córdoba) for granting the use of their computational resources.

References

1. Dimasi JA, Grabowski HG, Hansen RW (2016) Innovation in the pharmaceutical industry: new estimates of R&D costs. J Health Econ 47:20–33
2. Keseru GM, Makara GM (2009) The influence of lead discovery strategies on the properties of drug candidates. Nat Rev Drug Discov 8:203–212
3. Lipkin MJ, Stevens AP, Livingstone DJ et al (2008) How large does a compound screening collection need to be? Comb Chem High Throughput Screen 11:482–493
4. Phatak SS, Stephan CC, Cavasotto CN (2009) High-throughput and in silico screenings in drug discovery. Expert Opin Drug Discov 4:947–959
5. Baron R, Mccammon JA (2013) Molecular recognition and ligand association. Annu Rev Phys Chem 64:151–175
6. Jorgensen WL (2009) Efficient drug lead discovery and optimization. Acc Chem Res 42:724–733
7. Jorgensen WL (2012) Challenges for academic drug discovery. Angew Chem Int Ed Engl 51:11680–11684

8. Shoichet BK (2004) Virtual screening of chemical libraries. Nature 432:862–865
9. Spyrakis F, Cavasotto CN (2015) Open challenges in structure-based virtual screening: receptor modeling, target flexibility consideration and active site water molecules description. Arch Biochem Biophys 583:105–119
10. Cavasotto CN, Orry AJ (2007) Ligand docking and structure-based virtual screening in drug discovery. Curr Top Med Chem 7:1006–1014
11. Pagadala NS, Syed K, Tuszynski J (2017) Software for molecular docking: a review. Biophys Rev 9:91–102
12. Cavasotto CN, Aucar MG, Adler NS (2019) Computational chemistry in drug lead discovery and design. Int J Quantum Chem 119: e25678
13. Gilson MK, Zhou HX (2007) Calculation of protein-ligand binding affinities. Annu Rev Biophys Biomol Struct 36:21–42
14. Gohlke H, Klebe G (2002) Approaches to the description and prediction of the binding affinity of small-molecule ligands to macromolecular receptors. Angew Chem Int Ed Engl 41:2644–2676
15. Mobley DL, Gilson MK (2017) Predicting binding free energies: Frontiers and benchmarks. Annu Rev Biophys 46:531–558
16. Baron R, Mccammon JA (2008) (Thermo) dynamic role of receptor flexibility, entropy, and motional correlation in protein-ligand binding. ChemPhysChem 9:983–988
17. Kitchen DB, Decornez H, Furr JR et al (2004) Docking and scoring in virtual screening for drug discovery: methods and applications. Nat Rev Drug Discov 3:935–949
18. Amadasi A, Surface JA, Spyrakis F et al (2008) Robust classification of "relevant" water molecules in putative protein binding sites. J Med Chem 51:1063–1067
19. Cozzini P, Fornabaio M, Mozzarelli A et al (2006) Water: how to evaluate its contribution in protein-ligand interactions. Int J Quantum Chem 106:647–651
20. Richards WG (1977) Quantum pharmacology. Butterworths, London; Boston
21. Mucs D, Bryce RA (2013) The application of quantum mechanics in structure-based drug design. Expert Opin Drug Discov 8:263–276
22. De Benedetti PG, Fanelli F (2014) Multiscale quantum chemical approaches to QSAR modeling and drug design. Drug Discov Today 19:1921–1927
23. Blomberg MR, Borowski T, Himo F et al (2014) Quantum chemical studies of mechanisms for metalloenzymes. Chem Rev 114:3601–3658
24. Forti F, Cavasotto CN, Orozco M et al (2012) A multilevel strategy for the exploration of the conformational flexibility of small molecules. J Chem Theory Comput 8:1808–1819
25. Juarez-Jimenez J, Barril X, Orozco M et al (2015) Assessing the suitability of the multilevel strategy for the conformational analysis of small ligands. J Phys Chem B 119:1164–1172
26. Barbault F, Maurel F (2015) Simulation with quantum mechanics/molecular mechanics for drug discovery. Expert Opin Drug Discov 10:1047–1057
27. Cavasotto CN, Adler NS, Aucar MG (2018) Quantum chemical approaches in structure-based virtual screening and lead optimization. Front Chem 6:188
28. Crespo A, Rodriguez-Granillo A, Lim VT (2017) Quantum-mechanics methodologies in drug discovery: applications of docking and scoring in lead optimization. Curr Top Med Chem 17:2663–2680
29. Lodola A, De Vivo M (2012) The increasing role of QM/MM in drug discovery. Adv Protein Chem Struct Biol 87:337–362
30. Ehrlich S, Goller AH, Grimme S (2017) Towards full quantum-mechanics-based protein-ligand binding affinities. ChemPhysChem 18:898–905
31. Ryde U, Söderhjelm P (2016) Ligand-binding affinity estimates supported by quantum-mechanical methods. Chem Rev 116:5520–5566
32. Yilmazer ND, Korth M (2016) Prospects of applying enhanced semi-empirical QM methods for 2101 virtual drug design. Curr Med Chem 23:2101–2111
33. Rognan D (2011) Docking methods for virtual screening: principles and recent advances. In: Sotriffer C (ed) Virtual screening. Principles, challenges and practical guidelines. Wiley-VCH Verlag, Weinheim, Germany, pp 153–176
34. Sotriffer CA (2015) Protein-ligand docking: from basic principles to advanced applications. In: Cavasotto CN (ed) In silico drug discovery and design: theory, methods, challenges, and applications. CRC Press, Taylor & Francis Group, Boca Raton, FL, pp 155–188
35. Guedes IA, Pereira FSS, Dardenne LE (2018) Empirical scoring functions for structure-based virtual screening: applications, critical aspects, and challenges. Front Pharmacol 9:1089
36. Cavasotto CN (2012) Binding free energy calculations and scoring in small-molecule docking. In: Luque FJ, Barril X (eds) Physico-

chemical and computational approaches to drug discovery. Royal Society of Chemistry, London, pp 195–222
37. Li J, Fu A, Zhang L (2019) An Overview of Scoring Functions Used for Protein-Ligand Interactions in Molecular Docking. Interdiscip Sci 11(2):320–328
38. Wang Z, Sun H, Yao X et al (2016) Comprehensive evaluation of ten docking programs on a diverse set of protein-ligand complexes: the prediction accuracy of sampling power and scoring power. Phys Chem Chem Phys 18:12964–12975
39. Su M, Yang Q, Du Y et al (2019) Comparative assessment of scoring functions: the CASF-2016 update. J Chem Inf Model 59:895–913
40. Warren GL, Andrews CW, Capelli AM et al (2006) A critical assessment of docking programs and scoring functions. J Med Chem 49:5912–5931
41. Slater O, Kontoyianni M (2019) The compromise of virtual screening and its impact on drug discovery. Expert Opin Drug Discov 14:619–637
42. Cho AE, Guallar V, Berne BJ et al (2005) Importance of accurate charges in molecular docking: quantum mechanical/molecular mechanical (QM/MM) approach. J Comput Chem 26:915–931
43. Zhang D, Li H, Wang H et al (2016) Docking accuracy enhanced by QM-derived protein charges. Mol Phys 114:3015–3025
44. Raha K, Merz KM Jr (2004) A quantum mechanics-based scoring function: study of zinc ion-mediated ligand binding. J Am Chem Soc 126:1020–1021
45. Dewar MJS, Zoebisch EG, Healy EF et al (1985) Development and use of quantum mechanical molecular models. 76. AM1: a new general purpose quantum mechanical molecular model. J Am Chem Soc 107:3902–3909
46. Dixon SL, Merz KM Jr (1996) Semiempirical molecular orbital calculations with linear system size scaling. J Chem Phys 104:6643–6649
47. Raha K, Merz KM Jr (2005) Large-scale validation of a quantum mechanics based scoring function: predicting the binding affinity and the binding mode of a diverse set of protein-ligand complexes. J Med Chem 48:4558–4575
48. Fong P, Mcnamara JP, Hillier IH et al (2009) Assessment of QM/MM scoring functions for molecular docking to HIV-1 protease. J Chem Inf Model 49:913–924
49. Alvarez LD, Veleiro AS, Burton G (2015) Exploring the molecular basis of action of ring D aromatic steroidal antiestrogens. Proteins 83:1297–1306
50. Anisimov VM, Cavasotto CN (2011) Quantum mechanical binding free energy calculation for phosphopeptide inhibitors of the Lck SH2 domain. J Comput Chem 32:2254–2263
51. Anisimov VM, Ziemys A, Kizhake S et al (2011) Computational and experimental studies of the interaction between phosphopeptides and the C-terminal domain of BRCA1. J Comput Aided Mol Des 25:1071–1084
52. Anisimov VM, Cavasotto CN (2011) Hydration free energies using semiempirical quantum mechanical Hamiltonians and a continuum solvent model with multiple atomic-type parameters. J Phys Chem B 115:7896–7905
53. Klamt A, Schüürmann G (1993) COSMO: a new approach to dielectric screening in solvents with explicit expressions for the screening energy and its gradient. J Chem Soc Perkin Trans 2:799–805
54. Cavasotto CN, Palomba D (2015) Expanding the horizons of G protein-coupled receptor structure-based ligand discovery and optimization using homology models. Chem Commun 51:13576–13594
55. Kim M, Cho AE (2016) Incorporating QM and solvation into docking for applications to GPCR targets. Phys Chem Chem Phys 18:28281–28289
56. Chaskar P, Zoete V, Rohrig UF (2014) Toward on-the-fly quantum mechanical/molecular mechanical (QM/MM) docking: development and benchmark of a scoring function. J Chem Inf Model 54:3137–3152
57. Grosdidier A, Zoete V, Michielin O (2007) EADock: docking of small molecules into protein active sites with a multiobjective evolutionary optimization. Proteins 67:1010–1025
58. Elstner M, Porezag D, Jungnickel G et al (1998) Self-consistent-charge density-functional tight-binding method for simulations of complex materials properties. Phys Rev B 58:7260–7268
59. Brooks BR, Brooks CL, Mackerell AD et al (2009) CHARMM: the biomolecular simulation program. J Comput Chem 30:1545–1614
60. Zoete V, Schuepbach T, Bovigny C et al (2016) Attracting cavities for docking. Replacing the rough energy landscape of the protein by a smooth attracting landscape. J Comput Chem 37:437–447
61. Chaskar P, Zoete V, Röhrig UF (2017) On-the-Fly QM/MM docking with attracting cavities. J Chem Inf Model 57:73–84

62. Pecina A, Meier R, Fanfrlik J et al (2016) The SQM/COSMO filter: reliable native pose identification based on the quantum-mechanical description of protein-ligand interactions and implicit COSMO solvation. Chem Commun 52:3312–3315
63. Lepšík M, Řezáč J, Kolář M et al (2013) The Semiempirical quantum mechanical scoring function for in silico drug design. ChemPlusChem 78:921–931
64. Stewart JJP (2007) Optimization of parameters for Semiempirical methods V: modification of NDDO approximations and application to 70 elements. J Mol Model 13:1173–1213
65. Řezáč J, Hobza P (2011) A halogen-bonding correction for the semiempirical PM6 method. Chem Phys Lett 506:286–289
66. Řezáč J, Hobza P (2012) Advanced corrections of hydrogen bonding and dispersion for Semiempirical quantum mechanical methods. J Chem Theory Comput 8:141–151
67. Marenich AV, Cramer CJ, Truhlar DG (2009) Universal solvation model based on solute electron density and on a continuum model of the solvent defined by the bulk dielectric constant and atomic surface tensions. J Phys Chem B 113:6378–6396
68. Kolar M, Fanfrlik J, Lepsik M et al (2013) Assessing the accuracy and performance of implicit solvent models for drug molecules: conformational ensemble approaches. J Phys Chem B 117:5950–5962
69. Pecina A, Haldar S, Fanfrlik J et al (2017) SQM/COSMO scoring function at the DFTB3-D3H4 level: unique identification of native protein-ligand poses. J Chem Inf Model 57:127–132
70. Ajani H, Pecina A, Eyrilmez SM et al (2017) Superior performance of the SQM/COSMO scoring functions in native pose recognition of diverse protein–ligand complexes in cognate docking. ACS. Omega 2:4022–4029
71. Pecina A, Brynda J, Vrzal L et al (2018) Ranking power of the SQM/COSMO scoring function on carbonic anhydrase II-inhibitor complexes. ChemPhysChem 19:873–879
72. Jones G, Willett P, Glen RC et al (1997) Development and validation of a genetic algorithm for flexible docking. J Mol Biol 267:727–748
73. Bowers KJ, Chow E, Xu H et al (2006) Scalable algorithms for molecular dynamics simulations on commodity clusters. In: ACM/IEEE conference on supercomputing (SC06). ACM, Tampa, Florida
74. Trott O, Olson AJ (2010) AutoDock Vina: improving the speed and accuracy of docking with a new scoring function, efficient optimization, and multithreading. J Comput Chem 31:455–461
75. Morris GM, Goodsell DS, Halliday RS et al (1998) Automated docking using a Lamarckian genetic algorithm and an empirical binding free energy function. J Comput Chem 19:1639–1662
76. Morris GM, Huey R, Lindstrom W et al (2009) AutoDock4 and AutoDockTools4: automated docking with selective receptor flexibility. J Comput Chem 30:2785–2791
77. Pu C, Yan G, Shi J et al (2017) Assessing the performance of docking scoring function, FEP, MM-GBSA, and QM/MM-GBSA approaches on a series of PLK1 inhibitors. Medchemcomm 8:1452–1458
78. Ajani H, Jansa J, Kopruluoglu C et al (2018) Imidazo[1,2-c]pyrimidin-5(6*H*)-one as a novel core of cyclin-dependent kinase 2 inhibitors: synthesis, activity measurement, docking, and quantum mechanical scoring. J Mol Recognit 31:e2720
79. Friesner RA, Banks JL, Murphy RB et al (2004) Glide: a new approach for rapid, accurate docking and scoring. 1. Method and assessment of docking accuracy. J Med Chem 47:1739–1749
80. Halgren TA, Murphy RB, Friesner RA et al (2004) Glide: a new approach for rapid, accurate docking and scoring. 2. Enrichment factors in database screening. J Med Chem 47:1750–1759
81. Burger SK, Thompson DC, Ayers PW (2011) Quantum mechanics/molecular mechanics strategies for docking pose refinement: distinguishing between binders and decoys in cytochrome C peroxidase. J Chem Inf Model 51:93–101
82. Cavasotto CN, Singh N (2008) Docking and high throughput docking: successes and the challenge of protein flexibility. Curr Comput Aided Drug Des 4:221–234
83. Cozzini P, Kellogg GE, Spyrakis F et al (2008) Target flexibility: an emerging consideration in drug discovery and design. J Med Chem 51:6237–6255

Chapter 18

QM Calculations in ADMET Prediction

Alfonso Pozzan

Abstract

In recent years, there has been an increase in the application of quantum mechanics (QM) methods to describe properties related to the ADMET profile of small molecules. The application of these methods allows calculating useful descriptors and physiochemical properties contributing to ADMET prediction. Considering that QM methods are the only one that describe the electronic state of a molecules, such methods are particularly useful for studying the metabolism of drugs; furthermore, the introduction of mixed QM and molecular mechanics (QM/MM) is also increasing the understanding of drug interaction with cytochromes from a mechanistic point of view. Finally, combining the increase number of experimental data with machine learning algorithms and QM-derived descriptors allowed the creation of an end-user software capable of affecting the drug discovery process.

Key words Quantum mechanics, ADMET prediction, Metabolism prediction, In silico ADMET

1 Introduction

Every medicinal chemists working in drug discovery knows very well the importance of in vitro and in vivo studies of absorption, distribution, metabolism, excretion, and toxicology (ADMET) in order to identify molecules suitable to enter into preclinical and clinical studies.

ADMET properties combined with pharmacokinetics and pharmacodynamics (PK/PD) are the key properties to optimize in order to increase the possibility of a molecule to become an approved drug. This is why it is so extremely important to include, during the hit to lead (H2L) and the lead optimization (LO) phases, a series of experimental and theoretical studies leading to the synthesis of better molecules, not only in terms of target affinity, but also for their ADMET profile.

In Fig. 1, it is possible to observe a typical screening cascade utilized in drug discovery while progressing a molecule from the hit to the candidate selection phase. It should be worth considering the number of ADMET and PK/PD studies following the initial ones

Alexander Heifetz (ed.), *Quantum Mechanics in Drug Discovery*, Methods in Molecular Biology, vol. 2114,
https://doi.org/10.1007/978-1-0716-0282-9_18,

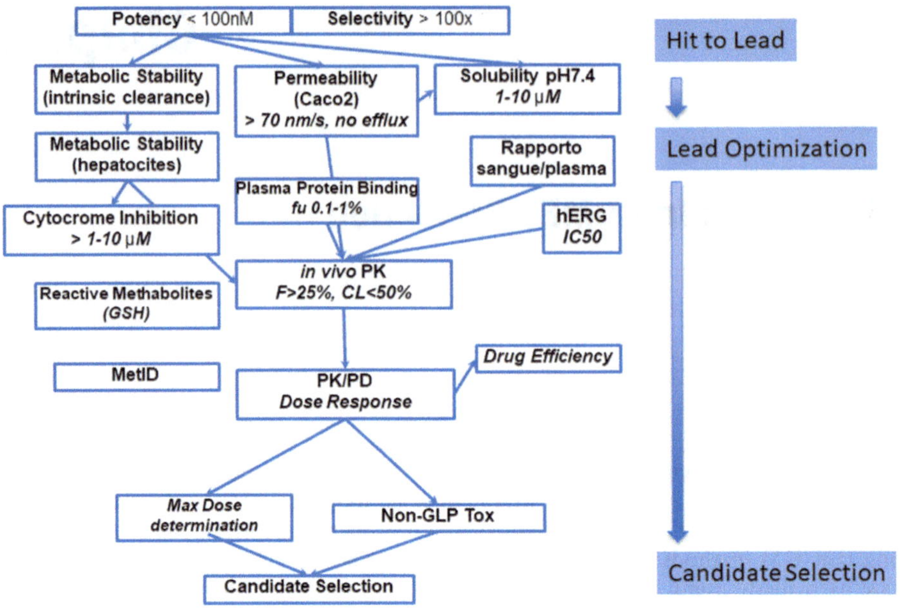

Fig. 1 Example of a typical screening cascade of a drug discovery project. It is worth noticing the number of assays concerning ADMET studies

addressing affinity and selectivity for the biological target. Despite the high number of ADMET and PK/PD studies, the attrition post candidate selection remains quite high. In a recent paper [1], four major pharmaceutical companies have joined efforts and analyzed the cause of attrition in drug development. The conclusions were as follows: control of physiochemical properties with lipophilicity is one of the key properties to consider. For example, it was observed that compounds failing in clinical Phase I were significantly more lipophilic (as assessed by calculated LogP) than those progressing into Phase II. At the same time, the authors recognize that a statistical application of good ranges of physiochemical properties could not address completely the high failure rates in drug development nor provide specific answer for improving the ADMET properties of a distinctive class of molecules. As many of such failures depend from specific drugs and biological macromolecules interactions, it is only with a better understanding of such mechanisms that we will progress, toward a design of better candidate molecules. From this perspective, quantum mechanics (QM) calculations allow to study molecules at the atom level, leading directly to molecular structure and energy, as well as atoms bonding, reactivity, and other observables. QM is based on the principles of chemistry and physics and, within an ideal spectrum of computational chemistry tools, sits at one extreme. QM methods have intrinsically the highest level of theory to describe, in fine details, the physiochemical properties of small molecules, their mechanism of interaction with biological targets, and their reactivity. In this review, we will discuss how the implementation of QM methods in drug discovery can be a useful companion tool for ADMET predictions. The word "companion" is necessary, because major advances in the field of in silico ADMET predictions have been possible thanks to the increased number of experimental data combined with various modeling techniques. QM methodologies, on the other hand, despite being more complex and computationally intensive, are constantly growing. In a recent review [2], it was highlighted how relevant some approaches are becoming within the drug discovery process. It is then the role of the experienced computational chemists to reach a compromise between accuracy and productivity and decide when and what QM method to apply in their research.

2 QM Methods

In principle, QM theory should be able to provide precise quantitative descriptions of molecular structures and their physiochemical properties. However, due to mathematical and computational complexities, this is still something that needs to be approximated. According to quantum chemistry theory, an electron bound to an atom cannot possess any arbitrary energy or occupy any position in space. These characteristics can be determined by solving the Schrödinger eq. QM attempts to solve such equation using various levels of approximation. The simplest are the semiempirical methods like MNDO [3], AM1 [4], PM3 [5], PM6 [6], and RM1 [7]. Such methods where present in the MOPAC [8] and AMPAC [9] FORTRAN codes and originated from the group of Michael Dewar [10]. Today, such approaches are extremely popular and some of them are available within various computational chemistry packages. Semiempirical methods are very fast (with today computers even extremely fast) thanks to the use of the zero differential overlap approximation. The approximation involves ignoring two-electron repulsion integrals via replacing with an ad hoc empirical parametrization. For the same reason, semiempirical methods can be applied to molecules as large as whole proteins. As part of the QM family, the next level in accuracy is represented by the ab initio methods. In such calculations, the Hamiltonian (in QM, a Hamiltonian is an operator corresponding to the sum of the kinetic energies and potential energies for all the particles in the system) provides a complete representation of all interactions between the nuclei and electrons in a molecule following different levels of orbital approximation and overall accuracy. One of the simplest ab initio methods is the Hartree–Fock (HF) self-consistent field (SCF) method. In such approximation, each electron wavefunction moves in an average potential generated by the remaining many-electron wavefunction, but without an explicit knowledge of where they are. For this reason, the electron repulsion overestimated as electrons with opposite spin can overinteract each other. Increased accuracy can be generally obtained using various methods aimed to consider electrons correlations. Such methods include the Møller–Plesset (MP) perturbation theory, configuration interaction (CI), and coupled-cluster (CC) methods [11]. In recent years, density functional methods (DFT) following the Hohenberg–Kohn theorem [12] have become more popular due to the fact that they are faster than traditional HF methods. In DFT, the many-electron wavefunction is completely bypassed in favor of the electron density. Because the level of theory to define the energy function for the electron density is still not completely understood, HF methods were considered more reliable than DFT. This however has recently changed and DFT techniques are now becoming extremely popular.

While not strictly pure QM calculations, the so-called hybrid methods combine molecular mechanics (MM) and QM. Such hybrid QM/MM calculations [13] combine quantum chemical and MM models, preserving the integrity of the overall energy function. Hybrid methods are approximately as expensive as the corresponding optimization of the QM region alone (the MM component is negligible in this context), making such optimizations feasible for large systems and ideally suited to studying the reactive portion of proteins or nucleic acids. A concise review covering the *state of the art* of ab initio quantum chemical methodologies has been published by Friesner [14] in 2005. The review provides enough background and examples without entering into the mathematical complexity of the theory and can be a valuable source of information for the nonexperts in the field.

3 Importance of Physiochemical Properties in ADMET Prediction

The importance of physiochemical properties in ADMET predictions has probably one of the longest history in computational chemistry. One of the most cited physiochemical property is the calculated logP (cLogP). LogP calculation was first reported by Hansch, Iwasa, and Fujita back in the '50s/'60s [15–17] and was then found of extremely importance in a variety of ADMET properties. In the Lipinski paper [18] that signed the beginning of the "rule of 5" era, simple descriptors like molecular weight and counts of oxygens, nitrogens, and hydrogens bonds were found to be very important for absorption and permeation. Leeson and Springthorpe [19] in 2007 further detailed the analysis of simple physiochemical properties alerting the scientific community about the risks of progressing candidates with cLogP > 4.5–5. This was further quantified with the introduction of concepts like the lipophilicity ligand efficiency (LLE) being LLE = pIC50 (or pKi)—cLogP (or ClogD) and a recommendation that clinical candidates should have LLE > 5 to reduce the risk of failure. In 2010, Young et al. [20, 21] introduced two simple parameters, the solubility forecast index (SFI) and the property forecast index (PFI), that is, SFI = cLogP + (number of aromatic rings) and PFI = Chromatographyc − $\log D_{pH7.4}$ + (number of aromatic rings), respectively. Also in this study, it was found that 67% of the analyzed candidate molecules have good solubility, lower plasma protein binding (albumin), lower interaction with cytochromes, lower intrinsic clearance, lower affinity for hERG, and reduced promiscuity if PFI < 5. Only permeability was found to improve for higher values of PFI.

Successive retrospective analyses on clinical candidates lead to improvement and refinement of the initial Lipinski rule. For example, Gleeson et al. formulated the rule 4/400. According to this rule, the optimal molecular weight should be <400 Da and the cLogP < 4 in order to reduce ADMET liabilities [22]. Veber et al. [23] instead found that good bioavailability could be obtained if the number of rotatable bonds in a molecule is less than 10 and the polar surface area (PSA) is lower than 140 $Å^2$.

As a further example, within our organization, it is common practice to map the physiochemical properties of molecules and compare (map) with the one derived from specific class of molecules such as CNS drugs, FDA-approved oral drugs, and anti-infectives. These maps help determine "where we are" in terms of a multidimensional physiochemical space and identify possible areas to focus the medicinal chemistry activities in order to improve a particular developability profile.

In Fig. 2, a simple example shows how principal component analysis (PCA) combined with simple calculated physiochemical properties helped developing molecules with peripheral or central CNS activity.

4 QM-Derived Molecular Descriptors Used in ADMET Predictions

4.1 Frontier Orbitals

Considering the importance that physiochemical properties have in ADMET prediction, what is the current contribution that QM provides in this field? If we look at the early applications of QM, the first publications back in the 1980s concern with the use of molecular orbital (MO) energies as predictors of molecule toxicity. Early works performed by Klopman et al. [24] on a small set of nitroarenes, and later by Debnath et al. [25] on an enlarged set of 188 molecules, indicate that QM-derived energies of the lowest unoccupied MO (E_{LUMO}) could be an extremely accurate physiochemical parameter that correlates with the experimental mutagenicity data (Ames test using S. typhimurium TA98 or TA1538). It should be noted that while Klopman et al. used a direct correlation with E_{LUMO}, in the work performed by Debnath et al., adding

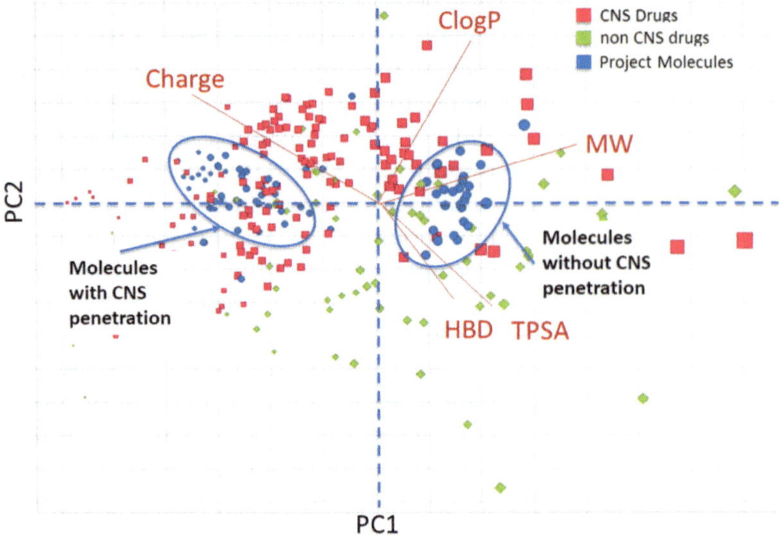

Fig. 2 PCA relative to a small set of simple physiochemical properties computed for a set of project molecules and two set of reference drugs (CNS and non-CNS). A graphical visualization can help the team in identify the best chemical space and in this class in developing two different chemical series with and without CNS penetration

cLogP and applying QSAR modeling via multiple regression analysis (MRA) could extend further the range of applicability of such methods. MO energies like the energies of the highest occupied MO (HOMO) or the LUMO are very popular quantum chemical descriptors. According to the frontier MO theory of chemical reactivity, the formation of transition states depends on the interaction between the HOMO and LUMO. The energy of the HOMO is linked to the ionization potential and characterizes the tendency of the molecule toward attack by electrophiles, while the energy of the LUMO is linked to the electron affinity and characterizes the susceptibility of the molecule toward attack by nucleophiles. Other definitions relating LUMO and HOMO energies are the hardness and softness. Hard nucleophiles have a low-energy HOMO; soft nucleophiles have a high-energy HOMO; similarly, hard electrophiles have a high-energy LUMO; and soft electrophiles have a low-energy LUMO. In Subheading 13, an example where HOMO energies were used to rank reactivity toward glutathione (GSH) is reported [26].

4.2 Charges, Polarizability, and Polarity Indexes

Local charges, electrostatic potential, dipole moment, and molecule polarizability are among the most popular physiochemical descriptors derived from QM calculations that once combined with various machine-learning methods are widely used to derive ADMET models. At QM level, charges are derived from electron densities and centered on each atom of the molecule. Atomic charges can be used as chemical descriptors encoding reactivity indices and directly linked to each specific atoms of the molecule. Moreover, partial charge distribution can be used to derive other descriptors like dipole moment or polarization indexes. Following the work by Klopman and Debnath on nitro derivatives [24, 25], Niazi et al. [27] published a QSAR model based on QM-derived descriptors where partial charges calculated for some specific atoms represent one of the key descriptors to model the toxicity of a series of nitro derivatives. Other QM descriptors included in the analysis were the molecule softness, the molecular polarizability, the dipole moment, and the electrophilicity index.

5 QM-Derived LogP Calculations

We have seen previously how lipophilicity plays an important role in ADMET predictions and how high lipophilicity is detrimental for progressing clinical candidates. The prediction of lipophilicity is generally obtained as an increment system adding contributions of fragments or atom types. Such methods are trained on a set of experimental data and are generally additive in nature. Very popular algorithms utilized consist are the atom-type based AlogP [28], MlogP [29], and XlogP [30] or the fragment-based cLogP [31]. Although these approaches are scientifically attractive, there are several problems with them; in particular for very flexible molecule with high molecular weight, such methods can provide results that are not very accurate. In a recent review by Shultz [32], analyzing two decades of post "rule of 5" FDA-approved drugs, the cLogP for such extra size molecules is generally overestimated. The hypothesis is that such molecules are flexible enough to bury polar atoms and collapse hydrophobic groups that fragment methods cannot take into account.

Few attempts have been made to predict and study lipophilicity using QM methods. This represents an attractive approach since QM methods can provide information concerning the polarizability and ionization state of a molecule as well as the probability to form hydrogen bonds. All these aspects are key factors characterizing the partitioning of a molecule between different phases but extremely challenging from a theoretical point of view. The need to consider conformational ensembles and solvation effects poses some limitation in QM terms. Nevertheless, in 2004, using ab initio MO-derived parameter in combination with accessible surface area of solute molecules allowed the calculation of a QM-derived cLogP [33]. According to this method, the LogP is defined using the following equation:

$$\log P_{oct/w} = a\Delta E_{oct/w} + b\text{ASA}_w + k$$

The $\Delta E_{oct/w}$ term represents the difference in the electrostatic-derived solvation energy between *n*-octanol and water, while the ASA_w represents the solvent accessible "surface area" of the molecule. An accurate electrostatic solvation energy was estimated using ab initio MO self-consistent reaction Field (SCRF) calculations within the conductor-like screening model (COSMO) [34, 35]. The set considered for the analysis and validation of the model was relatively small in size and made with molecules with no more than 30 heavy atoms. Among those, 69 molecules were hydrophobic, 34 contain one hydrogen bond acceptor, and the remaining 50 were amphiprotics. The results obtained proved to be extremely accurate for non-hydrogen bonders and less accurate for hydrogen bond acceptors and amphiprotic molecules. While the overall performance of this methods was extremely good when compared to traditional fragment-based approaches, the real scope of this research was improving the mechanistic understanding behind the log*P* calculation. The conclusions were a mixed combination of being successful and too optimistic at the same time, suggesting that a more sophisticated solvation model and a higher level of ab initio theory was required to improve the performance of the method. Twelve years later, as part of the SAMPL5 challenge [36], among the 76 different methods proposed for predicting the cyclohexane/water distribution coefficient, the top performer was a QM-derived method, proposed by Andreas Klamt et al. [37], using COSMO-RS, an evolution of the COSMO model. These results confirm that QM-based prediction of cLogP could play a relevant role in the near future.

6 Electronic State Determination

Ab initio QM-derived charges and spin-density distribution were used in early studies performed by Seagall et al. [38–40] for understanding the mechanism of drug metabolism performed by the cytochromes P450 superfamily. These studies however were limited to static models derived from experimentally obtained X-ray complex without drug metabolism predictions. Few years later, Jones et al. [41], reported an electronic model for predicting the regioselectivity in P450-mediated hydrogen atom abstraction. The methodology was based on semiempirical AM1 ground-state energies computed for the parent compound, and the product radicals therefore were quite fast and simple to implement. This paper was the first attempt to combine predictive models for P450-mediated aromatic and aliphatic oxidation and provided reasonable good correlation ($Q2_{CV,20\%out} = 0.79$ for the combined model) with experimental data of enzymes and substrates showing little preference in orientation, such as CYP1A2, CYP2B1, CYP2B4, CYP2E1, and CYP3A4. Despite the fact that only the reactivity component of P450-mediated oxidation was considered, the authors indicated that such models could be used for designing better molecules and possibly combined with homology or pharmacophore models representing the active site of the P450 cytochrome.

7 Reactivity Studies

Studies that involve reactivity, primarily with cytochromes P450, are certainly the area where QM calculations can provide a unique contribution to better understand and predict the site of metabolism of organic molecules. Early works in this area where published in the '80s mostly using semiempirical methods like AM1. Among the first studies reported in literature, the one proposed by Korzekwa et al. [42] focused on ranking molecules based on the energy of the radical species resulting from proton abstraction. These studies did not involve any structural information derived from the cytochrome P450, but purely described the proton abstraction. The assumption was that the ability of a drug to be metabolized would be a function of both its binding characteristics (reversible affinity for the active site) and the tendency of the molecule to undergo oxidation (electronic factors). One important aspect of this study was the utilization of the *p*-nitroso-phenoxy radical (PNR) as the radical species with the most appropriate transition-state geometry (Fig. 3) mimicking the energetic of a P450 enzyme involved in nonaromatic substrates oxidation.

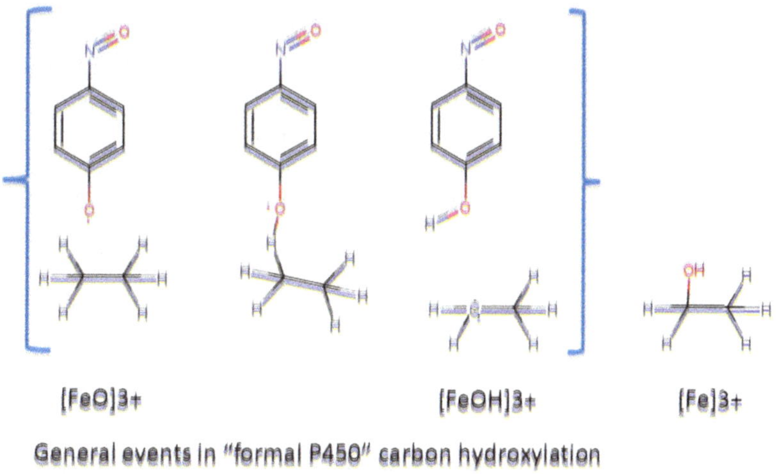

Fig. 3 A representative hydrogen abstraction reaction utilized by Korzekwa et al. [42] for simulating the active site of P450. Despite the extreme simplicity of such model, the use of PNR moiety allowed to rank correctly the oxidation of various classes of substrates

While this represent an oversimplification of the entire oxidative process, the model was able to correctly rank 20 different molecules, in agreement with the experimental rank order of various C–H cytochrome P450-mediated oxidations. Following the success of this model [43–45], the same group of scientist developed a second one accounting for oxidation of aromatic substrates [41]. In this case, the authors hypothesize that the aromatic oxidation would likely occur by addition of a triplet-like oxygen, a successive formation of a tetrahedral intermediate [46], and finally a rearrangement to epoxides, ketones, or phenols metabolites. Unfortunately, the PNR system used for the aliphatic oxidation did not generate the expected transition state with a tetrahedral geometry. Therefore, a different system employing a methoxide radical was used. Despite the two different radical systems, it was possible to normalize the respective energies and combine them into a single "global model." Such combined model, at the time of publication, was the first one capable to predict P450-mediated oxidations for both aromatic and aliphatic species and represent an important milestone in the field. Following the initial success obtained with such reactivity models, further approaches along the same concepts were investigated and reported in literature. One recent example is represented by SMARTCyp [47]. SMARTCyp is a reactivity-based P450 metabolism model that have gained popularity in the last few years because it is extremely fast and easy to use. The overall idea was to generate a broad collection of moieties representing as much as possible a broad set of molecule site of metabolism. This concept was put into practice via the optimization of 139 transition states using high-level ab initio methods. A simplified heme model was built (without side chains) including the porphyrinic central iron and an SCH_3^- and O_2^- in the axial positions.

On such heme model, transition states were then optimized using DFT methods with a mixed combination of basis sets: the *p* function enhanced double-ζ basis set of Schäfer et al. [48] for the iron and the 6-31G(d) basis set for all the other atoms [49–51]. Final energies where then computed using more accurate basis sets. The Schafer basis set for the iron was further enhanced with the inclusion of s, p, d, and f functions [52] while the 6-31G (d) used for the optimization was enhanced to the 6-311++G (2d,2p) basis set [53, 54] for all the other atoms.

The energies from all the 139 transition-state calculation were then encoded into a set of SMARTS to be used as a lookup table to extract predicted transition-state energies. As this is a simple 2D substructure search, SMARTCyp is extremely fast in generating a prediction. However, it should be noted that there is no reactivity associated for atoms not matching any SMARTS. One of the aspects of this method is that, given the simple heme representation, it should refer to a "generic" CYP-mediated reactivity prediction. On the other hand, the authors pointed out to the fact that one of the major isoforms responsible for the metabolism (the 3A4) is known to have a very large and flexible active site. For this reason, the SMARTCyp approach could be considered a surrogate model for 3A4-mediated substrate oxidation. Moving into the performance of such approach, in a data set of 394 molecules with experimental data, 65% correct predictions were obtained considering the highest reactive position. This value increased to 81% considering the top three reactive position. These can be considered good results that can help address potential metabolic stability issues and improve design strategies. If one of the limitations of SMARTCyp concerns the lack of 3D receptor information, the advantage of such approach is that new transition states can be added following new research in this area and without the need to reparametrize the entire model.

8 Improving Metabolic Prediction Combining QM and Experimental Data

Continuous method development and increased throughput in the various assays related to metabolic stability have progressively enlarged the amount of experimental data that can be used to train, develop, and validate in silico models. One approach relaying on experimental data is the CypScore model [55]. The model uses a data set containing 844 structures and 2336 metabolic transformations, carefully selected from literature and covering the principal P450 oxidation reactions (aliphatic hydroxylation, N-dealkylation, O-dealkylation, aromatic hydroxylation, double bond oxidation, N-oxidation, and S-oxidation). On such data, a number of QM-derived descriptors were used to build a CypScore model for each molecule's atom using multiple linear regression. In this way, the authors were able to encode the influence of the molecule electron distribution into a specific atomic position. Among the various descriptors, the most important where the atomic accessible solvent area, the atomic charge, the local ionization energy, the local electron affinity, the electrostatic potential, the C–H bond order, and the atom polarizability. Some of these parameters were relevant for all seven P450 oxidation reactions, while others were more specific for one or the other. One of the key aspects of CypScore is that the data used to train the model reflect the distribution of the most common P450 isoforms in human, rat, and mouse and therefore contain an "implicit" factor that should make the results comparable with the outcome of an experimental in vitro Met-ID assay. The assumption was proven by the authors, during the validation phase performed on a set of 39 well-characterized molecules from 14 LO programs where it was possible to predict at least one of the major metabolites in 87% of the molecules and even possible to predict all the major metabolites for 62% of the molecules.

Another notable example combining QM-derived descriptors with machine learning approaches is the one recently proposed by Finkelmann et al. [56], where the authors trained a machine learning classifier on a curated set of P450 metabolism data and used QM-derived descriptors. These novel descriptors include three general properties aimed to capture the behavior of a putative P450 reactive site. Such descriptors include the electronic and steric environment as well as the relative location within the molecule structure. More specifically, the electronic environment was described by the radial partial charge distribution and the steric environment by the radial mass distribution, while for describing the atom's relative location and topological environment, the authors used a histogram-type descriptor that recursively counts the occurrence of each interatomic distance (paths) for all atoms in a molecule. Such descriptors were calculated using fast density functional tight binding (DFTB) theory, while charges were derived from the natural population analysis (NPA) and Charge Model 5 (CM5). For machine learning, among the various methods that were used, random forest seemed to provide the best overall performances. The model was then trained on a published set [57] containing atom-based binary reactivity labels (1 for a reactive atom and 0 for a nonreactive one). The best model used 517 features and DFT charges and resulted in a correlation coefficient (fully cross-validated) between predicted and observed of 0.63. Furthermore, a retrospective analysis performed on two published metabolic stability optimization campaigns showed very good prediction in identifying most, if not all, of the experimentally determined site of metabolism. A more extensive evaluation was also carried out by one big pharma company (Bayer) using a benchmark of 25 compounds synthetized as part of 24 different drug discovery projects and all with Met-ID data from human hepatocytes. In this validation, the CypScore method is used as benchmark. Overall, the new approach was superior to CypScore for 17 cases, while for five compounds several false positives were observed in both methods. Overall, the model produced useful predictions on drug candidates structurally dissimilar from the one in the training to the point that it was considered suitable to guide drug discovery compound optimization process.

9 QM/MM Methods for Drug Metabolism Prediction

So far, we have seen how good QM can be in predicting the reaction component of a P450-mediated oxidation. At the same time, we have seen how the lack of structural information can lead to false positives. A simplistic explanation could imply different ligand orientation or steric effects [58] or that the electronic properties of a ligand could be modulated by the protein [59]. It is somehow obvious that the next step would try to combine QM with docking. This is somehow what the hybrid methods should be doing and combined QM/MM approaches are getting more and more important for modeling reactions in biomolecular systems. As part of this method, QM handles the catalytic/ligand part of the system, while the structural aspects concerning the entire protein are treated using MM force fields. Coupling two different methodologies allow building systems containing several thousand atoms; moreover, the rapid development of QM/MM in recent years has made possible to perform calculations to predict activation barriers for P450 reactions with accuracy near to 1 kcal/mol [60–62]. With this level of accuracy, site of metabolism prediction could really become an extremely important tool for drug discovery.

The first QM/MM study of human cytochrome P450 in complex with two nonsteroidal analgesic drugs (diclofenac and ibuprofen) was published by Bathelt et al. in 2005 [63]. The study relies on the X-ray structures of 3A4, 2C9, and 2B4 P450 isoforms and uses a combined QM/MM calculations at DFT/B3LYP: CHARMM27 level. Combining the structural information with QM/MM, the authors were able to build an electronic and geometric model describing the oxidizing species (Compound **I**)[1] involved in the metabolism of the two drugs. The results showed that Compound **I** is surprisingly similar in all the different P450 enzymes and independent of inclusion of heme substituents, truncation of the thiolate ligand, DFT settings, and protonation state of one catalytic histidine. Instead, the main source of variation was observed for the delocalization state of one of the three unpaired electrons of the axial thiolate of the heme. The spin density on the sulfur atom vary between 0.2 and 0.5 and would be the result of small changes in the environment of the heme group either due to different isoform or different conformation. Therefore, such results would indicate that the electronic properties of Compound **I** in the different human CYP450s are not distinguishable, which implies that observed differences in substrate selectivity are not caused by differences in their electronic properties.

Following the first QM/MM study on P450, from the same research group, a successive paper concerned with a specific model for benzene hydroxylation in P450 2C9 isoform was developed. The mechanism of benzene hydroxylation was analyzed in detail using both "side-on" and "face-on" modes of benzene binding with respect to the heme as well as different conformations of the enzyme substrate complexes. The results obtained indicated that the QM/MM-derived energy barriers for benzene hydroxylation

[1] "Compound I" is a conventional name referring to heme-based oxidizing species involving the formation of high valent ferryl intermediates of the heme iron.

are comparable with the experimental rate constants of benzene metabolism. The rearrangement pathways from the initial complex were modeled, to form an epoxide, a ketone, or an N-protonated porphyrin. Face-on geometry seems to lead to epoxide or keto products with equal probability. Epoxide formation would be favored in the side-on orientation. The N-protonated porphyrin species could form from the side-on conformation and confirmed the proposed role of this species as a proton shuttle.

10 Ab Initio Molecular Dynamics of P450

Molecular dynamics (MD) at ab initio level of theory represents one of the most accurate models that one could imagine for studying P450-mediated metabolism. This type of calculations are still quite demanding even with today's state-of-the-art computing resources. The reason behind is that, in order to calculate accurate reaction patterns with sufficient accuracy, the level of ab initio theory needs to be quite high and this needs to be performed for each step of a molecular dynamic simulation. Nerveless, in 2015, one of the first ab initio MD studies was proposed [64] concerning the aliphatic hydroxylation reaction and using Car–Parrinello [65] dynamics coupled with DFT ab initio theory. The initial model was obtained from a truncated representation of Compound **I** from the P450cam [66] X-ray structure. MD simulations were run for 1.2 ps in both high (300 K) and low (1 K) temperature conditions indicating that the hydroxyl–iron species have a short lifetime, between 300 fs and 3600 fs, depending on the enzyme active site architecture. While such studies helped to get more insights on the physical mechanism underlying substrates hydroxylation, it looks like ab initio MD is not yet feasible to fit into the timelines of a drug discovery process.

11 Prediction of Reactivity with GSH

One of the aspects related to drug toxicity is the generation of reactive metabolites that form in the liver. Most of such metabolites derive from the P450 oxidation but some of them can bound covalently to protein or nucleic acids. As most of these reactive metabolites form in the liver, drug-induced liver injury (DILI) has become the primary reason for withdrawal of approved drugs. For this reason, during the LO, it is common to add to the screening cascade the in vitro measure of reactivity versus GSH. While the prediction of P450-mediated metabolism is being extensively studied, very few authors have specifically addressed the GSH reactivity. One of the first exhaustive work in this area was published by Hughes et al. [67], using a very similar approach reported for the XenoSite method [57], that is, combining neural networks with various molecular descriptors containing also few reactivity ones derived from QM calculations. Quite surprisingly, while single QM-derived descriptors can discriminate between reactive and nonreactive atoms, the combination of various non-QM descriptors (mostly topological) outperformed the single descriptors. For example, using a data set of 1213 GSH-reactive and 271 GSH-nonreactive molecules, the QM self-polarizability could discriminate reactive from nonreactive molecules with 75.3% accuracy. The neural network-derived atom reactivity score (ARS) showed an improved 90.8% accuracy. The model was also tested on two external data sets and in both cases statistically significant predictions were observed. On top of the atom-based reactivity model, the authors built also a second one based on whole molecule descriptors. In this case, the most important descriptors were the reactivity of the most reactive atom (highest ARS) followed by the energy of the LUMO orbital.

12 Conclusions

QM calculations have the ability to provide an accurate representation of small molecules and biomolecules even if this comes at the expenses of more complexity and increasing computing time. In this review, we have discussed different approaches that use QM to predict ADMET properties with a particular emphasis for reactivity-driven drug metabolism, an area where other methods struggle. While this seems a limitation, we should remember that the vast majority of failures in clinical phases are due to hepatic toxicity induced by drugs metabolites. For this reason, any effort in this field would make a huge impact in the drug discovery process. Another area where QM can play a key role is in combination with MM (QM/MM methods) leading to a better understanding of the mechanisms of cytochrome P450 substrate oxidation also from the protein interaction/recognition side.

We should mention also the extremely good results obtained by mixing machine learning methods trained with experimental measures and, but not exclusively, QM-derived physiochemical descriptors. Among them, QM-derived electrostatic potential and solvation energies seem to play a central role also for ab initio-derived logP calculations. As a final remark, it should be also noted that QM is constantly growing and thanks to the diffusion of freely available, high-quality ab initio software, it will likely become a more prominent tool in the list of the computational medicinal chemists. Therefore, modern QM approaches to ADMET prediction will play a more direct role during the drug discovery process.

13 Case Study: Prediction of GSH Reactivity

In one of our projects, one detrimental aspect that we faced was the formation of reactive metabolites. These were characterized via GSH trapping after incubation in liver microsomes. Metabolite identification studies revealed the formation of an epoxide in one part of the molecule that had to be a (substituted) pyridine. This in order to maintain potency and selectivity. From the medicinal chemistry point of view, the strategies to mitigate such issue were to (1) add electron withdrawing groups, (2) add an extra nitrogen in the aromatic ring in order to reduce the electron density, and (3) change the nature of the linker in order to reduce the resonance effects to the aromatic ring. In order to prioritize the order in which such molecules could be synthetized, an attempt was made to predict the epoxide formation following a simple perturbation theory suggested by Klopman et al. for predicting the metabolism of aromatic hydrocarbons [68]. Oxidation of a double bond in aromatic molecules under catalysis by P450 was treated as an electrophilic reaction in which the LUMO orbital of an electrophilic oxygen atom interacts with the HOMO orbital of the aromatic compound. In this approach, the hypothesis is that the metabolites predominantly formed are those whose energy are the most favorite using the following equation:

$$\mathrm{QI}_A = \sum_i \frac{\left(C_{Ai} C_0^{\mathrm{LUMO}} \beta\right)^2}{E_i \beta - E_{\mathrm{P450}\beta}}$$

where QI_A is the perturbation energy for the substrate A, C_{Ai} is the coefficient of the occupied (HOMO, HOMO-n) atomic orbital i of atom A, $C_0{}^{\mathrm{LUMO}}$ (assumed constant = 1) is the coefficient of the oxygen atom in the LUMO orbital of P450, E_{i} is the energy of the occupied MO I of the substituted pyridine in β units, and E_{P450} (assumed to be 0.1) [2] is the energy of the LUMO orbital of P450 in β units. It should be noted that, assuming E_{P450} and $C_0{}^{\mathrm{LUMO}}$ constant, the perturbation energy would depend only from the substrate. QM calculations where then performed at the RB3LYP/6-31G(d) level for each of the substituted pyridine or pyrimidines (Figs. 4 and 5).

A good separation between molecules reacting or not with GSH seems to follow the rank of the HOMO energies (Table 1) which was used as simple descriptors to prioritize molecules before synthesis.

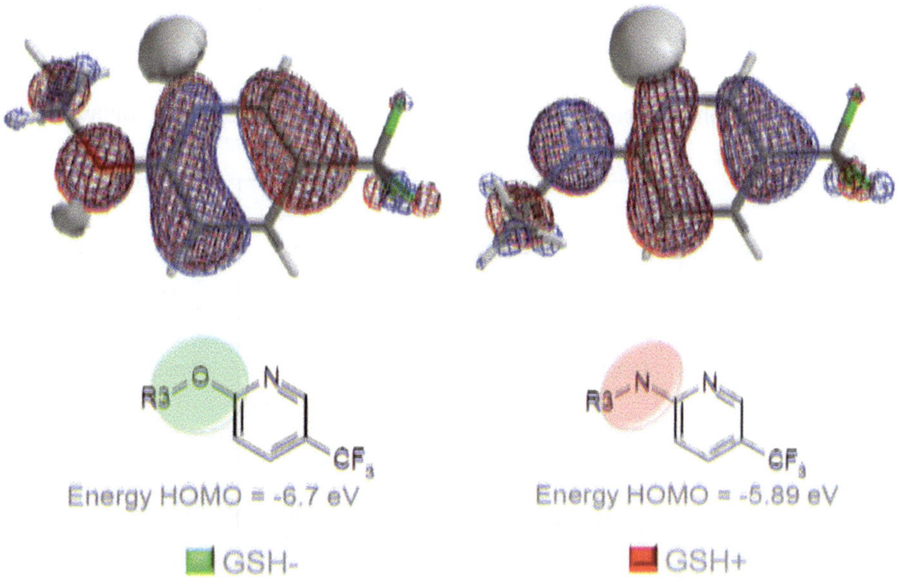

Fig. 4 HOMO orbitals and energies computed for the pyridine fragment where the oxidation occurs. The oxygen introduced at the linker level lowers the HOMO energy, that is, the whole system becomes "less reactive" toward attack by electrophiles according to the Klopman theory [68]

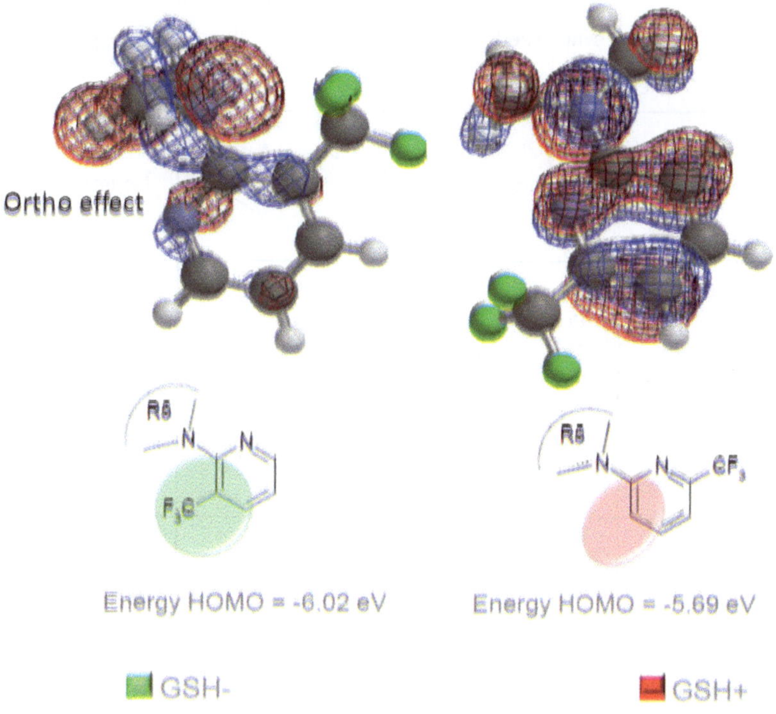

Fig. 5 HOMO orbitals and energies computed for the pyridine fragment where the oxidation occurs. The introduction of a methyl in position 3 helped reduce the reactivity in two ways: one related to a stabilization of the HOMO energy and the other related to the different localization of the orbital due to the steric hindrance and loss of conjugation. Both effects could reduce the reactivity toward the formation of oxidized species by P450 resulting in a different reactivity observed experimentally by the GSH trapping

Table 1
A good separation between molecules reacting or not with the GSH seems to follow the rank of the HOMO energies

Structure	GSH	HOMO energy [eV]	Structure	GSH	HOMO energy [eV]
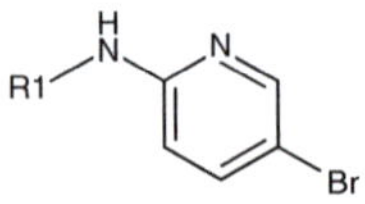	Positive	−5.64	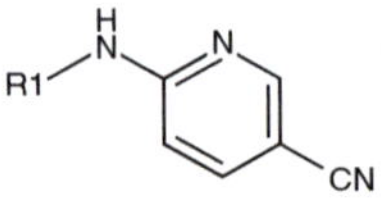	Positive	−6.04
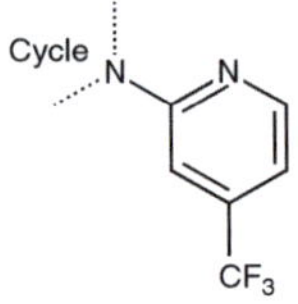	not synthetized	−5.66	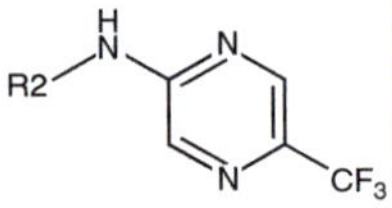	Positive	−6.25
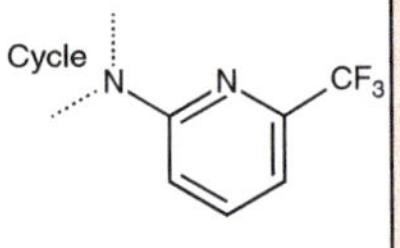	Positive	−5.69	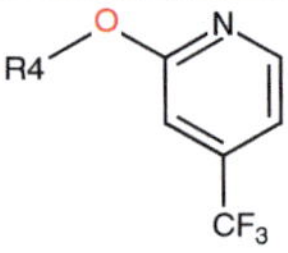	Negative	−6.61
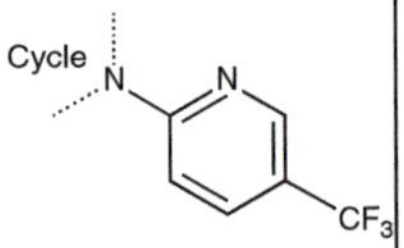	not synthetized	−5.74	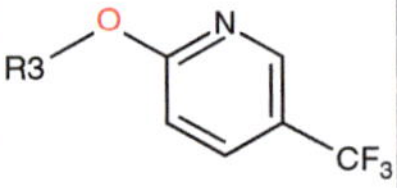	Negative	−6.70
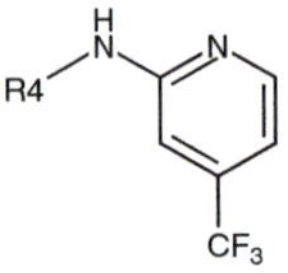	Positive	−5.83	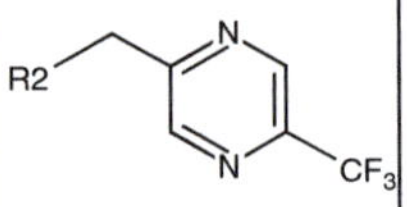	Negative	−7.02
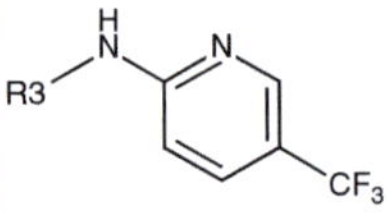	Positive	−5.89	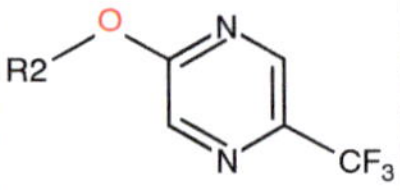	Negative	−7.08
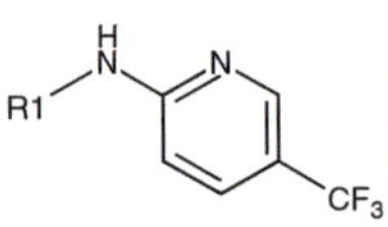	Positive	−5.89	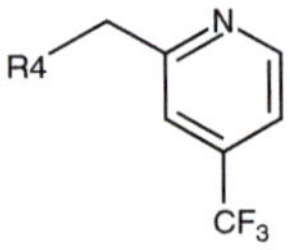	not synthetized	−7.09
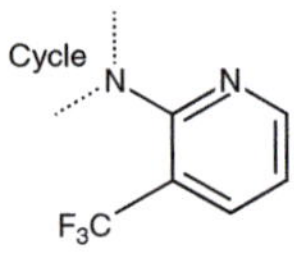	Negative	−6.02	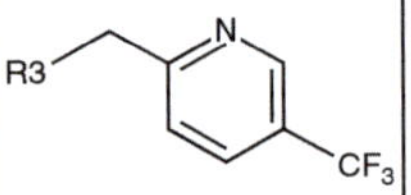	not synthetized	−7.10

These simple descriptors helped to prioritize molecules before synthesis and mitigate the formation of reactive metabolites

References

1. Waring MJ, Arrowsmith J, Leach AR et al (2015) An analysis of the attrition of drug candidates from four major pharmaceutical companies. Nat Rev Drug Discov 14:475–486
2. Arodola OA, Soliman MES (2017) Quantum mechanics implementation in drug-design workflows: does it really help? Drug Des Devel Ther 11:2551–2564
3. Dewar MJS, Thiel W (1977) Ground states of molecules. 38. The MNDO method. Approximations and parameters. J Am Chem Soc 99:4899–4907
4. Dewai MJS, Storch DM (1985) Comparative tests of theoretical procedures for studying chemical reactions 1. J Am Chem Soc 107:3898–3902
5. Stewart JJP (1989) Optimization of parameters for semiempirical methods I. method. J Comput Chem 10:209–220
6. Stewart JJP (2007) Optimization of parameters for semiempirical methods V: modification of NDDO approximations and application to 70 elements. J Mol Model 13:1173–1213
7. Rocha GB, Freire RO, Simas AM et al (2006) RM1: a reparameterization of AM1 for H, C, N, O, P, S, F, Cl, Br, and I. J Comput Chem 27:1101–1111
8. Stewart JJP (1990) MOPAC: A semiempirical molecular orbital program. J Comput Aided Mol Des 4:1–103
9. Inc. S (2019) AMPAC 10, www.semichem.com. Accessed 15 Aug 2019
10. Dewar M (2019). https://en.wikipedia.org/wiki/Michael_J._S._Dewar. Accessed 15 Aug 2019
11. Saebo S (1993) Local treatment of electron correlation. Annu Rev Phys Chem 44:213–236
12. Hohenberg P, Kohn W (1964) Inhomogeneous electron gas. Phys Rev 136:B864–B871
13. Field MJ, Bash PA, Karplus M (1990) A combined quantum mechanical and molecular mechanical potential for molecular dynamics simulations. J Comput Chem 11:700–733
14. Friesner RA (2005) Ab initio quantum chemistry: methodology and applications. Proc Natl Acad Sci U S A 102:6648–6653
15. Hansch C, Leo A, Unger SH et al (1973) Aromatic substituent constants for structure-activity correlations. J Med Chem 16:1207–1216
16. Fujita T, Iwasa J, Hansch C (1964) A new substituent constant, ir, derived from partition coefficients. J Am Chem Soc 86:5175–5180
17. Iwasa J, Fujita T, Hansch C (1965) Substituent constants for aliphatic functions obtained from partition coefficients. J Med Chem 8:150–153
18. Lipinski CA, Lombardo F, Dominy BW et al (1997) Experimental and computational approaches to estimate solubility and permeability in drug discovery and development settings. Adv Drug Deliv Rev 23:3–25
19. Leeson PD, Springthorpe B (2007) The influence of drug-like concepts on decision-making in medicinal chemistry. Nat Rev Drug Discov 6:881–890
20. Hill AP, Young RJ (2010) Getting physical in drug discovery: a contemporary perspective on solubility and hydrophobicity. Drug Discov Today 15:648–655
21. Young RJ, Green DVS, Luscombe CN et al (2011) Getting physical in drug discovery II: the impact of chromatographic hydrophobicity measurements and aromaticity. Drug Discov Today 16:822–830
22. Gleeson MP (2008) Generation of a set of simple, interpretable ADMET rules of thumb. J Med Chem 51:817–834
23. Veber DF, Johnson SR, Cheng H et al (2002) Molecular properties that influence the oral bioavailability of drug candidates. J Med Chem 45:2615–2623
24. Klopman G, Tonucci DA, Holloway M et al (1984) Relationship between polarographic reduction potential and mutagenicity of nitroarenes. Mutat Res 126:139–144
25. Debnath AK, Debnath G, Hansch C et al (1991) Structure-activity relationship of mutagenic aromatic and Heteroaromatic nitro compounds. Correlation with molecular orbital energies and hydrophobicity. J Med Chem 34:786–797
26. Hastner H (2011) Mitigation of reactive metabolism formation in lead optimization: a case study. Paper presented at the drug Methabolism discussion, Robinson College,Cambridge (UK) 14–16 September 2011
27. Niazi A, Jameh-Bozorghi S, Nori-Shargh D (2008) Prediction of toxicity of nitrobenzenes using ab initio and least squares support vector machines. J Hazard Mater 151:603–609
28. Ghose AK, Crippen GM (1986) Atomic physicochemical parameters for three-dimensional structure-directed quantitative structure-activity relationships I. partition coefficients as a measure of hydrophobicity. J Comput Chem 7:565–577

29. Moriguchi I, Hirono S, Liu Q et al (2011) Simple method of calculating Octanol/water partition coefficient. Chem Pharm Bull (Tokyo) 40:127–130
30. Cheng T, Zhao Y, Li X et al (2007) Computation of octanol-water partition coefficients by guiding an additive model with knowledge. J Chem Inf Model 47:2140–2148
31. Leo AJ, Hoekman D (2000) Calculating log P (oct) with no missing fragments; the problem of estimating new interaction parameters. Perspect Drug Discov Des 18:19–38
32. Shultz MD (2019) Two decades under the influence of the rule of five and the changing properties of approved oral drugs. J Med Chem 62:1701–1714
33. Chuman H, Mori A, Tanaka H et al (2004) Analyses of the partition coefficient, log P, using ab initio MO parameter and accessible surface area of solute molecules. J Pharm Sci 93:2681–2697
34. Tapia O, Goscinski O (1975) Self-consistent reaction field theory of solvent effects. Mol Phys 29:1653–1661
35. Klamt A, Schüürmann G (1993) COSMO: a new approach to dielectric screening in solvents with explicit expressions for the screening energy and its gradient. J Chem Soc Perkin Trans 2:799–805
36. Bannan CC, Burley KH, Chiu M et al (2016) Blind prediction of cyclohexane–water distribution coefficients from the SAMPL5 challenge. J Comput Aided Mol Des 30:927–944
37. Klamt A, Eckert F, Arlt W (2010) COSMO-RS: an alternative to simulation for calculating thermodynamic properties of liquid mixtures. Annu Rev Chem Biomol Eng 1:101–122
38. Hospital RH (1997) Ab initio molecular modeling in the study of drug. Eur J Drug Metab Pharmacokinet 44:1–13
39. Segall MD, Payne MC, Ellis SW et al (1998) An ab initio approach to the understanding of cytochrome P450-ligand interactions. Xenobiotica 28:15–20
40. Segall MD, Payne MC, Ellis SW et al (1998) First principles calculation of the activity of cytochrome P450. Phys Rev E 57:4618–4621
41. Jones JP, Mysinger M, Korzekwa KR (2002) Computational models for cytochrome P450: a predictive electronic model for aromatic oxidation and hydrogen atom abstraction. Drug Metab Dispos 30:7–12
42. Korzekwa KR, Jones JP, Gillette JR (1990) Theoretical studies on cytochrome P-450 mediated hydroxylation: a predictive model for hydrogen atom abstractions. J Am Chem Soc 112:7042–7046
43. Grogan J, DeVito SC, Pearlman RS et al (1992) Modeling cyanide release from nitriles: prediction of cytochrome P450 mediated acute nitrile toxicity. Chem Res Toxicol 5:548–552
44. Harris JW, Jones JP, Martin JL et al (1992) Pentahaloethane-based chlorofluorocarbon substitutes and halothane: correlation of in vivo hepatic protein Trifluoroacetylation and urinary Trifluoroacetic acid excretion with calculated enthalpies of activation. Chem Res Toxicol 5:720–725
45. Yin H, Anders MW, Korzekwa KR et al (1995) Designing safer chemicals: predicting the rates of metabolism of halogenated alkanes. Proc Natl Acad Sci 92:11076–11080
46. Korzekwa KR, Swinney DC, Trager WF (1989) Isotopically labeled chlorobenzenes as probes for the mechanism of cytochrome P-450 catalyzed aromatic hydroxylation. Biochemistry 28:9019–9027
47. Rydberg P, Gloriam DE, Zaretzki J et al (2010) SMARTCyp: a 2D method for prediction of cytochrome P450-mediated drug metabolism. ACS Med Chem Lett 1:96–100
48. Schäfer A, Horn H, Ahlrichs R (1992) Fully optimized contracted Gaussian basis sets for atoms Li to Kr. J Chem Phys 97:2571–2577
49. Hehre WJ, Ditchfield K, Pople JA (1972) Self-consistent molecular orbital methods. XII. Further extensions of gaussian-type basis sets for use in molecular orbital studies of organic molecules. J Chem Phys 56:2257–2261
50. Hariharan PC, Pople JA (1973) The influence of polarization functions on molecular orbital hydrogenation energies. Theor Chim Acta 28:213–222
51. Francl MM, Pietro WJ, Hehre WJ et al (1982) Self-consistent molecular orbital methods. XXIII. A polarization-type basis set for second-row elements. J Chem Phys 77:3654–3665
52. Rulíšek L, Jensen KP, Lundgren K et al (2006) The reaction mechanism of iron and manganese superoxide dismutases studied by theoretical calculations. J Comput Chem 27:1398–1414
53. McLean AD, Chandler GS (1980) Contracted Gaussian basis sets for molecular calculations. I. Second row atoms, Z=11-18. J Chem Phys 72:5639–5648
54. Krishnan R, Binkley JS, Seeger R et al (1980) Self-consistent molecular orbital methods. XX. A basis set for correlated wave functions. J Chem Phys 72:650–654
55. Hennemann M, Friedl A, Lobell M et al (2009) Cypscore: qunantitiative predication of reactivity toward cytochromes P450 based on

semiempirical molecular orbil theory. ChemMedChem 4(4):657–669

56. Finkelmann AR, Göller AH, Schneider G (2017) Site of metabolism prediction based on ab initio derived atom representations. ChemMedChem 12:606–612
57. Zaretzki J, Matlock M, Swamidass SJ (2013) XenoSite: accurately predicting CYP-mediated sites of metabolism with neural networks. J Chem Inf Model 53:3373–3383
58. de Groot MJ, Kirton SB, Sutcliffe MJ (2010) In Silico methods for predicting ligand binding determinants of cytochromes P450. In: Frontiers in medicinal chemistry—(Volume 3), pp 615–652
59. Harvey JN, Bathelt CM, Mulholland AJ (2006) QM/MM modeling of compound I active species in cytochrome P450, cytochrome C peroxidase, and ascorbate peroxidase. J Comput Chem 27:1352–1362
60. Bathelt CM, Ridder L, Mulholland AJ et al (2003) Aromatic hydroxylation by cytochrome P450: model calculations of mechanism and substituent effects. J Am Chem Soc 125:15004–15005
61. Lonsdale R, Harvey JN, Mulholland AJ (2010) Inclusion of dispersion effects significantly improves accuracy of calculated reaction barriers for cytochrome P450 catalyzed reactions. J Phys Chem Lett 1:3232–3237
62. Bathelt CM, Mulholland AJ, Harvey JN (2008) QM/MM modeling of benzene hydroxylation in human cytochrome P450 2C9 †. J Phys Chem A 112:13149–13156
63. Bathelt CM, Zurek J, Mulholland AJ et al (2005) Electronic structure of compound I in human isoforms of cytochrome P450 from QM/MM modeling. J Am Chem Soc 127:12900–12908
64. Elenewski JE, Hackett JC (2015) Ab initio dynamics of the cytochrome P450 hydroxylation reaction. J Chem Phys 142:064307
65. Car R, Parrinello M (1985) Unified approach for molecular dynamics and density-functional theory. Phys Rev Lett 55:2471–2474
66. Schlichting I, Berendzen J, Chu K et al (2000) The catalytic pathway of cytochrome p450cam at atomic resolution. Science 287:1615–1622
67. Hughes TB, Miller GP, Swamidass SJ (2015) Site of reactivity models predict molecular reactivity of diverse chemicals with glutathione. Chem Res Toxicol 28:797–809
68. Klopman G, Tu M, Fan BT (1999) META 4. Prediction of the metabolism of polycyclic aromatic hydrocarbons. Theor Chem Accounts 102:33–38

Chapter 19

Design and SAR Analysis of Covalent Inhibitors Driven by Hybrid QM/MM Simulations

Alessio Lodola, Donatella Callegari, Laura Scalvini, Silvia Rivara, and Marco Mor

Abstract

Quantum mechanics/molecular mechanics (QM/MM) hybrid technique is emerging as a reliable computational method to investigate and characterize chemical reactions occurring in enzymes. From a drug discovery perspective, a thorough understanding of enzyme catalysis appears pivotal to assist the design of inhibitors able to covalently bind one of the residues belonging to the enzyme catalytic machinery. Thanks to the current advances in computer power, and the availability of more efficient algorithms for QM-based simulations, the use of QM/MM methodology is becoming a viable option in the field of covalent inhibitor design. In the present review, we summarized our experience in the field of QM/MM simulations applied to drug design problems which involved the optimization of agents working on two well-known drug targets, namely fatty acid amide hydrolase (FAAH) and epidermal growth factor receptor (EGFR). In this context, QM/MM simulations gave valuable information in terms of geometry (i.e., of transition states and metastable intermediates) and reaction energetics that allowed to correctly predict inhibitor binding orientation and substituent effect on enzyme inhibition. What is more, enzyme reaction modelling with QM/MM provided insights that were translated into the synthesis of new covalent inhibitor featured by a unique combination of intrinsic reactivity, on-target activity, and selectivity.

Key words Quantum mechanics/molecular mechanics (QM/MM), Covalent inhibitor, Drug design, SAR, FAAH, EGFR

1 Introduction

The recognition process involving a drug and a target leads to the formation of a molecular complex capable of triggering a series of biochemical events that culminate in the insurgence of a therapeutic effect. The amplitude of this favorable pharmacological response is primarily dependent on the fraction of the molecular target occupied by the drug [1]. In condition of equilibrium, thermodynamics provides the driving force for drug binding, i.e., the protein–ligand complex is formed because it possesses a lower free energy than the protein and the ligand taken on their unbound

Alexander Heifetz (ed.), *Quantum Mechanics in Drug Discovery*, Methods in Molecular Biology, vol. 2114, https://doi.org/10.1007/978-1-0716-0282-9_19,

state. However, in condition of nonequilibrium such as in cells or in living organisms, not only thermodynamics matters [2]. In tissues, where small molecules are subject to metabolic transformations and drug targets are exposed to high concentration of their physiological substrates, the kinetic stability of the protein–ligand complex becomes important for observing the pharmacological response [3]. Specifically, the residence time of the small molecule at its target should be long enough to trigger those biochemical steps which are required for the insurgence of the pharmacological effect [4]. If this is not the case, hardly a small molecule will exert a biological effect when administered in vivo [5]. A practical way to increase the residence time of a small molecule at its target is to install on it a reactive group able to form a stable covalent bond with the receptor. If this new covalent bond is protected by solvolysis or, more generally, if it possesses high chemical stability, the drug target will likely result occupied for its entire life, with a significant advantage in terms of efficacy and duration of action [6]. Irreversible binding through covalent addition to the target is probably the most powerful strategy to obtain a sustained response in vivo since release from inhibition requires the resynthesis of the engaged target [7].

Two different strategies are currently applied to identify covalent inhibitors for a given target [8]. Both rely on the presence of a nucleophile, either member of a catalytic machinery or simply with a structural role, in the binding pocket of the protein of interest. In the first approach, briefly cited earlier, a scaffold of a reversible inhibitor already featured by a good affinity for the target is identified and an electrophile warhead is inserted on its structure [9]. If the electrophile is positioned close to the nucleophile with a geometry not too far from that of the hypothetical *transition state* for the reaction, a covalent bond can be formed within a reasonable time interval. A second approach used to identify a covalent inhibitor is based on the screening of libraries of electrophilic compounds of relatively small size. Once a covalent modifier is identified (either using mass spectrometry or biochemical assays), an optimization phase takes place. During this task, the nonreactive portion of the inhibitor (often called driver portion) is progressively modified to improve the stereo-electronic complementarity between the target and the inhibitor with the final aim of improving the potency [10]. Whether a covalent inhibitor is identified through rational design or in a screen, other runs of optimization are required to properly balance potency and selectivity. Again, this task can be done by modifying the driver portion of the inhibitor or altering the intrinsic reactivity of the warhead toward target nucleophile [11].

In the context of discovery and optimization of covalent agents, computer simulations played and are going to play a significant role both in academia and industrial settings [12, 13]. Ligand-based and structure-based computer-aided drug design techniques [14] have been successfully used to design new covalent inhibitors without explicitly accounting for their chemical mechanism of action [15, 16]. Most of these applications work under the hypothesis that driver portions can be optimized independently from the inhibitor warhead [17]. This condition is expected to be valid only when modifications in the driver portion do not have an impact on the dynamic of the covalent reaction or on the stability of the reaction product. In all the other cases, a useful computational method should take into account the effect of chemical substitution both on the recognition and on the reactivity phases [18]. Only recently, a number of computational approaches aimed at driving the design and optimization of covalent inhibitors have been reported in the literature [19, 20]. Among these, hybrid quantum mechanics/molecular mechanics (QM/MM) method [21] is probably the only one that allows to characterize the mechanism of action of a covalent inhibitor (see, for instance, references [22–27]) accounting for the accommodation of the driver portion and for the chemical reaction occurring between a nucleophile on the given target and the inhibitor warhead [28]. This is possible as the QM/MM approach combines the accuracy of QM methods in describing chemical transformation with the ability of MM force fields to provide a reasonable description of conformational energetics and nonbonded interactions in condensed systems [29].

In this chapter, after summarizing the basics of the QM/MM approach and the theoretical strategy to compute energy barriers in the context of enzyme-catalyzed reaction, we will report two case studies where these methods have been applied to elucidate the mechanism of action of reference inhibitors on drug targets of current pharmacological or clinical interest. These are fatty acid amide hydrolase (FAAH [30]), one of the key enzymes involved in the metabolism of endocannabinoids and neuromodulatory fatty acid ethanolamides [31] and epidermal growth factor receptor (EGFR [32]), a tyrosine kinase receptor controlling cell proliferation and tissue renewal. Eventually, we show how the QM/MM approach can be used in perspective drug design.

2 QM/MM Method

2.1 Definition of the Methodology

In QM/MM methodology, QM and a MM theory are coupled together to define a convenient QM/MM hybrid Hamiltonian that, applied in combination to geometry optimization, molecular dynamics (MD), or Monte Carlo (MC) sampling, is able to describe chemical reactions occurring in molecular systems composed by thousands of atoms [33, 34]. Other QM-based approaches have been generated to investigate chemical reactions in condensed phase, such as the empirical valence bond (EVB) method developed by Warshel [35], in which reactions are described through mixing states corresponding to classical valence-bond structures describing reactants, intermediates, and products, and Car–Parrinello MD (CPMD [36]), where the valence electrons of molecular structures are described by plane-wave basis sets rather than atom-centered ones. For more information on these methodologies, the reader is invited to look at more specialized reviews [37, 38].

In the QM/MM approach reported here, whose theory was firstly described by Warshel and Levitt in 1976 [39] and then refined by Karplus and coworkers [40], the system is divided in two regions of interests, namely, *reactive region* and *boundary region*. The *reactive region* of the active site is described at QM level, either based on the Hartree–Fock formalism or on the density functional theory (DFT) [41]. A practical choice is to include in this QM region all the atoms participating to the breakage and forming of covalent bonds. A molecular mechanics force field is instead used to describe the *boundary region*, i.e., the rest of the system not directly involved in the chemical transformation, but which influences the *reactive region* through nonbonded interactions [42]. The interaction between the *reactive region* (QM) and the *boundary region* (MM) has to be taken into account at any stage of the calculation. This is accomplished with the use of the following expression (Eq. 1) which includes three key terms [43]:

$$H = H_{\mathrm{QM}} + H_{\mathrm{QM/MM}} + H_{\mathrm{MM}} \tag{1}$$

where the H_{QM} is the Hamiltonian of *reactive region* treated at QM level, $H_{\mathrm{QM/MM}}$ is the Hamiltonian that couples *reactive* and *boundary region*, and H_{MM} is the Hamiltonian of *nonreactive region*, treated at classical level. In principle, the quantum region can be described by any QM method. For instance, semiempirical (SE) methods derived from Hartree–Fock theory, such as AM1 and PM3, can be used with the advantage of allowing very large QM systems (up to hundreds of atoms) to be treated [44]. However, these SE methods have known issues which may give a suboptimal description of the reaction energetics. More reliable, but highly expensive, calculations can be performed using electron correlation methods. These include MP2 perturbation methods and coupled-cluster theory approaches [45]. A computationally less demanding alternative is represented by the use of DFT methods. In this context, the self-consistent charge density functional tight binding (SCC-DFTB) approach [46], including its more recent DFTB3 variant [47], has been proposed as a fair alternative to expensive ab initio methods [48], at least for free-energy simulations where an extensive sampling of the molecular system under study is required to obtain meaningful information on the reaction mechanism. Although DFTB-based approaches are affected by systematic errors [49] (i.e., with proton affinities [50], H-bond energetics [51], and dispersion forces [52]), they have often provided reasonable results [53], in fair agreement with calculations performed at higher level of theory [54] or with experimental data [55, 56].

Specialized protein force fields including CHARMM [57] and AMBER [58] are routinely used to describe atoms belonging to the *boundary* regions, thus accounting for the H_{MM} Hamiltonian. With limitations due to the empirical nature of the mathematical functions employed to describe bonding and nonbonding interactions among atoms and the local validity of parameters used to build and calibrate these functions, MM force fields can provide a reasonable description of the conformational energetics of macromolecules and of nonbonded interactions in large systems [59]. For these reasons, they are applied in nearly all available QM/MM implementations.

As anticipated earlier, a QM/MM hybrid scheme must combine the quantum and classical parts to obtain a meaningful Hamiltonian. In general, the $H_{\mathrm{QM/MM}}$ term accounts for the QM/MM interaction energy, which involves (1) electrostatic interactions and (2) van der Waals interactions between atoms of the QM and of the MM regions. Based on how the QM and MM parts are coupled and therefore on how electrostatic interactions are computed, three different QM/MM approaches can be identified [60]. In the first one, known as *mechanical coupling*, the QM Hamiltonian is not influenced by the point charges of the MM system. Instead, a force field is used to account for the interactions between the QM and MM regions. Atom types are assigned also to the QM atoms, and the QM/MM interaction energy between these two regions is calculated by simply summing the electrostatic energy and the Van der Waals contribution, simply using the Coulomb law and the Lennard-Jones potential, respectively.

The second one is based on the *electrostatic coupling* approach, where the electrostatic interaction between the MM and the QM regions is obtained through the inclusion of classic charges belonging to MM atoms directly in the QM Hamiltonian, i.e., by incorporating MM atomic charges directly in one-electron integrals. MM atomic partial charges also interact with the nuclei of the atoms in the QM system. The electrostatic interaction of QM electrons with MM point charges moves from the $H_{\mathrm{QM/MM}}$ term of the *mechanical coupling* scheme (where it was described with a force field) to the quantum Hamiltonian, H_{QM} according to Eq. 2 [60]:

$$\hat{H}^{\mathrm{el}}_{\mathrm{QM(MM)}} = -\sum_{i}^{N} \sum_{J\in\Phi}^{L} \frac{q_J}{|\boldsymbol{r}_i - \boldsymbol{R}_J|} + \sum_{\alpha\in I+L}^{M} \sum_{J\in\Phi}^{L} \frac{q_J Q_\alpha}{|\boldsymbol{R}_\alpha - \boldsymbol{R}_J|} \quad (2)$$

where the q_J are the MM point charges separated by the R_J distance, Q_α are the nuclear charges of the QM atoms at R_α distance, and r_i describes electron positions. The indices i, J, and α run over the N electrons, L point charges, and M QM nuclei, respectively.

This electrostatic coupling is probably the most popular embedding scheme used today, at least in the context of biomolecular simulations [61], as it allows to account for the electrostatic influence of MM atoms on the QM region at a reasonable computational cost. The inclusion of this effect is likely to be important for many enzymes, given the polar nature of several enzyme active sites. The third approach is based on the *polarization coupling* scheme, in which classical charges of the MM atoms are not fixed but change on the base of the electrostatic potential generated by the QM core. The QM and MM regions experience a mutual polarization process, which improves the description of the electrostatic forces [62] but at a greater computational cost.

In all the three coupling schemes described earlier, QM/MM van der Waals interactions (representing dispersion and exchange repulsion interactions between QM and MM atoms) are calculated by MM. This requires that MM van der Waals parameters have to be assigned to each QM atom. This procedure is critical as van der Waals interactions are important at short range where they affect both QM/MM interaction energies and reactant geometries [63]. A known limitation of most QM/MM implementations is that the same set of van der Waals parameters is used for the QM atoms during a simulation. The nature of the groups involved in a chemical reaction (treated by QM) is subject to change as the reaction proceeds and the van der Waals parameters assigned to the QM atoms should be modified accordingly [64]. The effect of using different van der Waals parameters in QM/ MM simulations has been investigated. It was concluded that thermodynamic quantities such as the potential of mean force (PMF) are minorly affected by the van der Waals parameters. The improvement of the consistency of QM/MM methods for predicting energetic properties should thus focus on other factors [21].

2.2 Exploration of Reaction Mechanism

In the case of small systems where the use of a hybrid QM/MM approach is not required, QM alone can be used to characterize the potential energy surface (PES) of a chemical process by means of geometry optimization, coupled with calculations of second derivatives. This approach allows to identify stationary points along a minimum-energy path (MEP) and to roughly estimate their free energy after correcting potential energy for zero-point vibration as well as for thermal and entropy effects [65].

In this condition, the application of the *transition state theory* can allow to compare the calculated activation barrier to that one derived from the experimentally measured reaction rate [66]. If the barrier for the proposed mechanism is considerably larger than the experimental one, such a mechanism should not be trusted. On the same line, a mechanism with a calculated barrier comparable to the apparent experimental barrier is more plausible.

This workflow requires direct calculation, storage, and manipulation of the Hessian matrix, which contains the topological information of the PES of interest. It becomes extremely difficult to handle with molecular systems with many degrees of freedom [67]. In the context of QM/MM simulation, the exploration of the PES for a given mechanism can be performed using the adiabatic mapping approach, which does not require calculation of second derivatives. According to this method, the energy of the system is calculated after optimizing the structure of the reactants at a series of harmonically restrained values of a reaction coordinate, describing the process of interest such as the distance between two atoms. This approach is only valid if one conformation of the protein can represent the state of the system at a particular value of the reaction coordinate [68]. This is expected to be true when negligible structural changes occur at the active site with the progression of the reaction [69]. Only in this very specific condition does the minimization of the potential energy along the specified reaction coordinate provide a reasonable approximation of the "enthalpic component" of the free-energy profile for the considered reaction [70].

In other situations, the conditions described earlier are not satisfied, and structural fluctuations of the protein active site including those of the substrate may represent essential movements of the catalytic process [71]. The use of multiple structures as a starting point for reaction modelling offers a partial solution to the limited structural flexibility allowed by the adiabatic mapping approach. For a given reaction path, multiple PESs can be explored and several barrier heights can be computed and compared [72]. Depending on the type of statistical distribution, a convenient averaging method can be used to obtain a representative barrier value [73]. In this condition, it is also possible to estimate the uncertainty of the computed barrier which is critical when different reaction mechanisms are compared [74]. Regardless of the statistical robustness of the computed potential energy barrier, QM/MM calculations based on energy minimization remains potentially erratic. For instance, adiabatic mapping overestimates energy barriers if atom movements connected with the reaction are not included in the definition of the reaction coordinate, essentially because minimization alone cannot resolve steric strains caused by the progression of the system along the given reaction coordinate [75]. Despite these limitations, adiabatic mapping is useful for preliminary exploration of the PES, including the generation of models of transition states and intermediates. Several pioneering QM/MM works on enzyme reaction modelling were conducted with this approach and still preserve their validity [21, 60, 68].

2.3 Calculation of Potential of Mean Force

Computational techniques that sample multiple configurations of the molecular system along a given path have the capability to contribute a better description of a chemical transformation [76]. These approaches account for the conformational space explored by reactants, TSs, and products, as well as for solvent reorganization, during the chemical transformation of interest [77], allowing to compute the variation of the free energy of the system as a function of the progression of the reaction. The resulting free-energy profile, which is computed along a well-defined reaction coordinate, represents the PMF of the simulated process. Its knowledge is critical for enzyme reaction modelling as it allows to directly retrieve the activation free energy for the chemical transformation of interest. Thanks to the application of the *transition state theory,* this computed quantity can be meaningfully compared to the experimental barrier for the enzyme-catalyzed reaction under study [78, 79].

In principle, QM/MM simulations coupled with MD might allow to calculate activation free energy. However, classical MD does not explore phase space regions with high energy, as for transition states or transient intermediates. MC methods overcome this limitation [80], but the number of QM/MM MC iterations necessary to obtain converged properties is extremely high, making the overall process computationally demanding.

Conformational sampling of processes describing chemical reactions requires enhanced sampling (ES) methods in which an external bias is applied to sample high-energy region. Umbrella sampling (US) is one of the methods capable to achieve this task [81]. In this technique, a biasing potential (the "umbrella" potential) is applied to force the system to remain at a specific value of a given reaction coordinate [82]. Furthermore, this bias makes sure that all regions along the reaction coordinate are sampled with similar probability [83].

QM/MM US simulations are performed along a chemical path connecting reactants and products (or stable intermediate). Very often, US simulation begins with QM/MM MD of the Michaelis complex with a bias potential applied to restrain the reaction coordinate to a value corresponding to the reactant state. In subsequent US simulations, the reference value of the restraint is increased by a small quantity to ensure the progression along the reaction coordinate while allowing the sampling of the available conformational space for a relatively short time interval, routinely 50–100 ps with the current computational capability. As for adiabatic mapping methods (see previous texts), the reaction coordinate is often defined in terms of distances involving bond lengths. In US simulations, a typical difference between two consecutive points (or windows) is in the range of 0.1–0.2 Å. The reaction coordinate values assumed at each window of the US simulation are recorded. At the end of the whole process, the effects of the restraining potentials are removed and combined by the weighted histogram analysis method (WHAM) which gives the unbiased PMF along the reaction coordinate [84]. To obtain a continuous PMF for the modelled reaction, it is important that neighboring US windows produce overlapping distributions in terms of dispersion of the explored value of the reaction coordinate. This condition can be met by conveniently tuning the magnitude of the spring constant and/or the step size of the reaction coordinate. As with other sampling methods, it is important to test the convergence of the PMF with respect to the length of the simulation. It should be stressed that while US allows to sample multiple configurations of the system along a given path, it will fail to correctly estimate the free-energy difference between the two states if a major rearrangement of the conformation of the protein occurs during the reaction especially if this conformational transition is not described by the reaction coordinate.

Sampling reaction paths with US is computationally demanding since the underlying simulation requires a very large number of QM/MM computations of the potential energy including gradients. Hence, US is used in combination with QM/MM approaches derived from SE Hamiltonian such as AM1, PDDG-PM3, and SCC-DFTB. These SE methods are featured by systematic errors that often lead to rather inaccurate energy barriers. The accuracy of QM/MM US barriers can be improved by applying a correction term derived from adiabatic mapping study performed at a higher level of theory. This practice is well documented in the literature and it has been shown to give reaction barriers in close agreement to the experimental ones [21, 60].

Other ES methods have been used in the context of QM/MM simulations. Nonequilibrium methods are viable alternatives to US sampling and can provide rather rapid (but coarse) estimation of the activation free energy of a chemical transformation. One example is steered MD (SMD) [85], a method in which an external force is applied to part of the system, e.g., the distance separating two atoms, to drive it along a predefined direction.

While specifically suited for investigating ligand unbinding from proteins with MM force field [86], SMD has been proposed as an efficient sampling method in the context of QM/MM simulations [87]. The dissipative work performed on the system (W) during a QM/MM SMD simulation is calculated by numerical integration of the exercised force (F_{ex}) at each value of the reaction coordinate (Eq. 3), where $dx = v dt$.

$$W_{[x(t)]} = \int_0^{x(t)} F_{ex}(t)\, dx(t) \tag{3}$$

QM/MM SMD simulations are usually carried out by sequentially pulling the system along the reaction coordinate using a harmonic potential whose center moves with constant velocity and whose Hooke's constant is few hundreds of kcal/mol $Å^{-2}$. At each QM/MM MD integration step, the equilibrium position of the harmonic potential is increased to a fixed amount. A pulling velocity in the range of 0.01–0.1 Å/ps should allow to reduce the effect of dissipative frictional work on the overall work (W) computed on the system.

SMD is useful for comparing alternative reaction mechanisms. While single SMD run can only provide a qualitative description of the change in the mechanical work performed along the reaction path, independent works calculated through multiple SMD simulations can be combined together applying the Jarzynski equality (Eq. 4), providing a means of recovering an equilibrium PMF from nonequilibrium events [88].

$$\left\langle e^{-\Delta F/k_B T} \right\rangle = \left\langle e^{-W/k_B T} \right\rangle \tag{4}$$

It should be pointed out that obtaining statistically converged and accurate PMF profiles is far from easy and may result in increased computational cost if the steering velocity and number of replicas are not properly selected [89].

Examples of QM/MM SMD simulations applied to elucidate reaction mechanism are represented by the seminal work of Adrian Roitberg on chorismate mutase [87] and by others both in the field of organic reaction in solution [90] and in that of enzyme-catalyzed processes [91]. Also metadynamics [92], a nonequilibrium ES approach primarily used to investigate binding modes for drugs [93, 94], conformation transitions of proteins [95], and unbinding events of small molecules [96] from enzyme [97] and receptors [98], has been largely used in the context of chemical reaction modelling. Several examples of its use in enzyme catalysis can be found in reference [99] as well as in more recent articles [100].

2.4 Path-Collective Variables in QM/MM Simulations

QM/MM mechanistic modelling of complex enzymatic reactions is often carried out by separating the catalytic process into a sequence of consecutive steps. This approach is viable only when stable intermediates can be identified along a given path in a way that the end point of a chemical step can be used as starting point for the subsequent one. Furthermore, the arbitrary separation of a complex chemical transformation in independent steps may prevent to capture the concerted nature of the enzyme-catalyzed reactions.

The path-collective variable (PCV) approach overcomes these issues allowing to model a multistep process in a single simulation run. When PCVs are used, a multidimensional problem is reduced to only two dimensions. The first one, usually called S, describes the *progress* of the chemical transformation, while the second one, *Z*, defines the *tolerance* versus the exploration of paths alternative to that tracked by *S* (vide infra) [101].

The variable *S* (Eq. 5) is defined as a multiparametric function dependent on the position of atoms involved in the chemical transformation with respect to a reference path connecting the reactants to the products. This path is a discrete set of arbitrary geometries (called nodes) representative of the chemical transformation of interest and can be generated by extracting a set of frames from a PES of the reaction of interest or from a series of consecutive SMD simulations in which the reactants are gradually transformed into the products using classical reaction coordinates [102].

$$S(R) = \frac{\sum_{i=1}^{P} i e^{-\lambda(R-R(i))^2}}{\sum_{i=1}^{P} e^{-\lambda(R-R(i))^2}} \tag{5}$$

In Eq. 5, *R* is a vector representing the cartesian coordinates of the atoms involved in the chemical transformation, $(R - R_{(i)})^2$ corresponds to the squared difference between the coordinates assumed by the system at a given time (*R*) and the coordinates of the *i-th* node ($R_{(i)}$) along the reference path, and λ is a tunable parameter that helps the transition from one node (*i*) to the following one (*i* + 1) and its value depends on the average distances among nodes of the reference path.

A second multiparametric variable (*Z*) is employed in combination with *S*. According to its definition (Eq. 6), *Z* is the distance from the input path and can be used to define a confinement boundary which prevents the exploration of configurations of the system which are too far from the nodes of the reference path.

$$Z(R) = -\frac{1}{\lambda} \ln\left(\sum_{i=1}^{P} e^{-\lambda(R-R(i))^2}\right) \tag{6}$$

However, as the most probable reaction path does not necessarily coincide with *Z* = 0, it is important that the bias acting on *Z* is sufficiently soft to allow some deviations from the reference path. QM/MM MD simulations performed on the *S/Z* space using US or SMD allow to calculate the PMF for the chemical transformation of interest and to verify if a lower free-energy pathway alternative to the reference one can be identified. If this is the case, a novel reference path needs to be defined. This is done by selecting a set of discrete frames from the MD trajectory along the minimum free-energy path applying the fitting procedure developed by Maragliano and Vanden-Eijnden [103]. Thus, a new QM/MM simulation in the *S/Z* space is performed along the nodes defined by the new path. This procedure is repeated iteratively until the minimum free-energy path does not change passing from a PCV simulation to a new one. The PCV approach briefly summarized here has been initially developed to describe conformational transitions in macromolecules [101]. More recently, it has been applied to model chemical reactions in solution [104] and within enzyme [105, 106].

3 QM/MM Studies on Relevant Drug Targets

3.1 FAAH

3.1.1 Catalytic Mechanism for Acylation and Substrate Selectivity

FAAH is a member of the amidase signature family responsible for the inactivation of biologically active fatty acid ethanolamides [30]. FAAH terminates the signal brought by the endocannabinoid anandamide, catalyzing its hydrolysis to arachidonic acid and ethanolamine. FAAH catalyzes the hydrolysis of other amides such as oleoylethanolamide and palmitoylethanolamide which are agonists of the peroxisome proliferator-activated receptor (PPAR) subtype alpha. Inhibition of FAAH by small molecules has emerged as a promising therapeutic strategy to treat central nervous system diseases, including anxiety, inflammation, and depression [107].

FAAH possesses a unique catalytic triad composed of two serine residues (Ser217 and Ser241) and one lysine (Lys142), rather than the serine–histidine–aspartate triad found in classical serine hydrolases. This Ser-Ser-Lys triad is responsible for the remarkable ability of this enzyme to hydrolyze amides and esters at a similar rate, by a mechanism in which acylation is the rate-limiting event [108]. As the maximum hydrolytic activity of FAAH is observed in mild-basic conditions ($pH = 9$), it is believed that Lys142 participates to catalysis in its neutral form, i.e., with Lys142 triggering a series of deprotonation events which culminate in the formation of Ser241 alcoholate anion, the key nucleophilic species of the first step of acylation [109]. Computational investigations performed at B3LYP/6-31+G(d)//PM3/CHARMM22 level by adiabatic mapping [110] and PDDG-PM3/OPLS level by FEP in combination with MC sampling [111] support this role for Lys142.

Recently, an SCC-DFTB/CHARMM27 QM/MM potential has been used to model the process of FAAH acylation both in the presence of oleamide (OA) and oleoylmethyl ester (OME), two model substrates for which experimental kcat have been measured in the same experimental conditions, allowing to test reliability of the employed QM/MM approach [55].

PES surfaces for FAAH acylation in the presence of OA and OME have been explored by adiabatic mapping, according to the mechanism proposed by McKinney and Cravatt [112]. In the first step of acylation, Lys142 activates Ser241 nucleophile which in turn attacks the substrate carbonyl carbon, with formation of an anionic tetrahedral intermediate (TI). In the second step, the positively charged Lys142 readily protonates the leaving group (LG), leading to its expulsion. The role of Lys142 on Ser241 activation

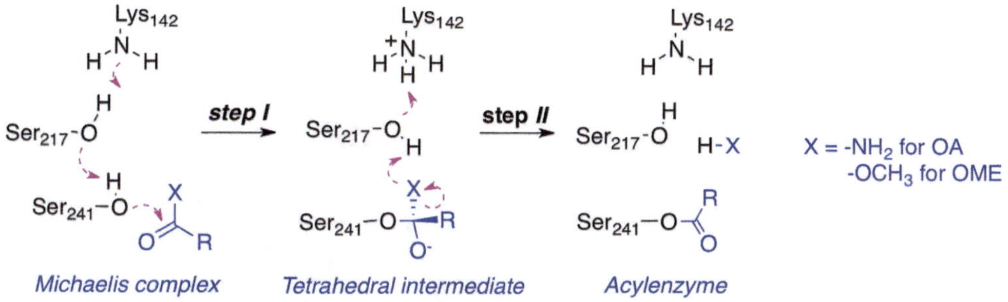

Fig. 1 Acylation mechanism of FAAH in the presence of OA and OME

and LG protonation is mediated by Ser217 which acts as key proton shuttle of the overall process (Fig. 1).

The exploration of the PESs for FAAH acylation at SCC-DFTB/CHARMM27 level indicates that the collapse of the TI is the rate-limiting step of the reaction for both OA and OME substrates, with calculated barriers of 19 and 21 kcal/mol at SCC-DFTB/CHARMM27 level. Similar findings have been obtained by Palermo et al. who modelled the hydrolysis of anandamide catalyzed by FAAH using a CPMD approach [113]. CPMD simulations identified protonation of the LG as the rate-limiting step of acylation with a free-energy barrier of 19 kcal/mol.

When SCC-DFTB/CHARMM27 energy values were averaged over multiple reaction paths ($n = 6$), the energy barriers became 21 ± 1 kcal/mol for OA and 25 ± 1 kcal/mol for OME [55], still in acceptable agreement with the experimentally deduced barriers of 16 kcal/mol for OA and 17 kcal/mol for OME, respectively. In qualitative term, the relevant experimental preference showed by FAAH in hydrolyzing OA at higher rate that OME is satisfactorily reproduced by SCC-DFTB/CHARMM27 calculations. Analysis of minimum potential energy path on the PES shows that protonation of the LG promoted by Lys142 emerged as the difficult step of the acylation for both OA and OME. The higher basicity of the nitrogen atom compared to that of oxygen in the TI configuration accounts for the lower barrier of LG protonation calculated for OA with respect to OME. This finding provides a theoretical explanation for the remarkable ability of FAAH to hydrolyze amides faster than esters.

3.1.2 Identification of Productive Binding Orientation for Covalent Inhibitors

A QM/MM approach has also been used to clarify the reaction of mechanism of a class of carbamic acid aryl esters designed to inhibit FAAH. The reference compound of the class, URB597 (the 3′-$CONH_2$ derivative of URB524) [114], was demonstrated to inhibit FAAH by carbamoylating Ser241 [115].

Molecular modelling investigation performed by our research group suggested URB524 and its derivatives, including URB597, can be docked within the FAAH catalytic site according to two

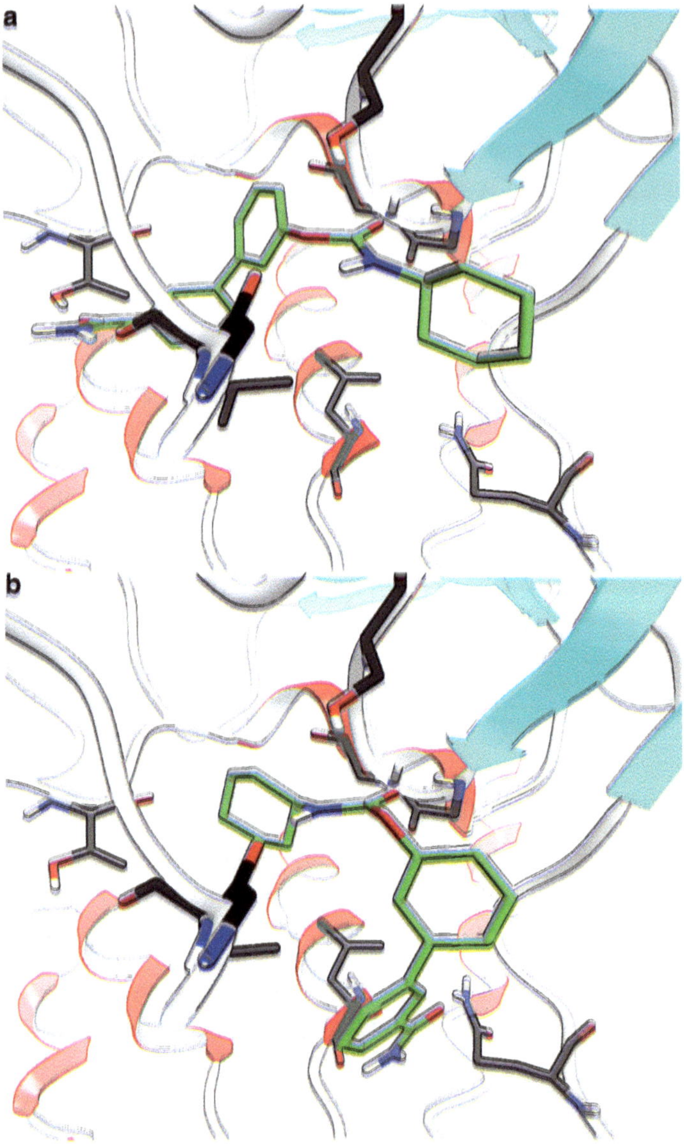

Fig. 2 Binding orientations for URB597 (green carbon atoms), docked within the active site of rat FAAH (black carton atom). Upper panel, binding orientation *I*; lower panel, binding orientation *II*

possible orientations, both placing the carbamoylating group of this inhibitor in proximity to Ser241 [116]. In the first (*binding orientation I,* Fig. 2a), the *m*-biphenyl moiety of URB524 occupied the so-called acyl-chain channel of FAAH, i.e., where the arachidonoyl portion of AEA is accommodated [117], while in the second one (*binding orientation II,* Fig. 2a), the *m*-biphenyl group lied in the cytosolic pocket of FAAH [118], i.e., where the ethanolamine head of AEA lays.

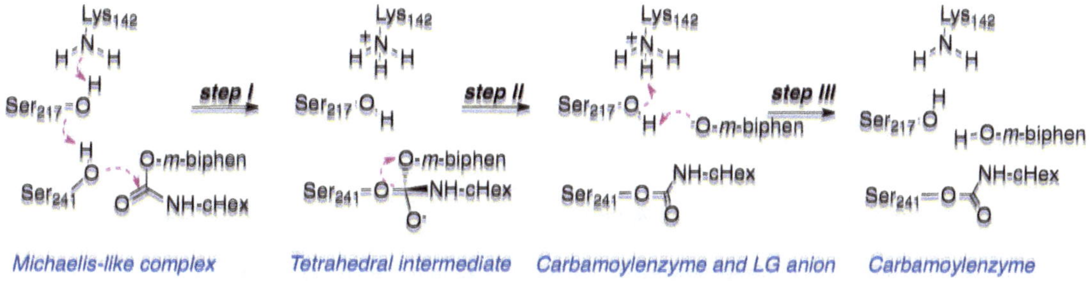

Fig. 3 Modelled mechanism of carbamoylation of FAAH Ser241 by QM/MM

A QM/MM approach was used to model the inhibitor binding process, applying a PM3/CHARMM22 potential corrected at B3LYP/6-31G+(d) level. PES surfaces of Ser241 carbamoylation were built for both binding orientations according to the reaction scheme depicted in Fig. 3.

Calculations showed that the carbamoylation in orientation *II* was energetically preferred (by nearly 15 kcal/mol at B3LYP/6-31 +G(d)//PM3/CHARMM22 level) over that one occurring in orientation *I*, proposing the second arrangement of URB524 at FAAH active site as the productive binding mode [119]. This prediction was later confirmed by the crystallographic resolution of the FAAH-URB597 structure [120]. Furthermore, in the mechanism occurring from binding orientation *II*, activation of Ser241 emerged as difficult step of the reaction (step I, Fig. 3). Nucleophilic attack performed by Ser241, which led to the formation of the TI, took place with a low barrier. Similarly, the expulsion of the biphenolate LG (step II, Fig. 3) occurred with a negligible barrier, leading to the formation of stable carbamoylated adduct. This mechanism of inhibition suggests a possible explanation on why the modulation of the electronic properties of the *O*-aryl moiety of carbamic acid aryl esters, through the insertion of electron-donor substituents in para position, has a negligible role on the inhibitory potency on FAAH [121].

3.1.3 Reversible Deacylation Versus Irreversible Decarbamoylation

URB597 has been shown by crystallography and by mass spectrometry to react with Ser241 to form a stable carbamoylated intermediate [120]. The carbamoylated form of FAAH is significantly more resistant to hydrolysis than the corresponding acylated form (produced by the reaction of the enzyme with OA or another substrate) and thus is expected to be responsible for time-dependent and persistent inhibition of FAAH in vivo.

A QM/MM approach based on adiabatic mapping calculations performed at SCC-DTFB/CHARMM27 level with B3LYP correction was used to elucidate and compare the energetics of FAAH deacylation starting from Ser241 acylated by OA versus that of FAAH decarbamoylation starting from Ser241 carbamoylated by

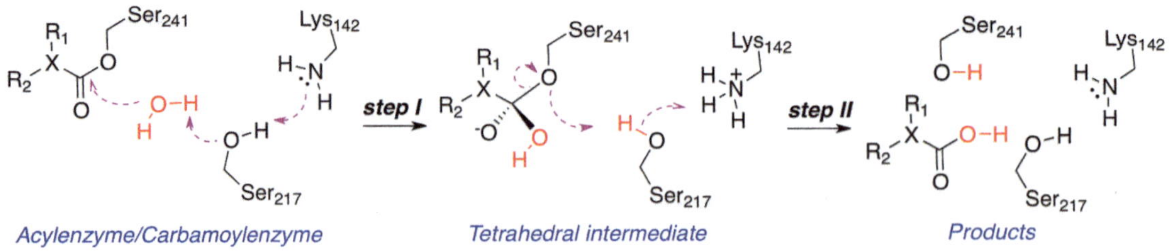

Fig. 4 Modelled mechanism of FAAH Ser241 deacylation/decarbamoylation by QM/MM

URB597 [22]. These two processes were modelled according to the reaction scheme reported in Fig. 4, where deacylation and decarbamoylation where promoted by the action of a "deacylating" water molecule identified in the X-ray structure of FAAH [120].

In the first step of the reaction, the key water molecule was activated by deprotonation, promoted by a combined action of Lys142 and Ser217. This event led to the formation of a hydroxide anion which in turn attacked the carbonyl carbon of the acylenzyme/carbamoylenzime leading to an anionic TI. In the second step of the reaction, protonation of Ser241 oxygen operated by Lys142 and Ser217 triggered the collapse of TI structure leading to expulsion of Ser241 side chain. At least in the case of the substrate, this second step was expected to restore the initial configuration of the catalytic triad. QM/MM calculations showed that the step I of FAAH deacylation occurs with a concerted mechanism, in which deprotonation of the water molecule and the nucleophilic attack at the carbonyl carbon were tightly coupled. The transition state (TS1) separating acylenzyme and TI configuration had an energy of nearly 17 kcal/mol above the reactant, while the TI itself was less stable than the reactant by 12.5 kcal/mol. Step II of the deacylation was characterized by the expulsion of Ser241 side chain from the TI, assisted by protonation reactions performed by a combined action of Ser217 and Lys142. The transition state (TS2) separating TI and reaction products (with formation of free Ser241 and oleic acid) had an energy of 14 kcal/mol, nearly 3 kcal/mol lower than that calculated for the formation of the TI. This suggested that TI formation was the rate-liming event for FAAH deacylation. Overall, reaction barriers at SCC-DFTB/CHARMM27 level are consistent with a fast deacylation and thus with an efficient FAAH catalysis in the presence of OA (Table 1) [112].

When it comes to decarbamoylation of Ser241, significantly higher energy barriers were found either for TS1 or for TS2 configurations. Formation and collapse of TI required 28.3 and 28.9 kcal/mol, respectively. As expected, decarbamoylation of FAAH resulted an unlikely event to happen. In the case of URB597, visual inspection of the TS geometry for TI formation (TI1) indicated that the N-cyclohexylcarbamoyl portion connected to Ser241 underwent a major rearrangement of its geometry

Table 1
SCC-DFTB/CHARMM27 potential energy values, expressed in kcal mol^{-1}, for key configurations identified along the deacylation path starting from acylated-(OA) or carbamoylated-FAAH

	OA	URB597
Acyl-Enz/Carb-Enz	0.0	0.0
TS1	16.6	28.3
TI	12.5	27.8
TS2	14.3	28.9
Products	8.2	9.6

during the reaction with the deacylating water molecule (Fig. 5, upper panel), while for the OA the acyl portion bound to Ser241 experienced a minimal variation of its geometry to form a covalent bond with the solvent (Fig. 5, lower panel). Active-site decomposition analysis [122] and gas-phase calculations indicate that significant distortion of the TS1 geometry observed in the case of URB594 reduces the electrostatic stabilization provided by FAAH active site during decarbamoylation accounting for the remarkable increase in the barrier for this process.

To test the validity of the computational approach, SCC-DFTB/CHARMM27 calculations were repeated using alternative starting conformations of FAAH covalent adducts. A total of eight FAAH structures extracted from a QM/MM MD trajectory were used to model deacylation or decarbamoylation at SCC-DFTB/CHARMM27 level. Although some differences were observed in the calculated energies, averaged values (Table 2) confirmed that Ser241 deacylation was significantly favored over Ser241 decarbamoylation. In light of this finding, this DFTB-based approach could be used to detect new irreversible inhibitors of FAAH acting on Ser241.

3.2 EGFR

3.2.1 Mechanism of Cys797 Alkylation by Acrylamide Inhibitors

EGFR is a transmembranal protein endowed with an extracellular EGF binding domain and a cytoplasmic domain which possesses a tyrosine kinase activity. EGFR activation promoted by EGF leads to receptor dimerization and autophosphorylation which in turn activates a signal transduction cascade promoting cell proliferation [32]. In non-small cell lung cancer (NSCLC), mutations in the kinase domain of EGFR have been observed in nearly the 50% of the cases. Of the known mutations, more than the 90% occurs in exon 19 or in exon 21: the first leading to deletion of the segment E746-A750 and the second resulting in arginine replacing leucine at position 858 (L858R) [123]. These mutations promote EGFR activation favoring the insurgence of NSCLC. The first generation of EGFR inhibitors included the 4-anilinoquinazoline gefitinib (Fig. 6) which elicited good responses in NSCLS patients

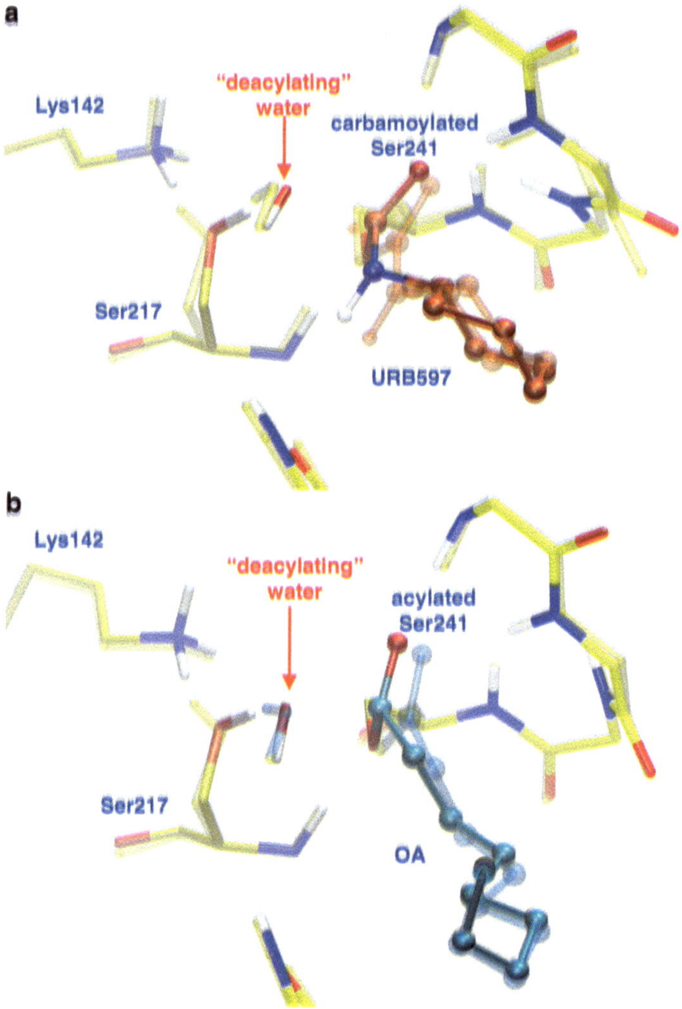

Fig. 5 Geometry of TS1 for FAAH decarbamoylation (upper panel) and FAAH deacylation (lower panel). FAAH carbon atoms are depicted in the yellow, URB597 in orange, and OA in cyan. The position of the N-cyclohexylcarbamoyl (upper panel) and oleoyl (lower panel) portion in the reactant state is displayed with a shaded ball and stick representing

Table 2
Average potential energy values ($n = 8$), expressed in kcal mol^{-1}, for key configurations identified along the deacylation path starting from acylated-(OA) or carbamoylated-FAAH

	OA	URB597
Acyl-Enz/Carb-Enz	0.0	0.0
TS1	19.1 ± 0.5	31.7 ± 0.9
TI	16.6 ± 0.9	31.1 ± 0.8
TS2	17.6 ± 0.7	31.0 ± 0.6
Products	8.9 ± 0.5	10.6 ± 0.6

[124]. However, most of them acquired drug resistance within 1-year treatment, which in several cases was due to the selection of T790 M mutation at the gatekeeper position [125]. The significant variation in the hydrophobic property of the "gatekeeper" residue is believed to be responsible for the reduced potency displayed by gefitinib and other first-generation inhibitors [126]. Second-generation of EGFR inhibitors such as afatinib (Fig. 6) demonstrated a good activity against T790M EGFR variants [127]. Thanks to the presence of an acrylamide warhead capable of alkylating Cys797, afatinib and other acrylamide-based analogues were able to circumvent ATP competition and thus to overcome the detrimental effect caused by the presence of a bulky and hydrophobic methionine.

The mechanism of EGFR wild-type inhibition operated by prototypical *N*-(4-anilinoquinazolin-6-yl)acrylamide has been recently elucidated on the basis of a computational investigation at SCC-DFTB/AMBER level using a PCV approach [128]. QM/MM simulations supported a mechanism (Fig. 7) in which an aspartate residue has a critical role in Cys797 alkylation. In details, Asp800 could firstly act as a base deprotonating Cys797-SH and then as an acid protonating the α position of the carbanion resulting from the nucleophilic attack of Cys797-S^- at α position of the acrylamide.

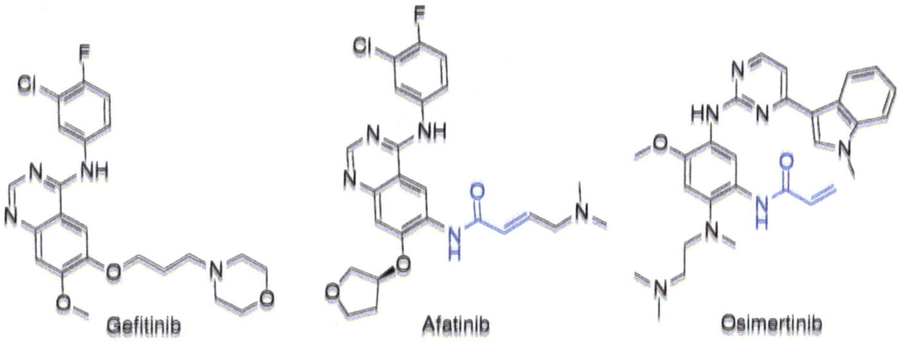

Fig. 6 Chemical structure of reference EGFR inhibitors. The acrylamide fragment in afatinib and osimertinib is highlighted in blue

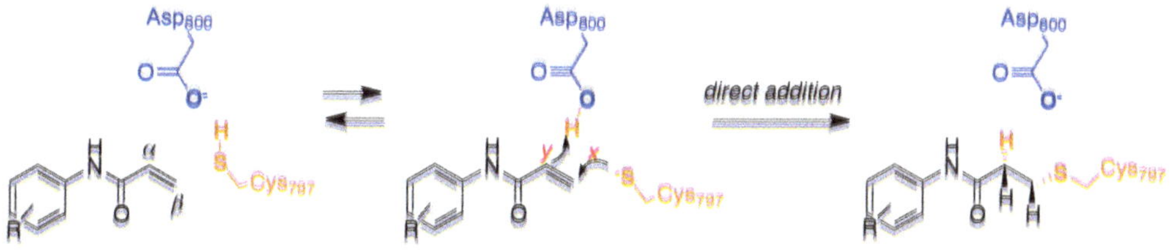

Fig. 7 Mechanism of Cys797 alkylation for acrylamide inhibitors of EGFR

QM/MM simulations indicated that the deprotonation of the thiol group of Cys797 by Asp800 carboxylate took place overcoming a free-energy barrier of 6 kcal/mol suggesting that this reaction was a fast event. Furthermore, the Cys797-S$^-$/Asp800-COOH pair was found more stable than the Cys797-SH/Asp800-COO$^-$ pair by approximately 4 kcal/mol. This suggested that the predominant form of Cys797 thiol side chain is anionic. While ab initio calculations (i.e., CCSD(T)) in gas phase indicate that SCC-DFTB underestimates the energetic cost for a cysteine residue to transfer a proton to an aspartate [129], our SCC-DFTB/AMBER results were consistent with two key experimental findings reported for EGFR such as (1) the rapid oxidation to sulfenic acid for Cys797-SH (which is favored by the presence of Cys-797 in its thiolate form) [130] and (2) titration experiments indicating that Cys797-SH has a pKa value approaching that one of a carboxylic acid [131]. We thus proposed that Cys797 participated to the reaction with its thiolate chain acting as a nucleophile. Cys797-S$^-$ approached the electrophile β carbon of the acrylamide fragment to form a new C-S bond. The acrylamide group of the inhibitor maintained a s-*cis* configuration during the entire alkylation process. The nucleophilic attack occurred in a concerted manner with the protonation of acrylamide α carbon performed by Asp800-COOH. These two events were required to overcome a free-energy barrier of nearly 15 kcal/mol to take place, in agreement with the fast alkylation of EGFR by *N*-(4-anilinoquinazolin-6-yl)acrylamide observed in a fluorescence-based assay [132].

The product of the reaction was a 3-(alkylsulfanyl)propenamide derivative which resulted significantly more stable than the reactants (~12 kcal/mol) in line with the spontaneous and irreversible alkylation of Cys797 [133, 134].

A similar mechanism of Cys797 alkylation has been recently reported for osimertinib [135], a third-generation EGFR inhibitor featured by a 2-aminopirimidine scaffold (Fig. 6). Osimertinib inhibits the activated form of EGFR regardless of the presence of T790M mutation, but compared to afatinib it spares wild-type EGFR thus avoiding severe side effects observed in patients treated with second-generation inhibitors [136].

In the QM/MM study involving osimertinib, 2D-US sampling simulation at a SCC-DFTB/AMBER level was performed along two intuitive reaction coordinates, describing nucleophilic attack by Cys797-S$^-$ and Cα protonation by Asp800-COOH, respectively. These simulations confirmed the presence of a concerted mechanism in which Cys-S$^-$ attack at the β carbon was tightly coupled with protonation at the α carbon of osimertinib acrylamide by Asp800-COOH leading to the formation of a highly stable alkylation product (Fig. 8). Recent calculations [129] indicated that, in the framework of a Michael-type reaction and in opposition to high level ab initio calculations, SCC-DFTB and other popular

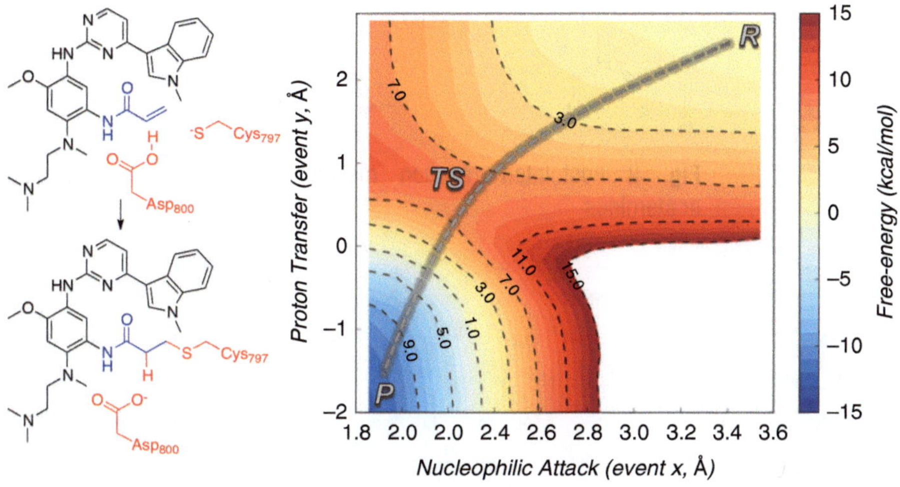

Fig. 8 Mechanism of Cys797 alkylation in EGFR T790 M by osimertinib

DFT functionals failed to identify a stable enolate intermediate as a thiolate sulfur approaches the β carbon of a prototypical acrylamide. To account for this drawback, we corrected SCC-DFTB PES at higher level. The resulting MP2/cc-pVTZ//SCC-DFTB/AMBER99SB PES for the Michael addition in EGFR indicated that the *ionic pathway*, corresponding to a stepwise reaction with formation of an enolate intermediate, and the *neutral pathway*, corresponding to the concerted reaction described earlier, have similar barriers (data not shown) suggesting that both pathways are accessible within the enzyme.

The presence of a low free-energy path for the reaction of alkylation by osimertinib (activation free energy 8 kcal/mol, reaction energy −12 kcal/mol) is likely due to the presence of a favorable environment in EGFR able to stabilize Cys797 in anionic form [131]. This suggests that the acrylamide itself could be effectively replaced by less electrophilic warheads thus helping the design of more selective and possibly safer inhibitors of EGFR.

3.3 Design of "Soft" Warheads Targeting Cys797 in EGFR

Warheads able to form a covalent bond with a specific cysteine while sparing other nucleophilic residues in cells may help the design of inhibitors featured by high selectivity and reduced side effects. Acetamides, if conveniently substituted, can meet this requirement possibly alkylating Cys797 by means of a S_N2 reaction (Fig. 9) while showing negligible reactivity with free cysteine in solution [137].

The reactivity of acetamide versus thiols can be modulated working on the LG. Reactivity studies with cysteine allowed us to identify (hetero)-aromatic thiols as poor LG substituents, thus suitable for our scope. Once inserted on an acetamide substituent, the resulting compounds were able to give the expected thioether

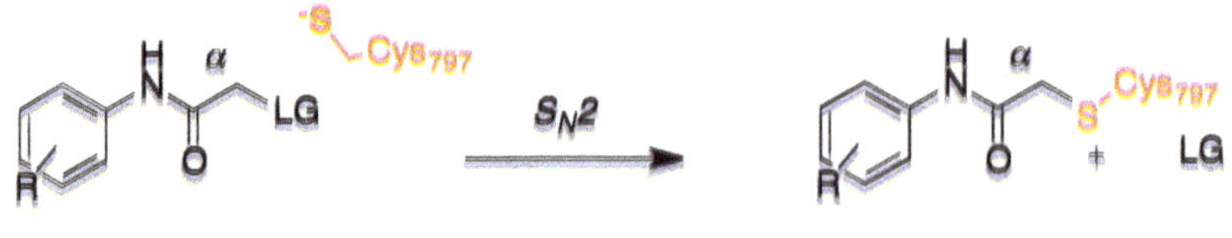

Fig. 9 Postulated reaction mechanism for *N*-(4-(anilino)quinazolin-6-yl) acetamides

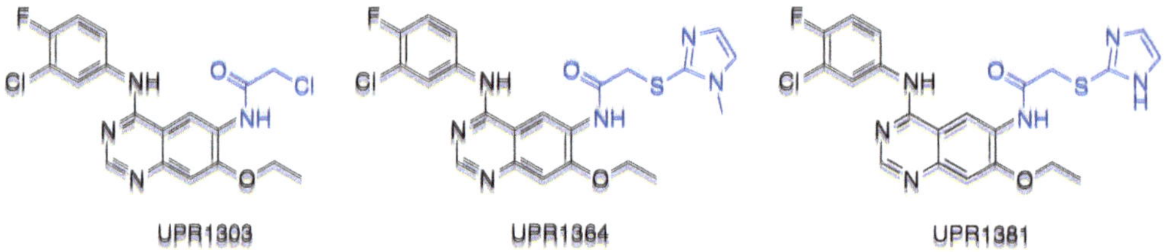

Fig. 10 Chemical structure of activated acetamides described here

product in the presence of cysteine but at a very low rate, i.e., with a half-life time ($t_{1/2}$) > 24 h. In the same experimental conditions, a reference chloroacetamide derivative displayed a $t_{1/2}$ of 1 h.

Thanks to the environment of EGFR, able to stabilize the thiolate form of Cys797, we speculated that (hetero)-aromatic thioacetamides, once installed on a high-affinity driver such as the 4-anilinoquinazoline scaffold, might be able to efficiently alkylate this specific cysteine. To support this hypothesis, we evaluated whether acetamides could undergo attack by Cys797 via a S_N2 reaction pathway using a QM/MM approach. The reaction mechanism was simulated for two 2-(imidazol-2-ylthio)acetamides (UPR1364 and UPR1381 Fig. 10) as well as for the highly reactive chloroacetamide UPR1303, employed as control.

The S_N2 reaction was modelled using PDDG/PM3/AMBER potential. The PDDG/PM3 Hamiltonian [138] was selected as it has been reported to give energy barriers for S_N2 reactions in qualitative agreement with ab initio calculations [139]. The Cys797/Asp800 pair was modelled in Cys-S-/Asp-COOH form. To simulate the nucleophilic substitution with UPR1364 and UPR1381, we applied an SMD approach applying a moving force along a reaction coordinate defined as the difference between the distance separating the acetamide Cα and the sulfur atom of the imidazol-2-ylthio group and the distance between the acetamide Cα and the sulfur atom of Cys797. In the case of UPR1303, the reaction coordinate was defined as the difference between the distance separating the acetamide Cα and the chlorine atom and the distance between the acetamide Cα and the sulfur atom of Cys797.

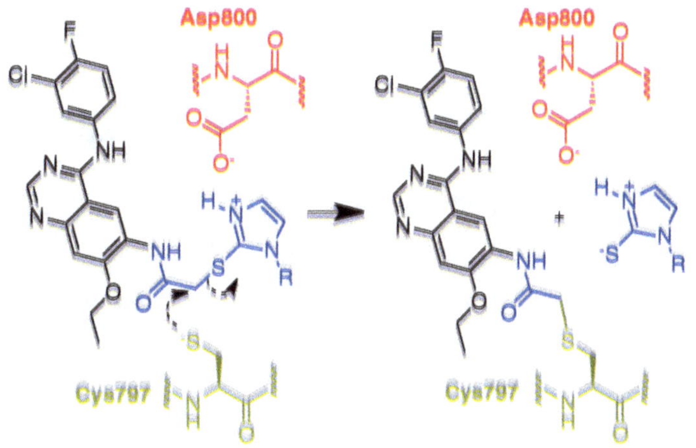

Fig. 11 Reactions modelled by QM/MM simulation for UPR1364 and UPR1381 with expulsion of a protonated imidazolylthiolate group. Asp800 was modelled as a carboxylate anion. R stands for $-CH_3$ (UPR1364) or -H (UPR1381)

Preliminary calculations showed that while for UPR1303 an alkylation product with low free energy could be identified, this was not the case for UPR1364 and UPR1381. We hypothesized that protonation of the imidazole could stabilize the imidazolylthio LG. In fact, the carboxylic acid of Asp800 was close enough to the distal nitrogen at the imidazole ring to protonate it. For compounds UPR1364 and UPR1381, Cys797 alkylation was modelled starting from the cationic species resulting from protonation of the nitrogen atom at the imidazole ring, with deprotonated carboxylate group of Asp800 (Fig. 11).

Free-energy profiles for Cys797 alkylation were obtained applying the Jarzynski equality using five independent work curves for each compound. The activation energy for Cys797 alkylation was 27.0 ± 0.7 kcal/mol for UPR1303, 40.4 ± 1.2 kcal/mol for UPR1364, and 34.2 ± 0.7 kcal/mol for compound UPR1381. The reaction energy was negative for UPR1303 and UPR1381 (−12.6 ± 0.7 and −10.2 ± 1.2 kcal/mol, respectively) indicating that Cys797 alkylation by these inhibitors was spontaneous. The reaction energy calculated for UPR1364 was positive (0.9 ± 1.0 kcal/mol) thus indicating that for this inhibitor the reaction was not favored.

Analysis of the geometries along the SMD trajectory indicated that the TS structure for alkylation of Cys797 by UPR1303 and UPR1381 was characterized by a trigonal bipyramidal geometry with the thiolate sulfur, the acetamide Cα, and the LG heteroatom forming an angle of nearly 180° (Fig. 12). This was not the case for UPR1364 where the close proximity between the N-methyl group emerging from the imidazole ring to the p-loop of EGFR hindered the geometry of the TS.

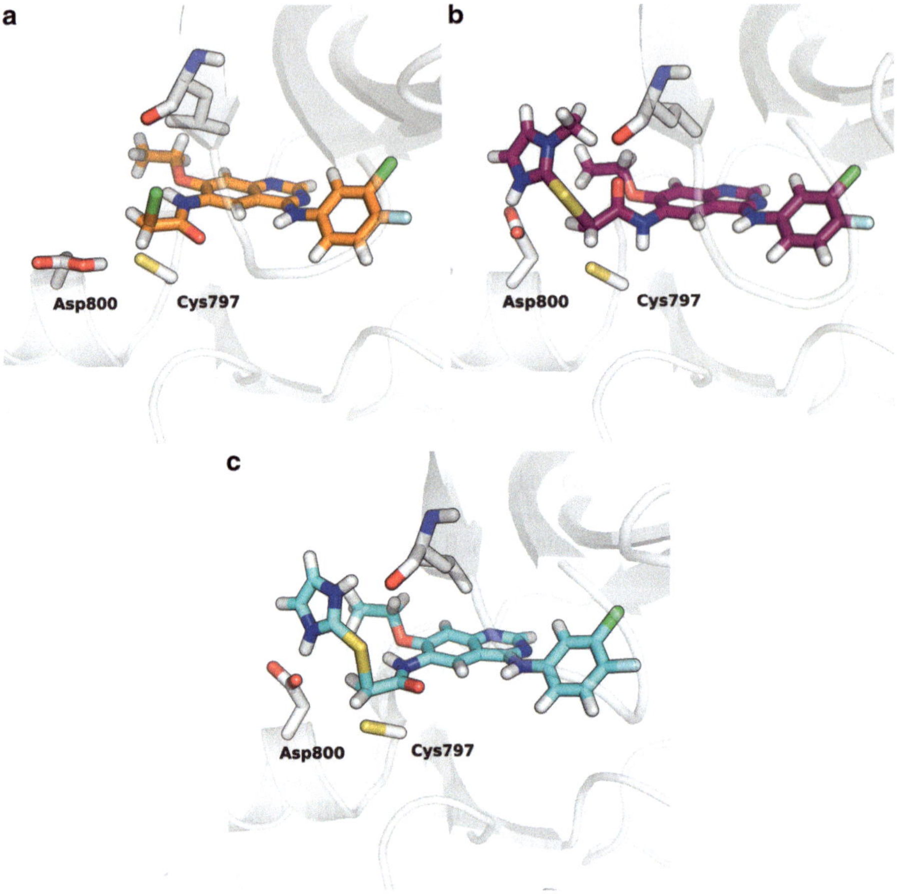

Fig. 12 Representative TS structures for Cys797 alkylation by UPR1303 (panel a, orange carbon atoms), UPR1381 (panel b, cyan carbon atoms), and UPR1364 (panel c, magenta carbon atoms)

QM/MM SMD simulations suggest that UPR1303 and UPR1381 can alkylate Cys797 leading to stable adducts. It is worth mentioning that these simulations were performed before the synthesis of the titled compounds. Computational results were essential for the decision to synthesize and test in biological assays UPR1381.

The ability of UPR1381 to inhibit EGFR L858R/T790M double mutant was assessed using a TR-FRET binding assay with and without preincubation. When a covalent modification occurs at an enzyme binding site, the detected IC_{50} value turns out to be dependent on the time of preincubation [140]. The IC_{50} value of UPR1381 displayed a significant reduction with 5-h preincubation, supporting a covalent mechanism of inhibition for this compound. In accordance with this finding, UPR1381 displayed long-lasting EGFR inhibition in A549 cells and inhibited the growth on H1975 cell line, which harbors L858R/T790M mutation, with a potency ($IC_{50} = 1.4\ \mu M$) comparable to that of a reference acrylamide-based inhibitor tested in the same condition ($IC_{50} = 0.70\ \mu M$).

Overall, our findings suggest that heteroaryl thioacetamide warheads can be used as alternative to acrylamides provided that their lower reactivity is counterbalanced by the presence of an appropriate scaffold able to place them in close proximity to the nucleophilic cysteine.

4 Conclusions

Recent advances in software and hardware development are making possible the use of QM/MM approaches to solve problems relevant for medicinal chemistry and chemical biology. Several research groups worldwide are currently trying to exploit results gathered from QM/MM simulations to drive the synthesis of covalent inhibitors in the context of both hit-to-lead and lead optimization processes. For instance, understanding substituent effect on reactivity and catalysis is critical to devise compounds featured by high potency and selectivity. Efforts are currently being taken to use QM-based calculations to predict the stability of the reaction product and possibly kinetic parameters associated to enzyme reactivation [25]. In this respect, calculating reaction barriers for deacylation by QM/MM at the scope of ranking β-lactam compounds bound to β-lactamases may help the design of new inhibitors possibly resistant to the action of this class of enzyme [141]. This is a key example, where a mechanistic understanding of the chemical reaction involving a drug target and a covalent modifier can suggest modifications that hardly other computational techniques may provide.

In drug design projects, there is the urgency to give indications to medicinal chemists as early as possible. In a similar context, QM/MM simulations have to be performed using SE methods to give answers in a reasonable amount of time. SE approaches are featured by systematic errors which often give reaction energetics far from those obtained with high-level ab initio approaches. On the other hand, SE methods are able to capture differences in terms of energy barriers for a set of inhibitors or substrates targeting the same enzyme [28, 41, 141]. This means that QM/MM simulations based on DFTB or PDDG/PM3 theory are helpful during the discovery process as they can prioritize the synthesis and testing of novel chemical entities.

Here, we have reported few representative examples of QM/MM studies performed on biological targets relevant for our drug discovery programs. In the case of FAAH, we showed the ability of QM/MM simulations to (1) predict the binding orientation of a reference inhibitor, (2) rationalize substituent effect on activity data, and (3) characterize the mechanism of enzyme reactivation discriminating a substrate from a covalent irreversible inhibitor. In the case of EGFR, QM/MM simulations allowed to

(1) characterize at atomic level the mechanism of inhibition of acrylamide-based inhibitors, (2) highlight the uncommon nucleophilicity possessed by the noncatalytic Cys797, and (3) assist the prospective design of novel acetamide warheads featured by low chemical reactivity. In our experience, QM/MM simulations can provide essential mechanistic answers that can help covalent drug design. We also believe that QM/MM methodology has the potentiality for becoming a standard and reference tool for computational chemists operating in field of medicinal chemistry and chemical biology. In these fields, one of the current topics is represented by the discovery of inhibitors targeting residues distinct from cysteine or serine [142]. Again, we believe that this kind of research might benefit from a wise use of QM/MM simulations.

Acknowledgements

Adrian Mulholland, (University of Bristol, Bristol, UK), Christo Christov (Michigan Tech University, Houghton, Michigan, US), and Davide Branduardi (Schrodinger Inc., London, UK) are kindly acknowledged for insightful discussions.

References

1. Copeland RA (2016) The drug-target residence time model: a 10-year retrospective. Nat Rev Drug Discov 15:87–95
2. Tonge P (2018) Drug-target kinetics in drug discovery. ACS Chem Neurosci 9:29–39
3. Nunez S, Venhorst J, Kruse CG (2012) Target-drug interactions: first principles and their application to drug discovery. Drug Discov Today 17:10–22
4. Copeland RA, Pompliano DL, Meek TD (2006) Drug-target residence time and its implications for lead optimization. Nat Rev Drug Discov 5:730–739
5. Vauquelin G, Charlton SJ (2010) Long-lasting target binding and rebinding as mechanisms to prolong in vivo drug action. Br J Pharmacol 161:488–508
6. Bauer RA (2015) Covalent inhibitors in drug discovery: from accidental discoveries to avoided liabilities and designed therapies. Drug Discov Today 20:1061–1073
7. Miyahisa I, Sameshima T, Hixon MS (2015) Rapid determination of the specificity constant of irreversible inhibitors (kinact/ki) by means of an endpoint competition assay. Angew Chem Int Ed Engl 54:14099–14102
8. Strelow JM (2017) A perspective on the kinetics of covalent and irreversible inhibition. SLAS Discov 22:3–20
9. Baillie TA (2016) Targeted covalent inhibitors for drug design. Angew Chem Int Ed Engl 55:13408–13421
10. Carmi C, Lodola A, Rivara S et al (2011) Epidermal growth factor receptor irreversible inhibitors: chemical exploration of the cysteine-trap portion. Mini Rev Med Chem 11:1019–1030
11. Singh J, Petter RC, Baillie TA et al (2011) The resurgence of covalent drugs. Nat Rev Drug Discov 10:307–317
12. Jorgensen WL (2004) The many roles of computation in drug discovery. Science 303:1813–1818
13. Jorgensen WL (2009) Efficient drug dead discovery and optimization. Acc Chem Res 42:724–733
14. London N, Miller RM, Krishnan S et al (2014) Covalent docking of large libraries for the discovery of chemical probes. Nat Chem Biol 10:1066–1072
15. Lonsdale R, Ward RA (2018) Structure-based design of targeted covalent inhibitors. Chem Soc Rev 47:3816–3830
16. De Cesco S, Kurian J, Dufresne C et al (2017) Covalent inhibitors design and discovery. Eur J Med Chem 138:96––114
17. Schmidt TC, Welker A, Rieger M et al (2014) Protocol for rational design of covalently

interacting inhibitors. ChemPhysChem 15:3226–3235
18. Zhang H, Jiang W, Chatterjee P et al (2019) Ranking reversible covalent drugs: from free energy perturbation to fragment docking. J Chem Inf Model 59:2093–2102
19. Scarpino A, Ferenczy GG, Keserű GM (2018) Comparative evaluation of covalent docking tools. J Chem Inf Model 58:1441–1458
20. Awoonor-Williams E, Walsh AG, Rowley CN (2017) Modeling covalent-modifier drugs. Biochim Biophys Acta 1865:1664–1675
21. Van der Kamp MW, Mulholland AJ (2013) Combined quantum mechanics/molecular mechanics (QM/MM) methods in computational enzymology. Biochemistry 52:2708–2728
22. Lodola A, Capoferri L, Rivara S et al (2013) Quantum mechanics/molecular mechanics modeling of fatty acid amide hydrolase reactivation distinguishes substrate from irreversible covalent inhibitors. J Med Chem 56:2500–2512
23. Chudyk EI, Limb MA, Jones C et al (2014) QM/MM simulations as an assay for carbapenemase activity in class a β-lactamases. Chem Commun (Camb) 50:14736–14739
24. Sgrignani J, Grazioso G, De Amici M et al (2014) Inactivation of TEM-1 by avibactam (NXL-104): insights from quantum mechanics/molecular mechanics metadynamics simulations. Biochemistry 53:5174–5185
25. Schirmeister T, Kesselring J, Jung S et al (2016) Quantum chemical-based protocol for the rational design of covalent inhibitors. J Am Chem Soc 138:8332–8335
26. Arafet K, Ferrer S, González FV et al (2017) Quantum mechanics/molecular mechanics studies of the mechanism of cysteine protease inhibition by peptidyl-2,3-epoxyketones. Phys Chem Chem Phys 19:12740–12748
27. Nutho B, Mulholland AJ, Rungrotmongkol T (2019) The reaction mechanism of zika virus NS2B/NS3 serine protease inhibition by dipeptidyl aldehyde: a QM/MM study. Phys Chem Chem Phys 21:14945–14956
28. Lodola A, De Vivo M (2012) The increasing role of qm/mm in drug discovery. Adv Protein Chem Struct Biol 87:337–362
29. De Vivo M (2011) Bridging quantum mechanics and structure-based drug design. Front Biosci 16:1619-–1633
30. Di Marzo V, Fontana A, Cadas H et al (1994) Formation and inactivation of endogenous cannabinoid anandamide in central neurons. Nature 372:686–691
31. Fegley D, Gaetani S, Duranti A et al (2005) Characterization of the fatty acid amide hydrolase inhibitor cyclohexyl carbamic acid 3′-carbamoyl-biphenyl-3-yl ester (URB597): effects on anandamide and oleoylethanolamide deactivation. J Pharmacol Exp Ther 313:352–358
32. Roskoski R Jr (2014) The ErbB/HER family of protein-tyrosine kinases and cancer. Pharmacol Res 79:34–74
33. Świderek K, Tuñón I, Moliner V et al (2015) Computational strategies for the design of new enzymatic functions. Arch Biochem Biophys 582:68–79
34. Amaro RE, Mulholland AJ (2018) Multiscale methods in drug design bridge chemical and biological complexity in the search for cures. Nat Rev Chem 2:0148
35. Warshel A, Weiss RM (1980) An empirical valence bond approach for comparing reactions in solutions and in enzymes. J Am Chem Soc 102:6218–6226
36. Car R, Parrinello M (1985) Unified approach for molecular dynamics and density-functional theory. Phys Rev Lett 55:2471–2474
37. Kamerlin SCL, Warshel A (2011) The empirical valence bond model: theory and applications. Wiley Interdiscip Rev Comput Mol Sci 1:30–45
38. Dal Peraro M, Ruggerone P, Raugei S et al (2007) Investigating biological systems using first principles car-parrinello molecular dynamics simulations. Curr Opin Struct Biol 17:149–156
39. Warshel A, Levitt M (1976) Theoretical studies of enzymic reactions: dielectric, electrostatic and steric stabilization of the carbonium ion in the reaction of lysozyme. J Mol Biol 103:227–249
40. Field MJ, Bash PA, Karplus M (1990) A combined quantum-mechanical and molecular mechanical potential for molecular-dynamics simulations. J Comput Chem 11:700–733
41. Mulholland AJ (2005) Modelling enzyme reaction mechanisms, specificity and catalysis. Drug Discov Today 10:1393–1402
42. Ryde U (2016) QM/MM calculations on proteins. Methods Enzymol 577:119–158
43. Warshel A (2003) Computer simulations of enzyme catalysis: methods, progress, and insights. Annu Rev Biophys Biomol Struct 32:425–443
44. Ainsley J, Lodola A, Mulholland AJ et al (2018) Combined quantum mechanics and molecular mechanics studies of enzymatic

reaction mechanisms. Adv Protein Chem Struct Biol 113:1–32
45. Claeyssens F, Harvey JN, Manby FR et al (2006) High-accuracy computation of reaction barriers in enzymes. Angew Chem Int Ed Engl 45:6856–6859
46. Elstner M, Porezag D, Jungnickel G et al (1998) Self-consistent-charge density-functional tight-binding method for simulations of complex materials properties. J Phys Rev B 58:7260–7268
47. Gaus M, Cui Q, Elstner M (2012) DFTB3: extension of the self-consistent-charge density-functional tight-binding method (SCC-DFTB). J Chem Theory Comput 7:931–948
48. Kromann JC, Christensen AS, Cui Q et al (2016) Towards a barrier height benchmark set for biologically relevant systems. Peer J 4: e1994
49. Christensen AS, Kubař T, Cui Q et al (2016) Semiempirical quantum mechanical methods for noncovalent interactions for chemical and biochemical applications. Chem Rev 116:5301–5337
50. Gaus M, Goez A, Elstner M (2013) Parametrization and benchmark of DFTB3 for organic molecules. J Chem Theory Comput 9:338–354
51. Domínguez A, Niehaus TA, Frauenheim T (2015) Accurate hydrogen bond energies within the density functional tight binding method. J Phys Chem A 119:3535–3544
52. Miriyala VM, Řezáč J (2017) Description of non-covalent interactions in SCC-DFTB methods. J Comput Chem 38:688–697
53. Gruden M, Andjeklović L, Jissy AK et al (2017) Benchmarking density functional tight binding models for barrier heights and reaction energetics of organic molecules. J Comput Chem 38:2171–2185
54. Elstner M (2006) The SCC-DFTB method and its application to biological systems. Theor Chem Accounts 116:316–325
55. Capoferri L, Mor M, Sirirak J et al (2011) Application of a SCC-DFTB QM/MM approach to the investigation of the catalytic mechanism of fatty acid amide hydrolase. J Mol Model 17:2375–2383
56. Jitonnom J, Limb MA, Mulholland AJ (2014) QM/MM free-energy simulations of reaction in serratia marcescens chitinase b reveal the protonation state of asp142 and the critical role of tyr214. J Phys Chem B 118:4771–4483
57. Brooks BR, Brooks CL 3rd, Mackerell AD Jr et al (2009) CHARMM: the biomolecular simulation program. J Comput Chem 30:1545–1614
58. Hornak V, Abel R, Okur A et al (2006) Comparison of multiple amber force fields and development of improved protein backbone parameters. Proteins 65:712–725
59. De Vivo M, Masetti M, Bottegoni G et al (2016) Role of molecular dynamics and related methods in drug discovery. J Med Chem 59:4035–4061
60. Senn HM, Thiel W (2009) QM/MM methods for biomolecular systems. Angew Chem Int Ed Engl 48:1198–1229
61. Boulanger E, Harvey JN (2018) QM/MM methods for free energies and photochemistry. Curr Opin Struct Biol 49:72–76
62. Ganguly A, Boulanger E, Thiel W (2017) Importance of mm polarization in qm/mm studies of enzymatic reactions: assessment of the QM/MM drude oscillator model. J Chem Theory Comput 13:2954–2961
63. Ridder L, Rietjens IMCM, Vervoort J et al (2002) Quantum mechanical/ molecular mechanical free energy simulations of the glutathione s-transferase (M1-1) reaction with phenanthrene 9,10-oxide. J Am Chem Soc 124:9926–9936
64. Riccardi D, Li GH, Cui Q (2004) Importance of Van der Waals interactions in QM/MM simulations. J Phys Chem B 108:6467–6478
65. Himo F (2006) Quantum chemical modeling of enzyme active sites and reaction mechanisms. Theor Chem Accounts 116:232–240
66. Marti S, Roca M, Andres J et al (2004) Theoretical insights in enzyme catalysis. Chem Soc Rev 33:98–107
67. Lonsdale R, Harvey JN, Mulholland AJ (2012) A practical guide to modelling enzyme-catalysed reactions. Chem Soc Rev 41:3025–3038
68. Sousa SF, Ribeiro AJM, Neves RPP et al (2017) Application of quantum mechanics/ molecular mechanics methods in the study of enzymatic reaction mechanisms. Wiley Interdiscip Rev Comput Mol Sci 7:e1281
69. Cerqueira NMFSA, Fernandes PA, Ramos MJ (2018) Protocol for computational enzymatic reactivity based on geometry optimisation. ChemPhysChem 19:669–689
70. Lodola A, Mulholland AJ (2013) Computational enzymology. Methods Mol Biol 924:67–89
71. Klahn M, Braun-Sand S, Rosta E et al (2005) On possible pitfalls in ab initio quantum mechanics/molecular mechanics minimization approaches for studies of enzymatic reactions. J Phys Chem B 109:15645–15650

72. Christov CZ, Lodola A, Karabencheva-Christova TG et al (2013) Conformational effects on the pro-s hydrogen abstraction reaction in cyclooxygenase-1: an integrated QM/MM and MD study. Biophys J 104: L5–L7
73. Ryde U (2017) How many conformations need to be sampled to obtain converged QM/MM energies? The curse of exponential averaging. J Chem Theory Comput 13:5745–5752
74. Lodola A, Sirirak J, Fey N et al (2010) Structural fluctuations in enzyme-catalysed reactions: determinants of reactivity in fatty acid amide hydrolase from multivariate statistical analysis of quantum mechanics/molecular mechanics paths. J Chem Theory Comput 6:2948–2960
75. Mulholland AJ (2001) The QM/MM approach to enzymatic reactions. Theor Comp Chem 9:597–653
76. Acevedo O, Jorgensen WL (2010) Advances in quantum and molecular mechanical (QM/MM) simulations for organic and enzymatic reactions. Acc Chem Res 43:142–151
77. Warshel A, Sharma PK, Kato M et al (2006) Electrostatic basis for enzyme catalysis. Chem Rev 106:3210–3235
78. Glowacki DR, Harvey JN, Mulholland AJ (2012) Taking Ockham's razor to enzyme dynamics and catalysis. Nat Chem 4:169–176
79. Swiderek K, Tunon I, Moliner V (2014) Predicting enzymatic reactivity: from theory to design. Wiley Interdiscip Rev Comput Mol Sci 4:407–421
80. Acevedo O, Jorgensen WL (2014) Quantum and molecular mechanical (QM/MM) Monte Carlo techniques for modeling condensed-phase reactions. Wiley Interdiscip Rev Comput Mol Sci 4:422–435
81. Roux B (1995) The calculation of the potential of mean force using computer simulations. Comput Phys Commun 91:275–282
82. Gao J, Ma S, Major DT et al (2006) Mechanisms and free energies of enzymatic reactions. Chem Rev 106:3188–3209
83. Kaestner J (2011) Umbrella sampling. Wiley Interdiscip Rev Comput Mol Sci 1:932–942
84. Kumar S, Rosenberg JM, Bouzida D et al (1992) The weighted histogram analysis method for free- energy calculations on biomolecules. i. the method. J Comput Chem 13:1011–1021
85. Izrailev S, Stepaniants S, Balsera M et al (1997) Molecular dynamics study of unbinding of the avidin-biotin complex. Biophys J 72:1568–1581
86. Murcia M, Jirouskova M, Li J et al (2008) Functional and computational studies of the ligand-associated metal binding site of beta3 integrins. Proteins 71:1779–1791
87. Crespo A, Martí MA, Estrin DA, Roitberg AE (2005) Multiple-steering QM-MM calculation of the free energy profile in chorismate mutase. J Am Chem Soc 127:6940–6941
88. Jarzynski C (1997) Non-equilibrium equality for free energy differences. Phys Rev Lett 78:2690–2693
89. Ramírez CL, Martí MA, Roitberg AE (2016) Steered molecular dynamics methods applied to enzyme mechanism and energetics. Methods Enzymol 578:123–143
90. Soto-Delgado J, Tapia RA, Torras J (2016) Multiscale treatment for the molecular mechanism of a Diels-Alder reaction in solution: a QM/MM-MD study. J Chem Theory Comput 12:4735–4742
91. Karabencheva-Christova TG, Torras J, Mulholland AJ et al (2017) Mechanistic insights into the reaction of chlorination of tryptophan catalyzed by tryptophan 7-halogenase. Sci Rep 7:17395
92. Laio A, Parrinello M (2002) Escaping free-energy minima. Proc Natl Acad Sci U S A 99:12562–125666
93. Gervasio FL, Laio A, Parrinello M (2005) Flexible docking in solution using metadynamics. J Am Chem Soc 127:2600–2607
94. Russo S, Callegari D, Incerti M et al (2016) Exploiting free-energy minima to design novel EphA2 protein-protein antagonists: from simulation to experiment and return. Chemistry 22:8048–8052
95. Scalvini L, Vacondio F, Bassi M et al (2016) Free-energy studies reveal a possible mechanism for oxidation-dependent inhibition of MGL. Sci Rep 6:31046
96. Cavalli A, Spitaleri A, Saladino G et al (2015) Investigating drug-target association and dissociation mechanisms using metadynamics-based algorithms. Acc Chem Res 48:277–285
97. Callegari D, Lodola A, Pala D et al (2017) Metadynamics simulations distinguish short- and long-residence-time inhibitors of cyclin-dependent kinase 8. J Chem Inf Model 57:159–169
98. Ibrahim P, Clark T (2019) Metadynamics simulations of ligand binding to GPCRs. Curr Opin Struct Biol 55:129–137
99. Ensing B, De Vivo M, Liu Z et al (2006) Metadynamics as a tool for exploring free energy landscapes of chemical reactions. Acc Chem Res 39:73–81

100. Raich L, Nin-Hill A, Ardèvol A et al (2016) Enzymatic cleavage of glycosidic bonds: strategies on how to set up and control a QM/MM metadynamics simulation. Methods Enzymol 577:159–183
101. Branduardi D, Gervasio FL, Parrinello M (2007) From a to B in free energy space. J Chem Phys 126:054103
102. Zinovjev K, Tuñón I (2014) Exploring chemical reactivity of complex systems with path-based coordinates: role of the distance metric. J Comput Chem 35:1672–1681
103. Maragliano L, Vanden-Eijnden EJ (2008) Single-sweep methods for free energy calculations. J Chem Phys 128:184110
104. Branduardi D, De Vivo M, Rega N et al (2011) Methylphosphate dianion hydrolysis in solution characterized by path collective variables coupled with DFT-based enhanced sampling simulations. J Chem Theory Comput 7:539–543
105. Lodola A, Branduardi D, De Vivo M et al (2012) A catalytic mechanism for cysteine N-terminal nucleophile hydrolases, as revealed by free energy simulations. PLoS One 7:e32397
106. Murillo-López J, Zinovjev K, Pereira H et al (2019) Studying the phosphoryl transfer mechanism of the E. coli phosphofructokinase-2: from X-ray structure to quantum mechanics/molecular mechanics simulations. Chem Sci 10:2882–2892
107. Lodola A, Castelli R, Mor M et al (2015) Fatty acid amide hydrolase inhibitors: a patent review (2009–2014). Expert Opin Ther Pat 25:1247–1266
108. McKinney MK, Cravatt BF (2003) Evidence for distinct roles in catalysis for residues of the serine-serine-lysine catalytic triad of fatty acid amide hydrolase. J Biol Chem 278:37393–37399
109. Lodola A, Mor M, Sirirak J et al (2009) Insights into the mechanism and inhibition of fatty acid amide hydrolase from quantum mechanics/molecular mechanics (QM/MM) modelling. Biochem Soc Trans 37 (Pt 2):363–367
110. Lodola A, Mor M, Hermann JC et al (2005) QM/MM modelling of oleamide hydrolysis in fatty acid amide hydrolase (FAAH) reveals a new mechanism of nucleophile activation. Chem Commun (Camb) 35:4399–4401
111. Tubert-Brohman I, Acevedo O, Jorgensen WL (2006) Elucidation of hydrolysis mechanisms for fatty acid amide hydrolase and its lys142ala variant via QM/MM simulations. J Am Chem Soc 128:16904–16913
112. McKinney MK, Cravatt BF (2005) Structure and function of fatty acid amide hydrolase. Annu Rev Biochem 74:411–432
113. Palermo G, Campomanes P, Cavalli A, Rothlisberger U, De Vivo M (2015) Anandamide hydrolysis in FAAH reveals a dual strategy for efficient enzyme-assisted amide bond cleavage via nitrogen inversion. J Phys Chem B 119:789–801
114. Mor M, Rivara S, Lodola A et al (2004) Cyclohexylcarbamic acid 3′- or 4′-substituted biphenyl-3-yl esters as fatty acid amide hydrolase inhibitors: synthesis, quantitative structure-activity relationships, and molecular modeling studies. J Med Chem 47:4998–5008
115. Alexander JP, Cravatt BF (2005) Mechanism of carbamate inactivation of FAAH: implications for the design of covalent inhibitors and in vivo functional probes for enzymes. Chem Biol 12:1179–1187
116. Lodola A, Rivara S, Mor M (2011) Application of computational methods to the design of fatty acid amide hydrolase (FAAH) inhibitors based on a carbamic template structure. Adv Protein Chem Struct Biol 85:1–26
117. Bracey MH, Hanson MA, Masuda KR et al (2002) Structural adaptations in a membrane enzyme that terminates endocannabinoid signaling. Science 298:1793–1796
118. Basso E, Duranti A, Mor M et al (2004) Tandem mass spectrometric data-FAAH inhibitory activity relationships of some carbamic acid O-aryl esters. J Mass Spectrom 39:1450–1455
119. Lodola A, Mor M, Rivara S et al (2008) Identification of productive inhibitor binding orientation in fatty acid amide hydrolase (FAAH) by QM/MM mechanistic modelling. Chem Commun (Camb) 2:214–216
120. Mileni M, Kamtekar S, Wood DC et al (2010) Crystal structure of fatty acid amide hydrolase bound to the carbamate inhibitor urb597: discovery of a deacylating water molecule and insight into enzyme inactivation. J Mol Biol 400:743–754
121. Lodola A, Capoferri L, Rivara S et al (2011) Understanding the role of carbamate reactivity in fatty acid amide hydrolase inhibition by QM/MM mechanistic modelling. Chem Commun (Camb) 47:2517–2519
122. Mulholland AJ, Richards G (1997) Acetyl-CoA enolization in citrate synthase: a quantum mechanical/molecular mechanical (QM/MM) study. Proteins 27:9–25
123. Sharma SV, Bell DW, Settleman J, Haber DA (2007) Epidermal growth factor receptor

mutations in lung cancer. Nat Rev Cancer 7:169–181

124. Pao W, Chmielecki J (2010) Rational, biologically based treatment of EGFR-mutant non-small-cell lung cancer. Nat Rev Cancer 10:760–774
125. Engelman JA, Janne PA (2008) Mechanisms of acquired resistance to epidermal growth factor receptor tyrosine kinase inhibitors in non-small cell lung cancer. Clin Cancer Res 14:2895–2899
126. Michalczyk A, Klüter S, Rode HB, Simard JR, Grütter C, Rabiller M, Rauh D (2008) Structural insights into how irreversible inhibitors can overcome drug resistance in EGFR. Bioorg Med Chem 16:3482–3488
127. Hirsh V (2011) Afatinib (BIBW 2992) development in non-small-cell lung cancer. Future Oncol 7:817–825
128. Capoferri L, Lodola A, Rivara S et al (2015) Quantum mechanics/molecular mechanics modeling of covalent addition between EGFR-cysteine 797 and N-(4-anilinoquinazolin-6-yl) acrylamide. J Chem Inf Model 55:589–599
129. Awoonor-Williams E, Isley W, Dale S et al (2019) Quantum chemical methods for modeling covalent modification of biological thiols. ChemRxiv. https://doi.org/10.26434/chemrxiv.8061845.v2
130. Truong TH, Carroll KS (2012) Redox regulation of epidermal growth factor receptor signaling through cysteine oxidation. Biochemistry 51:9954–9965
131. Truong TH, Ung PM, Palde PB et al (2016) Molecular basis for redox activation of epidermal growth factor receptor kinase. Cell Chem Biol 23:837–848
132. Klüter S, Simard JR, Rode HB et al (2010) Characterization of irreversible kinase inhibitors by directly detecting covalent bond formation: a tool for dissecting kinase drug resistance. Chembiochem 11:2557–2566
133. Schwartz PA, Kuzmic P, Solowiej J et al (2014) Covalent EGFR inhibitor analysis reveals importance of reversible interactions to potency and mechanisms of drug resistance. Proc Natl Acad Sci U S A 111:173–178
134. Carmi C, Galvani E, Vacondio F et al (2012) Irreversible inhibition of epidermal growth factor receptor activity by 3-aminopropanamides. J Med Chem 55:2251–2264
135. Callegari D, Ranaghan KE, Woods CJ et al (2018) L718Q mutant EGFR escapes covalent inhibition by stabilizing a non-reactive conformation of the lung cancer drug osimertinib. Chem Sci 9:2740–2749
136. Cross DA, Ashton SE, Ghiorghiu S et al (2014) AZD9291, an irreversible EGFR TKI, overcomes T790M-mediated resistance to EGFR inhibitors in lung cancer. Cancer Discov 4:1046–1061
137. Castelli R, Bozza N, Cavazzoni A et al (2019) Balancing reactivity and antitumor activity: heteroarylthioacetamide derivatives as potent and time-dependent inhibitors of EGFR. Eur J Med Chem 162:507–524
138. Repasky MP, Chandrasekhar J, Jorgensen WL (2002) PDDG/PM3 and PDDG/MNDO: improved semiempirical methods. J Comput Chem 23:1601–1622
139. Vayner G, Houk KN, Jorgensen WL et al (2004) Steric retardation of sn2 reactions in the gas phase and solution. J Am Chem Soc 126:9054–9058
140. Barf T, Kaptein A (2012) Irreversible protein kinase inhibitors: balancing the benefits and risks. J Med Chem 55:6243–6262
141. Hirvonen VHA, Hammond K, Chudyk EI et al (2019) An efficient computational assay for β-lactam antibiotic breakdown by class a β-lactamases. J Chem Inf Model 59:3365–3369
142. Gehringer M, Laufer SA (2019) Emerging and re-emerging warheads for targeted covalent inhibitors: applications in medicinal chemistry and chemical biology. J Med Chem 62:5673–5724

Chapter 20

What Next for Quantum Mechanics in Structure-Based Drug Discovery?

Richard A. Bryce

Abstract

There is significant potential for electronic structure methods to improve the quality of the predictions furnished by the tools of computer-aided drug design, which typically rely on empirically derived functions. In this perspective, we consider some recent examples of how quantum mechanics has been applied in predicting protein–ligand geometries, protein–ligand binding affinities and ligand strain on binding. We then outline several significant developments in quantum mechanics methodology likely to influence these approaches: in particular, we note the advent of more computationally expedient ab initio quantum mechanical methods that can provide chemical accuracy for larger molecular systems than hitherto possible. We highlight the emergence of increasingly accurate semiempirical quantum mechanical methods and the associated role of machine learning and molecular databases in their development. Indeed, the convergence of improved algorithms for solving and analyzing electronic structure, modern machine learning methods, and increasingly comprehensive benchmark data sets of molecular geometries and energies provides a context in which the potential of quantum mechanics will be increasingly realized in driving future developments and applications in structure-based drug discovery.

Key words Quantum mechanics, QM/MM docking, Semiempirical quantum mechanics, Structure-based drug design, Ligand strain, Binding free energies, Machine learning, Energy decomposition, Pair natural orbital theory, Density functional tight binding theory, Free energy perturbation

1 Introduction

Key to structure-based drug design is a molecular-level understanding of the interaction of a ligand with its target receptor. Most commonly, the ligand is a small organic molecule although rational design approaches are increasingly being applied to larger peptides and proteins [1, 2]. Similarly, the complexity of target structures has evolved from globular proteins to membrane-bound proteins, large protein assemblies and nonprotein targets such as DNA and RNA. Nevertheless, the underpinning physics of the interactions between a ligand and receptor remain the same, often featuring hydrogen bonds, salt bridges, halogen bonds and dispersive contributions, such as π–π stacking interactions. There may in addition

Alexander Heifetz (ed.), *Quantum Mechanics in Drug Discovery*, Methods in Molecular Biology, vol. 2114,
https://doi.org/10.1007/978-1-0716-0282-9_20,

be covalent interactions between the ligand and its receptor's residues or bound metal ions.

Computational methods are now integral across the drug discovery process [3] and are widely used as a cost-effective way to guide the design of new more potent ligands. These tools often depend on a suitably robust prediction of the affinity of the ligand–receptor interaction. Thus, in lead discovery, virtual screening may employ receptor–ligand docking based on an empirically-derived scoring function to predict the bound pose of a ligand in a target receptor binding pocket. In lead optimization, free energy perturbation (FEP) methods [4] are being deployed with a good level of success, guiding the design of increasing large changes in ligand chemical structure [5–7]. FEP approaches, based on molecular simulation methods, depend on an accurate empirical force field for the ligand and receptor [6]. Indeed, the accuracy of the empirical potential underlying the statistical mechanical calculations is crucial. As outlined earlier, the types of interaction between ligand and receptor can be diverse and challenging to capture in a general way with an empirical functional form [8]. However, the methods of quantum mechanics (QM) have the potential to provide more accurate descriptions of ligand–protein interactions without the need for extensive and sometimes bespoke parametrizations required for force fields.

Consequently, in recent years, there has been growing interest in evaluating the potential of quantum chemical methods for structure-based drug design [9, 10]. A major enabling factor for this direction has been the increased computational efficiency of more accurate QM approaches [11]. Improved performance has been achieved through algorithmic and theoretical developments alongside the evolution of computer architecture and capability. In our 2013 review on the role of QM in drug design [12], we considered the growing impact of QM methods on several activities encountered in computational medicinal chemistry, namely, protein–ligand docking, protein–ligand binding affinities and estimating the effect of ligand strain on binding. Developments in these areas subsequent to our article have been well covered by more recent reviews [11, 13–19]. In this perspective, we briefly highlight selected examples of these activities before considering recent developments in computational quantum chemistry with implications for the future use of QM in structure-based drug design.

2 Developments in QM for Drug Design

2.1 Ligand–Receptor Docking

The prediction of the bound geometry of a ligand in a receptor pocket via computational docking typically employs scoring functions based on force fields, simplified empirical models or statistical potentials. The large number of energy calculations required to evaluate each of the generated trial ligand–receptor orientations, although high in computational cost, could in principle be scored by a QM or hybrid quantum mechanical/molecular mechanical (QM/MM) potential. The benefit of directly using QM to guide sampling, rather than using QM to selectively rescore empirically generated protein–ligand orientations, would be to detect bound ligand poses otherwise overlooked by or inaccessible to empirical potentials, for example, due to strong electronic polarization effects or covalent bond formation.

In a step towards a fully integrated QM/MM docking approach, Rohrig and coworkers [20] employed their EADock dihedral space sampling (DSS) docking algorithm to generate ligand poses via a classical potential. The docking solutions were then clustered and filtered and a subset of 250 conformations rescored via a QM/MM potential: the ligand, zinc ion, and zinc-coordinating residues were modelled at the SCC-DFTB level [21] of semiempirical QM (SQM), while the remaining metalloprotein was modelled with the CHARMM force field [22] and the effect of water via the FACTS solvation model [23]. For 226 zinc-bound ligands, this QM/MM rescoring approach recovered the crystallographic pose in 66% of cases, although providing only a rather modest improvement of 5% over the MM implementation.

The use of QM, QM/MM or MM potentials can lead to rather rugged energy landscapes that are challenging to sample effectively. In order to smooth this landscape, an interesting approach was adopted by Rohrig et al. [24] Their Attracting Cavities algorithm [25] initially substitutes the protein active site for a smooth potential composed of attractive van der Waals centers and point charges; in this simplified cavity, a grid-based search of ligand poses is performed; the atomistic protein active site is reintroduced as a soft-core potential and then in full atomistic detail. Using this Attracting Cavities scheme [24], a similar performance to their earlier QM/MM docking study [20] was obtained for the 226 zinc-binding ligand set, of up to 73% complexes docked correctly, depending on specific implementation details [24]; in this case, the success rate improved by 10% or more over Attracting Cavities docking with an MM potential and outperformed empirically scored docking using GOLD [26], AutoDock [27] or AutoDock Vina methods [28].

The use of compute-intensive QM/MM energy evaluations of course introduces significant overhead, ranging from a factor of three slower when compared with Attractive Cavities docking with a classical potential to 1000 times slower when compared with the GOLD docking algorithm. However, such QM/MM schemes have the potential to study a wide range of systems without the need for involved bespoke parameterizations, for example, for docking at electronically complex metal sites such as iron, cobalt and nickel, which feature in important proteins related to human disease, [29] allowing for different oxidation, spin and coordination states.

2.2 Ligand–Receptor Binding Affinities

As for ligand pose prediction, the accuracy of ligand–protein binding affinities may be hindered by inadequate parameterization of the underlying empirical potential. Consequently, an extensive range of strategies to integrate QM methods directly or indirectly into computation of ligand–protein binding energies has been explored [13]. In the pursuit of ever larger QM subsystems in these calculations, the use of fragmentation-based methods has been of particular interest. This includes the fragment molecular orbital (FMO) approach [30], molecular fragmentation with conjugate caps (MFCC) [31] and the recently introduced molecules-in-molecules (MIM) method [32, 33].

In the MIM method, a multilayer scheme is applied: the primary subsystem is described at a high QM level and lower levels of QM capture environmental polarization effects and long range interactions. In particular, the MIM method is compatible with large basis sets which are important in capturing the subtleties of noncovalent ligand–protein interactions. The method is capable of predicting the total energy of molecules up to over 700 atoms, to within 0–2 kcal/mol of the full, unfragmented energy calculation. A three-layer calculation of protein–ligand interaction energies for 89 complexes was performed at the B97-D3(BJ)/6-311++G**: B97-D3(BJ)/6-31G*:PM6-D3 level, with QM regions ranging from 250 to 600 atoms. A good correlation r^2 of 0.7–0.9 with experimental affinities was found.

Progressing beyond single structure gas phase calculations, further improvements in agreement with experiment would be expected via the inclusion of solvation and entropy effects. Indeed, much work has been done on exploring such schemes within the context of computing QM and QM/MM protein–ligand affinities [13, 34, 35]. One possibility is to use a QM/MM potential directly within an FEP scheme, to compute the relative free energy of binding of two ligands to a receptor. In this case, a SQM potential is necessitated by the computational cost of the multiple molecular simulations required. Alternatively, a reference potential approach can be used, where FEPs from one ligand to another, bound and unbound, are computed at the MM level, and then the MM $\rightarrow$ QM/MM free energy is computed for the end points, that is, the two ligand–receptor complexes and the two ligands free in solution. Recent work on host–guest systems [36], using the PM6-DH+ level of theory [37] for the ligand, suggests that the reference potential approach provided an accuracy comparable to the direct QM/MM FEP scheme but at a factor of five reduced computational cost [36].

2.3 Ligand Strain on Receptor Binding

Also related to computing the accurate binding affinity of a ligand to its receptor is the associated internal energy change of the ligand on receptor binding. The relative conformational energy of a ligand in its bound pose, typically derived from an X-ray crystal structure, has been compared with its unbound conformation(s). Estimates from classical force fields have produced high estimates of strain energy: for example, Nicklaus et al. [38] calculated the strain energy of 27 bound ligands which ranged in magnitude up to 40 kcal/mol. Work by Perola and Charifson using the OPLS-AA and MMFF force fields found that, for a 150 ligand set, 40% of the compounds bound with a computed strain energy of over 5 kcal/mol and that, for a quarter of these ligands, the predicted strain energy exceeded 9 kcal/mol.

The ability of force fields to accurately calculate strain energies has therefore been questioned [39]. This may be a reasonable view given the difficulties of parametrizing force fields to capture the broad chemical diversity of ligand structures; this problem would instead seem well suited to a quantum chemical approach. A comparison of QM and MM methods for computing ligand strain energy considered a large set of relatively rigid FDA-approved drugs [40]. The study found inconsistent estimates of relative conformational energies via OPLS, MMFF and CHARMM force fields. By contrast, consistent correlation between predicted relative conformational energies was obtained using QM levels of theory such as M06-2X/6-31G* and the dispersion and counterpoise corrected B3LYP-gCP-D3/6-31G* model [40].

Nevertheless, in an interesting meta-analysis, Peach et al. [41] combined ligand strain energy data from several studies, concluding that on average strain energies from MM calculations were ~2–3 kcal/mol; this was significantly lower than that computed from SQM and DFT methods and comparable to MP2 estimates. However, maximum stain energy values at the MP2 level were within 1–2 kcal/mol of its average value, whereas for the force fields the maximum value exceed 40 kcal/mol [41]. Therefore, it would seem that QM does have a valuable role to play in computing strain energy on binding more accurately across ligand chemical space. However, additional care is required in estimating such contributions by QM or MM, with potential errors arising from the assignment of the X-ray structure of the bound ligand in the crystal and possible inaccuracies in its computed or measured unbound solution conformation(s).

We next turn to consider current developments in QM methods that are likely to further impact the practise of docking, binding and conformational energy calculations as well as other activities relevant computer-aided drug design.

3 Developments in QM Methods

3.1 Comprehensive Databases of Molecular Conformation and Interactions

In order to develop methods capable of modelling intermolecular interactions accurately, the availability of data sets of accurate geometries and energetics of bound molecules is vital. In particular, having chemical structures that represent the commonly encountered types of interaction in biological systems is helpful to the design and validation of widely applicable potentials for computing protein–ligand affinities. To compute benchmark interaction energies, methods based on systematic expansions of the many-particle Schrödinger equation in the framework of coupled-cluster theory can offer "chemical accuracy" (i.e., to within 1 kcal/mol).

One of the first benchmarks of this quality was the S22 set [42], which contains 22 small molecule dimer complexes modelling the types of hydrogen bonding and dispersive interactions found between amino acids and within base pairs. The energetics were computed at the estimated level of the coupled-cluster method with single, double and perturbative triple excitations, CCSD(T), in the complete basis set (CBS) limit. In essence, the CCSD(T)/CBS energies can be approximated by the MP2/CBS energy with a CCSD(T) correction via a moderately sized basis set. The S22 data set has been invaluable both in testing and in fitting computationally expedient potential energy functions, including empirical potentials [43] and semiempirical QM Hamiltonians [44]. Beyond bound equilibrium geometries, subsequent dissociation curves for the bimolecular complexes of S22 were reported by Molnar et al. [45] and Gráfová et al. [46], with the latter denoted as the S22x5 set.

Subsequently, the data set was expanded to 66 complexes, labelled the S66 set [47], with its corresponding dissociation curves in the S66x8 set [48]. A separate database, the X40x10 set, contains equilibrium and nonequilibrium geometries of halogenated compounds [49]. The Benchmark Energy and Geometry Database (BEGDB, http://www.begdb.com) [50] is a compilation of data sets including these aforementioned collections. This BEGDB compilation was used [51] to compare the accuracy in predicted binding energy and geometry of a range of semiempirical QM, density functional theory (DFT) and symmetry-adapted perturbation theory (SAPT) methods [52]. The study indicated that DFT methods with dispersion corrections, such as M06-2X [53] and B97-D3 functionals [54], and SAPT methods with a second order or greater expansion were able to yield dimer interaction energies to within 1 kcal/mol across over 1200 gas phase dimers. The errors of PM6- and PM7-based methods were higher and rather similar in performance to one another.

Other compilations of data sets exist, for example, the GMTKN30 (General Main Group Thermochemistry, Kinetics, and Noncovalent Interactions) database which is a collection of

30 different data sets [55]. Using a subset of GMTKN30 alongside reference calculations on open shell species, the PM6 Hamiltonian [56] was reparametrized to enable more accurate computation of the properties of organic diradical molecules while also slightly improving its performance for closed shell species [57]. In a similar vein, Grimme and coworkers used a range of data sets including S22 and S66x8 as a basis on which to derive parameters for the density functional tight binding semiempirical QM methods, GFN-xTB and GFN2-xTB [58, 59].

3.2 Efficient and Accurate QM Methods for Large Molecular Systems

While large data sets containing a diverse range of small molecules is desirable in testing the generality of a potential, the need to consider many body effects beyond bimolecular interactions has been highlighted [11]. However, the CCSD(T) method scales poorly, as N^7 for N basis functions, making computation of energetics for large molecules and multimolecular systems challenging. A convergence of recent developments in local correlation methods and pair natural orbital (PNO) theory have led to computationally tractable routes to computing energetics approaching CCSD(T) accuracy for systems of over 600 atoms [60–63]. PNO-based CCSD (T) computations can lie within 1 kcal/mol of CCSD(T) accuracy, for the cost equivalent to a hybrid DFT calculation. Its accuracy has been reported to be systematically improved to within 0.25 kcal/mol for a tenfold increase in computational expense [63].

Thus, the domain-based local PNO local correlation method, DLPNO-CCSD(T), was used to compute CCSD(T)/CBS quality energetics for the S30L set of 30 large non-covalently bound host–guest complexes (of up to 200 atoms each) [64]. These dimers are highly pertinent to ligand–receptor systems, including interactions such as hydrogen and halogen bonding, π–π stacking, nonpolar dispersion, CH–π and cation–dipolar interactions. The S30L set was subsequently used in evaluating the performance of the GFN-xTB and GFN2-xTB semiempirical QM methods [58, 59]. Interestingly, the mean signed error in energies for the S30L set was as low as −0.6 kcal/mol using the GFN2-xTB method, as compared with an error of −3.9 kcal/mol and −4.3 kcal/mol for the semiempirical QM methods, PM6-D3H4X and DFTB3-D3(BJ), respectively [59].

Recently, the energies of eight conformers for each of 37 organic molecules was determined at the DLPNO-CCSD(T)/cc-pVTZ level of theory, based on geometries obtained at the PBE-D3/def2-TZVP level [65]; this molecular database, denoted 37Conf8 by the authors, includes marketed drugs such as oseltamivir, cimetidine and gemcitabine. A range of SQM methods as well as the MMFF94 force field and PBE-D3 methods were benchmarked using the 37Conf8 database [65]. The tight binding methods DFTB3, GFN-xTB and GFN2-xTB were able to rank conformations nearly as well as the PBE-D3/def2-TZVP level of theory: these SQM methods gave a correlation r^2 of 0.8–0.9 and median absolute errors in conformational energy of below 2 kcal/mol. The MMFF94 force field also gave good agreement in ranking conformations of the 37 molecules but had a median error in energy of ~3 kcal/mol. The NDDO-based SQM methods did not perform as well as the tight binding density functional methods, either in ranking (r^2 of 0.5–0.7) or in median absolute error in conformational energy which was above 2 kcal/mol.

From this study, however, it was evident that MMF94 and most SQM methods, including GFN-xTB and GFN2-xTB, struggled to correctly model the conformational landscape of the cyclic ether, 18-crown-6, in the 37Conf8 set [65]. Sizable errors in the relative energies of certain conformers were found (~10–20 kcal/mol). This discrepancy is reminiscent of specific issues that were detected in modelling β-cyclodextrin conformers using NDDO-based SQM methods [66]; this problem arose from inaccurate conformational energetics of adjacent hydroxyl groups in the glucosyl residues of β-cyclodextrin. Conformational databases such as 37Conf8 will play an important role in seeking to identify and ultimately correct such deficiencies in potential energy models; in doing so, this will be a step towards addressing the vexed question of quantifying changes in ligand conformational energy on binding to a receptor (vide supra).

3.3 Machine Learning-Based Quantum Chemistry

Machine learning (ML) methods such as decision trees, support vector machines, kernel-based methods and neural networks are well suited to the analysis of large data sets including those of molecular data. Developments in algorithms from the field of computer science, coupled to the availability of large data sets, have led to a renaissance in the application of ML approaches. Given a set of inputs and descriptors, ML methods can learn to robustly classify objects, for example, to distinguish druglike from non-druglike compounds. However, the quality of the prediction depends on the diversity and extent of the training set. It has been pertinently noted [67] that the application of ML to chemical and pharmaceutical problems is not new but can be dated back to 1974, to a work by Cramer, Redl and Berkoff that applied a ML strategy to chemical substructure analysis [68]. Today, ML methods are embedded across the drug discovery process [3], from target identification via druggability analysis to distinguishing compounds based on their predicted pharmacokinetic and toxicological characteristics.

Modern ML methods are also powerful tools for fitting potential energy functions [69]. In a recent example of this [70], kernel ridge regression was used to reparametrize the molecular orbital-based OM2 semiempirical method: [71] refitting of OM2 used the geometries and computed atomization enthalpies of 5000 $C_7H_{10}O_2$ isomers drawn from the G4MP2 set. The new OM2 model led to a reduction in its mean absolute error in computed atomization enthalpies from 6.3 to 1.7 kcal/mol for the remaining 1095 isomers. In a similar way, a combined ML/genetic algorithm approach [72] was adopted to determine empirical force field parameters for simulating liquid methanol. Quantum chemical data at the MP2 level was used to derive a classical potential based on the framework of the AMOEBA polarizable multipole model [73]; the resulting model was found to outperform the original AMOEBA parametrization in reproducing condensed phase properties of methanol [72].

Interestingly, ML methods can be used to predict energy as a function of molecular geometries directly rather than parametrize terms for a predetermined functional form of a potential. For example, kernel ridge regression was used to generate a ML potential, trained on selected points from the potential energy surface of methyl chloride at the CCSD(T)/CBS level [74]. After training, the resulting ML potential was able to accurately predict the CCSD (T)/CBS energy of tens of thousands of input nuclear configurations in a matter of seconds. In another study [75], a deep neural network with three hidden layers was trained on geometries and energies of organic molecules at the ωB97X/6-31G* level [76]. This training set was a 57,951 molecule subset of the GDB-11 database of all possible molecules [77], containing only C, H, N and O atoms of up to eight heavy atoms. The resulting ML potential, called ANI-1, was able to predict DFT quality energies for larger molecular systems, with accuracy exceeding that of the PM6 and SCC-DFTB models.

ML potentials are capable of orders of magnitude increase in compute speed over the reference quantum chemical calculations. A significant caveat, however, lies in ensuring the training set is suitably diverse to span the desired chemical space, such that the machine learned predictions can interpolate effectively. Otherwise, the ML potential, with its lack of theoretical grounding, can lead to large outliers in predictions for chemical structures and conformations of molecules that are not well represented in the training set.

3.4 Tools for Interpretation of Electronic Structure

While ML-based potentials may show promise in offering accurate energies at a fraction of the cost of an electronic structure calculation, an issue that should not be overlooked is the interpretability of a model in providing the chemical insight essential in guiding molecular design. Using efficient and accurate methods based on the electronic wavefunction or electron density, it is possible to unpick the nature of this interaction. In medicinal chemistry terms, it is useful to discern what chemical groups are the main anchors for the observed or computed affinity between ligand and receptor, with a view to optimizing interactions further. Experimentally or computationally, schemes such as fluorine scanning or other probes of structure–activity can offer these insights. In silico, it is also possible to factor interaction energies into group contributions: for example, the FMO approach [30] was used to resolve the QM interaction of agonist ligands with human orexin-2 receptor into residue-based components [78]; the residue-wise energies correlated well with site-directed mutagenesis data. Furthermore, a pairwise interaction energy decomposition analysis (PIEDA) scheme was used to resolve the FMO interaction energy into relative contribution of electrostatics, exchange repulsion, charge transfer and dispersion. The computed interaction energy indicated that the agonist prefers a "U"-shaped rather than "L"-shaped pose [78]. For both poses, the PIEDA analysis indicated these interaction energies comprised approximately equally electrostatic and dispersive contributions. For 16 agonist ligands modelled in a "U"-shaped pose with orexin-2 receptor, the total FMO protein–ligand energies gave a correlation r^2 of 0.9 with experimental pEC_{50} measurements, despite the omission of solvation, strain and entropic effects [78].

Indeed, it is desirable to understand at the electronic level the physical nature of the driving force behind the binding: what are the relative roles of dispersive interactions, electrostatics and ligand strain energy. Besides PIEDA, methods of analysis include atoms in molecules (AIM) [79], natural bond orbital analysis (NBO) [80], SAPT energy decomposition [52], the local energy decomposition (LED) method [81] and absolutely localized molecular orbital/energy decomposition analysis (ALMO-EDA) [82]. These methods are useful in deciphering the presence and strength of interactions, such as hydrogen bonding or dispersion-dominated effects, and the role of other physical effects, such as polarization and charge transfer, in governing intermolecular interactions. Methods such as AIM can detect subtle intra- and intermolecular bonding features in ligand–receptor complexes, such as halogen bonding and nonclassical hydrogen bonds. There is scope for wider use of energy decomposition methods in guiding molecular design in medicinal chemistry.

4 Outlook for QM in Structure-Based Drug Discovery

There are a growing number of applications of quantum chemistry to prospective prediction of compounds targeting therapeutically important receptors. An interesting recent application of QM in structure-based drug design was in the use of density functional calculations to characterize transition state and intermediate structures of insulin-regulated aminopeptidase [83], a protein target linked to enhanced cognitive function. Pharmacophores were derived based on these structures and subsequently used to screen a compound library of over 5 million chemicals, identifying known and a previously unknown peptidase inhibitor. QM methods have also been used more directly in virtual screening: [84] a focused in silico library of ~10^4 compounds was filtered by calculation of the docked ligand–protein interaction energy at the PM7 level. For this, the protein active site was modelled by a cluster of small molecules representing key amino acid functional groups

[84]. Application of this protocol to jumonji demethylase JMJD3, alongside other filtering steps, led to identification of a number of novel low micromolar inhibitors spanning several chemical scaffolds.

Drug design projects such as these will be impacted by the significant ongoing developments in the accuracy and efficiency of QM approaches. New robust and computationally expedient SQM methods include GFN2-xTB [59], adding to the arsenal of density functional tight-binding based-approaches [21, 85]. On the horizon, one may also anticipate new ML-refitted SQM methods such as the reparametrized OM2 method [70]. Even faster computations can be afforded by potentials obtained directly by ML; while ensuring the broad transferability of such methods is a challenge, its promise is well illustrated by the performance of the ANI-1 deep neural network potential [75].

The existence of large sets of accurate quantum chemical data for relevant molecular systems are intrinsically required for both the development and validation of these new potentials; the advent of more efficient, high accuracy ab initio QM methods, such as DLPNO-CCSD(T), will fuel the generation of increasingly comprehensive and more diverse data sets of conformation and interactions of molecules. This in turn will underpin the development of computationally fast and accurate classical, SQM-, or ML-based potentials. High-accuracy ML-based potentials could be derived, either as suitably transferable to the system of interest or derived bespoke. The resulting ML-based potential may, for example, be integrated within a QM/MM scheme as part of an improved scoring function to guide protein–ligand docking. Possibly, even ab initio QM approaches could be used for rescoring selected poses or directly applied in docking via multilayer fragmentation schemes such as FMO [30] or MIM [32]. However, for accurate calculation of ligand–receptor interaction energetics in docking, virtual screening or compound optimization, the challenge of capturing solvent and entropic effects remains [85].

The ability to incorporate thermodynamic contributions to protein–ligand binding using QM or QM/MM potentials can in principle be obtained through statistical mechanical methods, such as FEP calculations [36]. Alternatively, a recent study [86] derived system-specific ligand–protein force fields for computing free energies of binding. An electrostatic model of the ligand–protein system was obtained from linear-scaling QM calculations of the complex, with partitioning of the resulting electron density using an AIM approach. Applying these bespoke force fields within an FEP scheme, the absolute binding free energies of six benzene derivatives to the L99A mutant of T4 lysozyme were calculated. The estimated free energies were within 0.8 kcal/mol of the experimental binding affinities, improving on reported estimates of 1.3 kcal/mol and 2.1 kcal/mol via the OPLS and AMBER force fields, respectively [86].

As illustrated by examples in this issue of Methods for Molecular Biology, and supported by reports in the wider medicinal chemistry literature [11, 13–19], there is a growing use of quantum chemical approaches to prospectively guide structure-based drug design projects. Concurrently, there are a number of exciting developments in quantum chemical methodology: the computational overhead associated with high level ab initio QM calculations of chemical accuracy is significantly reduced. Improved SQM methods are becoming available through fitting to large databases of high-quality QM energies and geometries, in some cases aided by ML approaches. ML is also being used to derive potentials of QM accuracy with very low computational overhead. Electronic decomposition methods can be used to analyze interactions at the QM level for insights into ligand design but also to derive accurate bespoke force fields. The methodological framework exists for these emerging QM-based approaches to be integrated in a range of ways into the hit identification and lead optimization stages of drug discovery. Given this context, it is likely that the rate at which quantum chemistry will impact structure-based drug design will accelerate, playing an ever-increasing role in the rational design of new medicines.

References

1. Kuroda D, Shirai H, Jacobson MP, Nakamura H (2012) Computer-aided antibody design. Protein Eng Des Sel 25:507–522
2. K Singh R, Lee J-K, Selvaraj C, Singh R, Li J, Kim S-Y, C Kalia V (2018) Protein engineering approaches in the post-genomic era. Curr Protein Pept Sci 19:5–15
3. Rognan D (2017) The impact of in silico screening in the discovery of novel and safer drug candidates. Pharmacol Ther 175:47–66
4. Zwanzig RW (1954) High-temperature equation of state by a perturbation method. I. Nonpolar gases. J Chem Phys 22:1420–1426
5. Wang L, Wu Y, Deng Y, Kim B, Pierce L, Krilov G, Lupyan D, Robinson S, Dahlgren MK, Greenwood J (2015) Accurate and reliable prediction of relative ligand binding potency in prospective drug discovery by way of a modern free-energy calculation protocol and force field. J Am Chem Soc 137:2695–2703
6. Steinbrecher TB, Dahlgren M, Cappel D, Lin T, Wang L, Krilov G, Abel R, Friesner R, Sherman W (2015) Accurate binding free energy predictions in fragment optimization. J Chem Inf Model 55:2411–2420
7. Ciordia M, Pérez-Benito L, Delgado F, Trabanco A s A, Tresadern G (2016) Application of free energy perturbation for the design of Bace1 inhibitors. J Chem Inf Model 56:1856–1871
8. Paton RS, Goodman JM (2009) Hydrogen bonding and π-stacking: how reliable are force fields? A critical evaluation of force field descriptions of nonbonded interactions. J Chem Inf Model 49:944–955
9. Peters MB, Raha K, Merz K (2006) Quantum mechanics in structure-based drug design. Curr Opin Drug Discov Devel 9:370
10. Raha K, Peters MB, Wang B, Yu N, Wollacott AM, Westerhoff LM, Merz KM Jr (2007) The role of quantum mechanics in structure-based drug design. Drug Discov Today 12:725–731
11. Christensen AS, Kubar T, Cui Q, Elstner M (2016) Semiempirical quantum mechanical methods for noncovalent interactions for chemical and biochemical applications. Chem Rev 116:5301–5337
12. Mucs D, Bryce RA (2013) The application of quantum mechanics in structure-based drug design. Expert Opin Drug Discovery 8:263–276
13. Ryde U, Soderhjelm P (2016) Ligand-binding affinity estimates supported by quantum-

mechanical methods. Chem Rev 116:5520–5566

14. Adeniyi AA, Soliman ME (2017) Implementing QM in docking calculations: is it a waste of computational time? Drug Discov Today 22:1216–1223
15. Cavasotto CN, Adler NS, Aucar MG (2018) Quantum chemical approaches in structure-based virtual screening and lead optimization. Front Chem 6:188
16. Yilmazer N, Korth M (2016) Recent progress in treating protein–ligand interactions with quantum-mechanical methods. Int J Mol Sci 17:742
17. Lepšík M, Řezáč J, Kolář M, Pecina A, Hobza P, Fanfrlík J (2013) The semiempirical quantum mechanical scoring function for in silico drug design. ChemPlusChem 78:921–931
18. Merz KM Jr (2014) Using quantum mechanical approaches to study biological systems. Acc Chem Res 47:2804–2811
19. Zhou T, Huang D, Caflisch A (2010) Quantum mechanical methods for drug design. Curr Top Med Chem 10:33–45
20. Chaskar P, Zoete V, Röhrig UF (2014) Toward on-the-fly quantum mechanical/molecular mechanical (QM/MM) docking: development and benchmark of a scoring function. J Chem Inf Model 54:3137–3152
21. Elstner M (2006) The SCC-DFTB method and its application to biological systems. Theor Chem Accounts 116:316–325
22. Brooks BR, Brooks CL III, Mackerell AD Jr, Nilsson L, Petrella RJ, Roux B, Won Y, Archontis G, Bartels C, Boresch S (2009) CHARMM: the biomolecular simulation program. J Comput Chem 30:1545–1614
23. Haberthür U, Caflisch A (2008) FACTS: fast analytical continuum treatment of solvation. J Comput Chem 29:701–715
24. Chaskar P, Zoete V, Röhrig UF (2016) On-the-fly QM/MM docking with attracting cavities. J Chem Inf Model 57:73–84
25. Zoete V, Schuepbach T, Bovigny C, Chaskar P, Daina A, Röhrig UF, Michielin O (2016) Attracting cavities for docking. Replacing the rough energy landscape of the protein by a smooth attracting landscape. J Comput Chem 37:437–447
26. Verdonk ML, Cole JC, Hartshorn MJ, Murray CW, Taylor RD (2003) Improved protein–ligand docking using GOLD. Proteins 52:609–623
27. Goodsell DS, Morris GM, Olson AJ (1996) Automated docking of flexible ligands: applications of autodock. J Mol Recognit 9:1–5
28. Trott O, Olson AJ (2010) Autodock Vina: improving the speed and accuracy of docking with a new scoring function, efficient optimization, and multithreading. J Comput Chem 31:455–461
29. Riccardi L, Genna V, De Vivo M (2018) Metal–ligand interactions in drug design. Nat Rev Chem 2:100
30. Fedorov DG, Nagata T, Kitaura K (2012) Exploring chemistry with the fragment molecular orbital method. PCCP 14:7562–7577
31. Zhang DW, Xiang Y, Zhang JZ (2003) New advance in computational chemistry: full quantum mechanical Ab initio computation of streptavidin– biotin interaction energy. J Phys Chem B 107:12039–12041
32. Thapa B, Beckett D, Erickson J, Raghavachari K (2018) Theoretical study of protein–ligand interactions using the molecules-in-molecules fragmentation-based method. J Chem Theory Comput 14:5143–5155
33. Thapa B, Beckett D, Jovan Jose K, Raghavachari K (2018) Assessment of fragmentation strategies for large proteins using the multilayer molecules-in-molecules approach. J Chem Theory Comput 14:1383–1394
34. Brahmkshatriya PS, Dobes P, Fanfrlik J, Rezac J, Paruch K, Bronowska A, Lepsík M, Hobza P (2013) Quantum mechanical scoring: structural and energetic insights into cyclin-dependent kinase 2 inhibition by Pyrazolo [1, 5-a] Pyrimidines. Curr Comput Aided Drug Des 9:118–129
35. Hayik SA, Dunbrack R Jr, Merz KM Jr (2010) Mixed quantum mechanics/molecular mechanics scoring function to predict protein–ligand binding affinity. J Chem Theory Comput 6:3079–3091
36. Olsson MA, Ryde U (2017) Comparison of QM/MM methods to obtain ligand-binding free energies. J Chem Theory Comput 13:2245–2253
37. Korth M (2010) Third-generation hydrogen-bonding corrections for semiempirical QM methods and force fields. J Chem Theory Comput 6:3808–3816
38. Nicklaus MC, Wang S, Driscoll JS, Milne GW (1995) Conformational changes of small molecules binding to proteins. Biorg Med Chem 3:411–428
39. Butler KT, Luque FJ, Barril X (2009) Toward accurate relative energy predictions of the bioactive conformation of drugs. J Comput Chem 30:601–610
40. Avgy-David HH, Senderowitz H (2015) Toward focusing conformational ensembles on bioactive conformations: a molecular

mechanics/quantum mechanics study. J Chem Inf Model 55:2154–2167

41. Peach ML, Cachau RE, Nicklaus MC (2017) Conformational energy range of ligands in protein crystal structures: the difficult quest for accurate understanding. J Mol Recognit 30: e2618
42. Jurečka P, Šponer J, Černý J, Hobza P (2006) Benchmark database of accurate (MP2 and CCSD(T) complete basis set limit) interaction energies of small model complexes, DNA base pairs, and amino acid pairs. PCCP 8:1985–1993
43. Bereau T, von Lilienfeld OA (2014) Toward transferable interatomic van der Waals interactions without electrons: the role of multipole electrostatics and many-body dispersion. J Chem Phys 141:034101
44. McNamara JP, Hillier IH (2007) Semi-empirical molecular orbital methods including dispersion corrections for the accurate prediction of the full range of intermolecular interactions in biomolecules. PCCP 9:2362–2370
45. Molnar LF, He X, Wang B, Merz KM Jr (2009) Further analysis and comparative study of intermolecular interactions using dimers from the S22 database. J Chem Phys 131:08B603
46. Gráfová L, Pitonak M, Rezac J, Hobza P (2010) Comparative study of selected wave function and density functional methods for noncovalent interaction energy calculations using the extended S22 data set. J Chem Theory Comput 6:2365–2376
47. Řezác J, Riley KE, Hobza P (2011) S66: a well-balanced database of benchmark interaction energies relevant to biomolecular structures. J Chem Theory Comput 7:2427–2438
48. Řezác J, Riley KE, Hobza P (2011) Extensions of the S66 data set: more accurate interaction energies and angular-displaced nonequilibrium geometries. J Chem Theory Comput 7:3466–3470
49. Rezac J, Riley KE, Hobza P (2012) Benchmark calculations of noncovalent interactions of halogenated molecules. J Chem Theory Comput 8:4285–4292
50. Řezáč J, Jurečka P, Riley KE, Černý J, Valdes H, Pluháčková K, Berka K, Řezáč T, Pitoňák M, Vondrášek J (2008) Quantum chemical benchmark energy and geometry database for molecular clusters and complex molecular systems (www.begdb.com): a users manual and examples. Collect Czechoslov Chem Commun 73:1261–1270
51. Li A, Muddana HS, Gilson MK (2014) Quantum mechanical calculation of noncovalent interactions: a large-scale evaluation of Pmx, Dft, and sapt approaches. J Chem Theory Comput 10:1563–1575
52. Jeziorski B, Moszynski R, Szalewicz K (1994) Perturbation theory approach to intermolecular potential energy surfaces of van der Waals complexes. Chem Rev 94:1887–1930
53. Zhao Y, Truhlar DG (2006) A new local density functional for Main-group thermochemistry, transition metal bonding, thermochemical kinetics, and noncovalent interactions. J Chem Phys 125:194101
54. Grimme S (2006) Semiempirical GGA-type density functional constructed with a long-range dispersion correction. J Comput Chem 27:1787–1799
55. Goerigk L, Grimme S (2010) Efficient and accurate double-hybrid-meta-GGA density Functionals evaluation with the extended GMTKN30 database for General Main group thermochemistry, kinetics, and noncovalent interactions. J Chem Theory Comput 7:291–309
56. Stewart JJ (2007) Optimization of parameters for Semiempirical methods V: modification of NDDO approximations and application to 70 elements. J Mol Model 13:1173–1213
57. Saito T, Kitagawa Y, Takano Y (2016) Reparameterization of PM6 applied to organic Diradical molecules. Chem Eur J 120:8750–8760
58. Grimme S, Bannwarth C, Shushkov P (2017) A robust and accurate tight-binding quantum chemical method for structures, vibrational frequencies, and noncovalent interactions of large molecular systems parametrized for all spd-block elements (Z= 1–86). J Chem Theory Comput 13:1989–2009
59. Bannwarth C, Ehlert S, Grimme S (2019) GFN2-xTB—an accurate and broadly Parametrized self-consistent tight-binding quantum chemical method with multipole electrostatics and density-dependent dispersion contributions. J Chem Theory Comput 15:1652–1671
60. Riplinger C, Sandhoefer B, Hansen A, Neese F (2013) Natural triple excitations in local coupled cluster calculations with pair natural orbitals. J Chem Phys 139:134101
61. Ma Q, Schwilk M, Köppl C, Werner H-J (2017) Scalable electron correlation methods. 4. Parallel explicitly correlated local coupled cluster with pair natural orbitals (PNO-CCSD-F12). J Chem Theory Comput 13:4871–4896
62. Liakos DG, Sparta M, Kesharwani MK, Martin JM, Neese F (2015) Exploring the accuracy limits of local pair natural orbital coupled-cluster theory. J Chem Theory Comput 11:1525–1539

63. Neese F, Atanasov M, Bistoni G, Maganas D, Ye S (2019) Chemistry and quantum mechanics in 2019: give us insight and numbers. J Am Chem Soc 141:2814–2824
64. Sure R, Grimme S (2015) Comprehensive benchmark of association (free) energies of realistic host–guest complexes. J Chem Theory Comput 11:3785–3801
65. Cavallo L, Minenkov Y, Sharapa D, Genaev A (2018) A robust and cost-efficient scheme for accurate conformational energies of organic molecules. ChemPhysChem 20(1):92–102
66. Jaiyong P, Bryce RA (2017) Approximate quantum chemical methods for modelling carbohydrate conformation and aromatic interactions: β-cyclodextrin and its adsorption on a single-layer graphene sheet. PCCP 19:15346–15355
67. Willett P (2011) Chemoinformatics: a history. Wiley Interdiscip Rev Comput Mol Sci 1:46–56
68. Cramer RD III, Redl G, Berkoff CE (1974) Substructural analysis. Novel approach to the problem of drug design. J Med Chem 17:533–535
69. Behler J (2016) Perspective: machine learning potentials for atomistic simulations. J Chem Phys 145:170901
70. Dral PO, von Lilienfeld OA, Thiel W (2015) Machine learning of parameters for accurate semiempirical quantum chemical calculations. J Chem Theory Comput 11:2120–2125
71. Weber W, Thiel W (2000) Orthogonalization corrections for semiempirical methods. Theor Chem Accounts 103:495–506
72. Li Y, Li H, Pickard FC IV, Narayanan B, Sen FG, Chan MK, Sankaranarayanan SK, Brooks BR, Roux B (2017) Machine learning force field parameters from ab initio data. J Chem Theory Comput 13:4492–4503
73. Ren P, Wu C, Ponder JW (2011) Polarizable atomic multipole-based molecular mechanics for organic molecules. J Chem Theory Comput 7:3143–3161
74. Dral PO, Owens A, Yurchenko SN, Thiel W (2017) Structure-based sampling and self-correcting machine learning for accurate calculations of potential energy surfaces and vibrational levels. J Chem Phys 146:244108
75. Smith JS, Isayev O, Roitberg AE (2017) Ani-1: an extensible neural network potential with Dft accuracy at force field computational cost. Chem Sci 8:3192–3203
76. Chai J-D, Head-Gordon M (2008) Systematic optimization of long-range corrected hybrid density functionals. J Chem Phys 128:084106
77. Fink T, Reymond J-L (2007) Virtual exploration of the chemical universe up to 11 atoms of C, N, O, F: assembly of 26.4 million structures (110.9 million stereoisomers) and analysis for new ring systems, stereochemistry, physicochemical properties, compound classes, and drug discovery. J Chem Inf Model 47:342–353
78. Heifetz A, Aldeghi M, Chudyk EI, Fedorov DG, Bodkin MJ, Biggin PC (2016) Using the fragment molecular orbital method to investigate agonist–orexin-2 receptor interactions. Biochem Soc Trans 44:574–581
79. Bader RF (1991) A quantum theory of molecular structure and its applications. Chem Rev 91:893–928
80. Reed AE, Curtiss LA, Weinhold F (1988) Intermolecular interactions from a natural bond orbital, donor-acceptor viewpoint. Chem Rev 88:899–926
81. Schneider WB, Bistoni G, Sparta M, Saitow M, Riplinger C, Auer AA, Neese F (2016) Decomposition of intermolecular interaction energies within the local pair natural orbital coupled cluster framework. J Chem Theory Comput 12:4778–4792
82. Mao Y, Horn PR, Head-Gordon M (2017) Energy decomposition analysis in an adiabatic picture. PCCP 19:5944–5958
83. Svensson F, Engen K, Lundback T, Larhed M, Skold C (2015) Virtual screening for transition state analogue inhibitors of Irap based on quantum mechanically derived reaction coordinates. J Chem Inf Model 55:1984–1993
84. Esposito C, Wiedmer L, Caflisch A (2018) In Silico identification of JMJD3 demethylase inhibitors. J Chem Inf Model 58:2151–2163
85. Christensen AS, Elstner M, Cui Q (2015) Improving intermolecular interactions in DFTB3 using extended polarization from chemical-potential equalization. J Chem Phys 143:084123
86. Cole DJ, de Vaca IC, Jorgensen WL (2019) Computation of protein–ligand binding free energies using quantum mechanical bespoke force fields. Med Chem Commun 10 (7):1116–1120

INDEX

Alexander Heifetz (ed.), *Quantum Mechanics in Drug Discovery*, Methods in Molecular Biology, vol. 2114, https://doi.org/10.1007/978-1-0716-0282-9,

O

P

Q

R

S

T

U

V

W

X

Z

FSC
www.fsc.org
MIX
Papier aus verantwortungsvollen Quellen
Paper from responsible sources
FSC® C105338

If you have any concerns about our products,
you can contact us on
ProductSafety@springernature.com

In case Publisher is established outside the EU,
the EU authorized representative is:
Springer Nature Customer Service Center GmbH
Europaplatz 3, 69115 Heidelberg, Germany

Printed by Libri Plureos GmbH
in Hamburg, Germany